A. Liese, K. Seelbach, C. Wandrey

Industrial Biotransformations

A. Liese, K. Seelbach, C. Wandrey

Industrial
Biotransformations

Weinheim · New York · Chichester · Brisbane · Singapore · Toronto

Chemistry Library

Dr. Andreas Liese
Institut für Biotechnologie
Forschungszentrum
Jülich GmbH
D-52425 Jülich
Germany

Dr. Karsten Seelbach
Corporate Process
Technology
Degussa-Hüls AG
D-45764 Marl
Germany

Prof. Dr. Christian Wandrey
Institut für Biotechnologie
Forschungszentrum
Jülich GmbH
D-52425 Jülich
Germany

The cover illustration shows the production plant of L-methionine. The picture was supplied by Professor W. Leuchtenberger, Degussa-Hüls AG. In the foreground the conversion of N-acetyl-L-methionine with acylase (EC 3.5.1.14) to L-methionine is shown.

Library of Congress Card No.: applied for

British Library Cataloguing-in-Publication Data:
A catalogue record for this book
is available from the Britsh Library

Die Deutsche Bibliothek – CIP Cataloguing-in-Publication Data:
A catalogue record for this publication is available from Die Deutsche Bibliothek
ISBN 3-527-30094-5

© WILEY-VCH Verlag GmbH, D-69469 Weinheim (Federal Republic of Germany), 2000

Gedruckt auf säurefreiem und chlorfrei gebleichtem Papier

Composition: Kühn & Weyh, Freiburg
Printing: betz-druck, Darmstadt
Bookbinding: Osswald, Neustadt/Wstr.

Printed in the Federal Republic of Germany.

Content

1 Introduction

The main incentive in writing this book was to gather information on one-step biotransformations that are of industrial importance. With this collection, we want to illustrate that more enzyme-catalyzed processes have gained practical significance than their potential users are conscious of. There is still a prejudice that biotransformations are only needed in cases where classical chemical synthesis fails. Even the conviction that the respective biocatalysts are not available and, if so, then too expensive, unstable and only functional in water, still seems to be widespread. We hope that this collection of industrial biotransformations will in future influence decision-making of synthesis development in such a way that it might lead to considering the possible incorporation of a biotransformation step in a scheme of synthesis.

We therefore took great pains in explicitly describing the substrates, the catalyst, the product and as much of the reaction conditions as possible of the processes mentioned. Wherever flow schemes were available for publication or could be generated from the reaction details, this was done. Details of some process parameters are still incomplete, since such information is only sparingly available. We are nevertheless convinced that the details are sufficient to convey a feeling for the process parameters. Finally, the use of the products is described and a few process-relevant references are made.

We would go beyond the scope of this foreword, should we attempt to thank all those who were kind enough to supply us with examples. Of course, we only published openly available results (including the patent literature) or used personally conveyed results with the consent of the respective authors. We are aware of the fact that far more processes exist and that by the time the book is published, many process details will be outdated. Nonetheless, we believe that this compilation with its overview character will serve the above-mentioned purpose. This awareness could be augmented if the reader, using his or her experience, would take the trouble of filling out the printed worksheet at the end of this book with suggestions that could lead to an improvement of a given process or the incorporation of a further industrial process into the collection.

Requesting our industrial partners to make process schemes and parameters more accessible did not please them very much. Even so, we are asking our partners once again to disclose more information than they have done in the past. In many instances, far more knowledge of industrial processes has been gained than is publicly available. Our objective is to be able to make use of these "well known secrets" as well. We would like to express our gratitude to all those who supplied us with information in a progress-conducive manner. Thanks also go to those who did not reject our requests completely and at least supplied us with a photograph in compensation for the actually requested information.

The book begins with a short historical overview of industrial biotransformations. Since the process order of the compilation is in accordance with the enzyme nomenclature system, the latter is described in more detail. We also include a chapter on reaction engineering to enable an easier evaluation of the processes.

The main part of the book, as you would expect, is the compilation of the industrial biotransformations. The comprehensive index will allow a facile search for substrates, enzymes and products.

We sincerely hope that this book will be of assistance in the academic as well as the industrial field, when one wants to get an insight into industrial biotransformations. We would be very thankful to receive any correction suggestions or further comments and contributions. At least we hope to experience a trigger effect that would make it worth while for the readership, the authors and the editors to have a second edition succeeding the first.

We are indebted to several coworkers for screening literature and compiling data, especially to Jürgen Haberland, Doris Hahn, Marianne Hess, Wolfgang Lanters, Monika Lauer, Christian Litterscheid, Nagaraj Rao, Durda Vasic-Racki, Murillo Villela Filho, Philomena Volkmann and Andrea Weckbecker.

We thank especially Uta Seelbach for drawing most of the figures during long nights, as well as Nagaraj Rao and the "enzyme group" (Nils Brinkmann, Lasse Greiner, Jürgen Haberland, Christoph Hoh, David Kihumbu, Stephan Laue, Thomas Stillger and Murillo Villela Filho).

And last but not least we thank our families for their support and tolerance during the time that we invested in our so called 'book project'.

2 History of Industrial Biotransformations – Dreams and Realities

DURDA VASIC-RACKI

Faculty of Chemical Engineering and Technology
University of Zagreb
HR-10000 Zagreb, Croatia

Throughout the history of mankind, microorganisms have been of tremendous social and economic importance. Without even being aware of their existence, man used them in the production of food and beverages already very early in history. Sumerians and Babylonians practised beer brewing **before 6000 B.C.**, references to wine making can be found in the Book of Genesis, and Egyptians used yeast for baking bread. However, the knowledge of the production of chemicals such as alcohols and organic acids by fermentation is relatively recent and the first reports in the literature appeared only in the **second half of the 19th century**. Lactic acid was probably the first optically active compound to be produced industrially by fermentation. It was accomplished in the USA in **1880** [1]. In **1921**, Chapman reviewed a number of early industrial fermentation processes for organic chemicals [2].

In the course of time, it was discovered that microorganisms could modify certain compounds by simple, chemically well-defined reactions which were further catalyzed by enzymes. Nowadays, these processes are called **"biotransformations"**. The essential difference between fermentation and biotransformation is that there are several catalytic steps between substrate and product in fermentation while there are only one or two in biotransformation. The distinction is also in the fact that the chemical structures of the substrate and the product resemble one another in a biotransformation, but not necessarily in a fermentation.

2.1 From the "flower of vinegar" to the recombinant *E. coli* – The history of microbial biotransformations

The story of microbial biotransformations is closely connected with vinegar production which dates back to some **2000 years B.C.**

Vinegar production is perhaps the oldest and best known example of microbial oxidation which may illustrate some of the important developments in the field of biotransformations by living cells (figure 1).

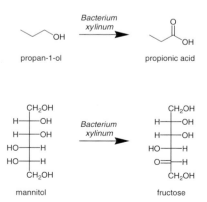

$$\text{ethanol} + O_2 \xrightarrow{\textbf{E}} \text{acetic acid} + H_2O$$

ethanol   oxygen   acetic acid   water

Fig. 1 Vinegar production (**E** = biocatalyst).

A prototype bioreactor with immobilized bacteria has been known in France since the **17th century**. The oldest bioreactor using immobilized living microorganisms, a so-called generator, was developed in **1823** [3,4]. Even today, acetic acid is still known as "vinegar" if it is obtained by oxidative fermentation of ethanol-containing solutions by acetic acid bacteria [5].

In **1858**, Pasteur [6] was the first to demonstrate the microbial resolution of tartaric acid. He performed fermentation of the ammonium salt of racemic tartaric acid, mediated by the mold *Penicillium glaucum*. The fermentation yielded (–)-tartaric acid (figure 2).

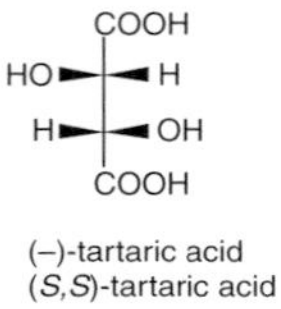

COOH
HO—H
H—OH
COOH

(–)-tartaric acid
(*S,S*)-tartaric acid

Fig. 2 Pasteur's product of the first resolution reaction.

This was also the first time that a method in which microorganisms degrade one enantiomer of the racemate while leaving the other untouched was used.

In **1862**, Pasteur [7] investigated the conversion of alcohol to vinegar and concluded that the pellicle, which he called "the flower of vinegar", "serves as a transport of air oxygen to a multitude of organic substances".

In **1886**, Brown confirmed Pasteur's findings and named the causative agent in vinegar production as *Bacterium xylinum*. He also found that it could oxidize propanol to propionic acid and mannitol to fructose (figure 3) [8].

propan-1-ol $\xrightarrow[\text{xylinum}]{\textit{Bacterium}}$ propionic acid

CH₂OH   CH₂OH
H—OH   H—OH
H—OH   *Bacterium*   H—OH
HO—H   *xylinum*   HO—H
HO—H   O=H
CH₂OH   CH₂OH

mannitol   fructose

Fig. 3 Reactions catalyzed by *Bacterium xylinum*, the vinegar biocatalyst.

In **1897**, Buchner [9] reported that cell-free extracts prepared by grinding yeast cells with sand could carry out alcoholic fermentation reactions in the absence of living cells. This initiated the usage of resting cells for biotransformations.

Neuberg and Hirsch [10] discovered in **1921** that the condensation of benzaldehyde with acetaldehyde in the presence of yeast forms optically active 1-hydroxy-1-phenyl-2-propanone (figure 4).

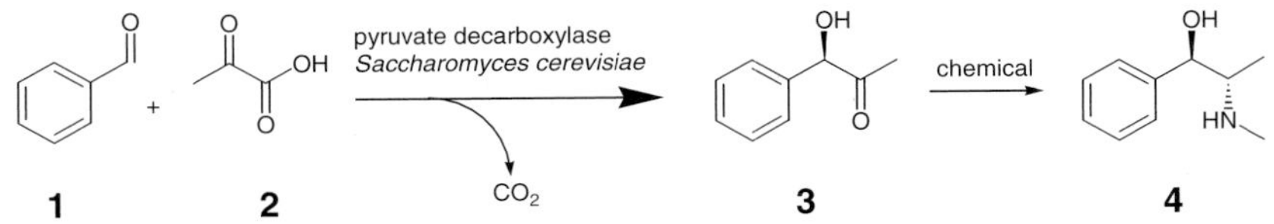

1 = benzaldehyde
2 = 2-oxo-propionic acid
3 = 1-hydroxy-1-phenylpropan-2-one
4 = 2-methylamino-1-phenylpropan-1-ol

Fig. 4 L-Ephedrine production.

The obtained compound was further chemically converted into D-(–)-ephedrine by Knoll AG, Ludwigshafen, Germany in **1930** (figure 5) [11].

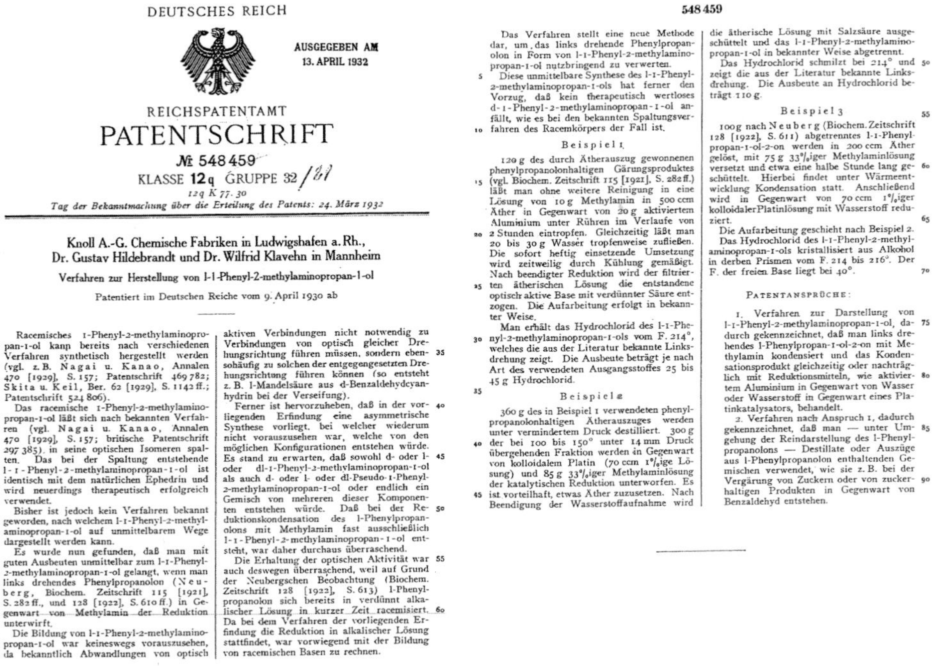

Fig. 5 Knoll's patent of **1930**.

The bacterium *Acetobacter suboxydans* was isolated in **1923** [12]. Its ability to carry out limited oxidation was used in a highly efficient preparation of L-sorbose from D-sorbitol (figure 6).

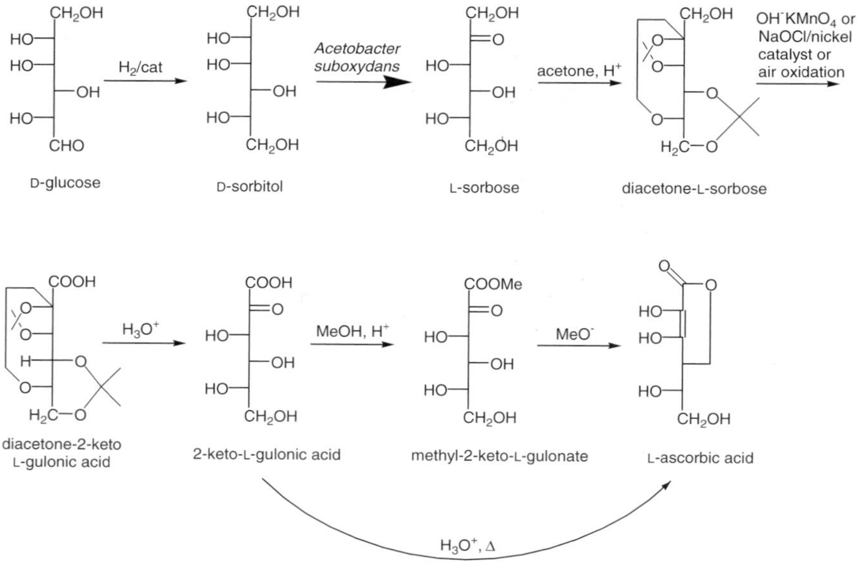

Fig. 6 Reichstein-Grüssner synthesis of vitamin C (L-ascorbic acid).

L-Sorbose became important in the **mid-1930's** as an intermediate in the Reichstein-Grüssner synthesis of L-ascorbic acid [13].

In **1953**, Peterson at al. [14] reported that *Rhizopus arrhius* converted progesterone to 11α-hydroxyprogesterone (figure 7), which was used as an intermediate in the synthesis of cortisone.

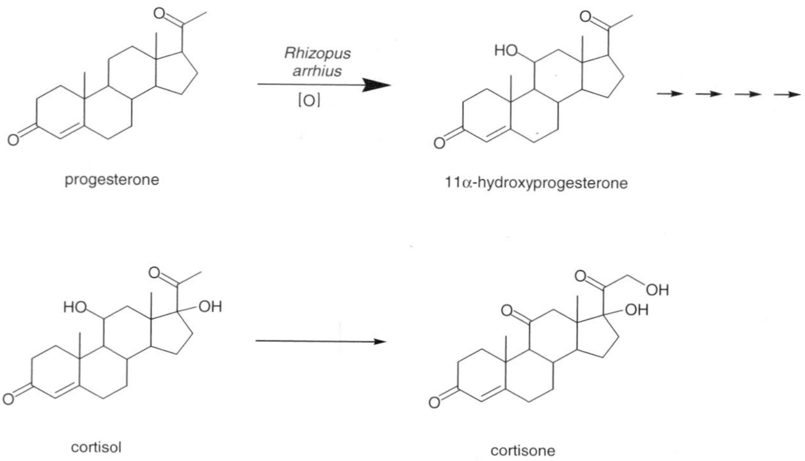

Fig. 7 Microbial 11α-hydroxylation of progesterone.

This microbial hydroxylation simplified and considerably improved the efficiency of the multi-step chemical synthesis of corticosteroid hormones and their derivatives. Although the chemical synthesis [15] (figure 8) from deoxycholic acid that was developed at Merck, Germany, was workable, it was recognized that it was complicated and uneconomical: 31 steps were necessary to obtain 1 kg of cortisone acetate from 615 kg of deoxycholic acid. The microbial 11α-hydroxylation of progesterone quickly reduced the price of cortisone from $200 to $ 6 per gram. Further improvements have led to a current price of less than $1 per gram [16].

In the **1950's**, the double helix structure and the chemical nature of RNA and DNA – the genetic code of heredity – were discovered. This discovery can be regarded as one of the milestones among this century's main scientific achievements. It led to the synthesis of recombinant DNA and gave a fillip to genetic engineering in the seventies'. Such developments quickly made the rDNA technology a part of industrial microbial transformations. Application of this technology for the production of small molecules began in **1983**. Ensley et al. [17] reported on the construction of a strain of *E.coli* that excreted indigo, one of the oldest known dyes. They found that the entire pathway for conversion of naphthalene to salicylic acid is encoded by genes of *Pseudomonas putida*. These genes can be expressed in *E.coli*. Their results led to the unexpected finding that a subset of these genes was also responsible for the microbial production of indigo. Moreover, they showed that indigo formation was a property of the dioxygenase enzyme system that forms *cis*-dihydrodiols from aromatic hydrocarbons. Finally, they proposed a pathway for indigo biosynthesis in a recombinant strain of *E. coli* (figure 9).

Genencor International is developing a commercially competitive biosynthetic route to indigo using recombinant *E.coli* that can directly synthesize indigo from glucose [18]. Anderson et al. in **1985** [19] reported the construction of a metabolically engineered bacterial strain that was able to synthesize 2-keto-L-gulonic acid (figure 10), a key intermediate in the production of L-ascorbic acid (vitamin C).

BASF, Merck and Cerestar are building a 2-keto-L-ketogulonic acid plant in Krefeld, Germany. The start up of operation is scheduled for **1999**. They developed a new fermentation route from sorbitol directly to the ketogulonic acid [20]. This method is probably similar to the method described in **1966** [21].

The Cetus Corporation (Berkeley, California, USA) bioprocess for converting alkenes to alkene oxides emerged in **1980** [22]. This bioprocess appeared to be very interesting, thanks to the possibility of replacing energy-consuming petrochemical processes.

There were high hopes that the development of recombinant DNA technology would speed up technological advances. Unfortunately, there is still a lot left to be done about the development and application of bioprocesses before the commercial production of low-value chemicals becomes feasible [23]. However, today even the traditional chemical companies like Dow Chemical, DuPont, Degussa-Hüls AG etc., pressurized by investors and technological advances, are trying to use microbial or enzymatic transformations in production. They are doing this to see whether natural feedstocks can bring more advantages than crude oil. One only needs to compare the cost of a barrel of oil with that of corn starch to see that the latter is quite cheaper [20].

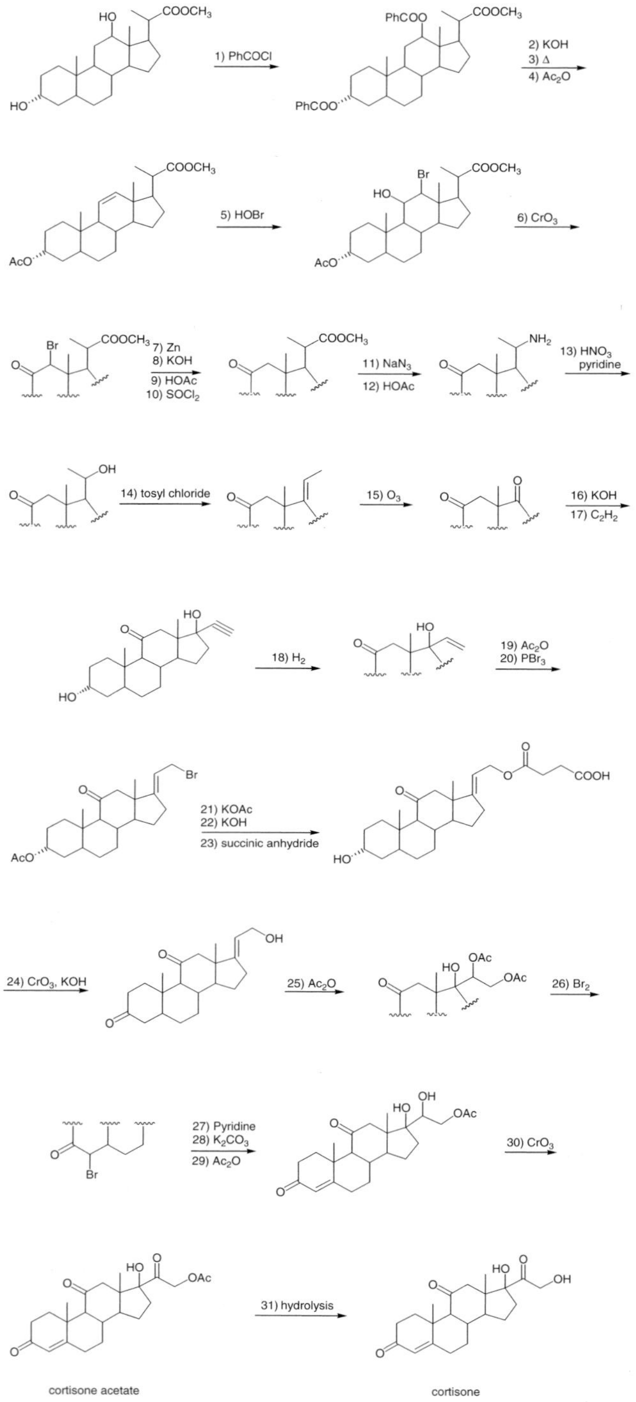

Fig. 8 Chemical synthesis of cortisone.

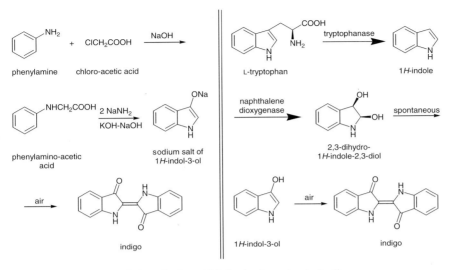

Fig. 9 Comparison of chemical and biological routes to indigo.

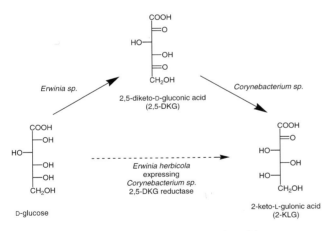

Fig. 10 Biosynthesis of 2-keto-L-gulonic acid.

Acrylamide is one of the most important commodity chemicals. Its global consumption is about 200,000 tonnes per year. It is used in the production of various polymers for use as flocculants, additives or for petroleum recovery. In conventional synthesis, copper salts are used as catalysts in the hydration of nitriles. However, this is rather disadvantageous as the preparation of the catalysts is quite complex. Additionally, it is difficult to regenerate the used catalyst and separate and purify the formed acrylamide. Furthermore, since acrylamides are readily polymerized, their production under moderate conditions is highly desirable. In contrast to the conventional chemical process, there is no need to recover unreacted acrylonitrile in the enzymatic process, because the conversion and yield of the enzymatic hydration pro-

9

cess are almost 100 %. The removal of the copper ions from the product is no longer necessary. Overall, the enzymatic process – being carried out below 10 °C under mild reaction conditions and requiring no special energy source – proves to be simpler and more economical. The immobilized cells are used repeatedly and a very pure product is obtained. The enzymatic process, which was first implemented in **1985**, is already producing about 6000 tons of acrylamide per year for Nitto [24,25]. The use of biocatalyst for the production of acrylamide may not be the first case in which biotransformation as a part of biotechnology was used in the petrochemical industry. However, it is the first successful example of the introduction of an industrial biotransformation process for the manufacture of a commodity chemical (figure 11).

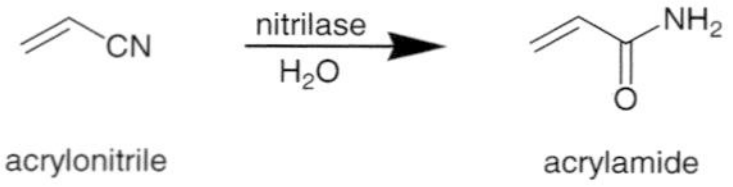

Fig. 11 Acrylamide synthesis.

Some representative industrial microbial transformations are listed in Table I.

Table I: Some representative industrial biotransformations catalyzed by whole cells.

Product	Biocatalyst	Operating since	Company
vinegar	bacteria	1823	various
L-2-methylamino-1-phenylpropan-1-ol	yeast	1930	Knoll AG, Germany
L-sorbose	*Acetobacter suboxydans*	1934	various
prednisolone	*Arthrobacter simplex*	1955	Schering AG, Germany
L-aspartic acid	*Escherichia coli*	1958	Tanabe Seiyaku Co., Japan
7-ADCA	*Bacillus megaterium*	1970	Asahi Chemical Industry, Japan
L-malic acid	*Brevibacterium ammoniagenes*	1974	Tanabe Seiyaku Co., Japan
D-*p*-hydroxyphenylglycine	*Pseudomonas striata*	1983	Kanegafuchi, Chemical Co., Japan
acrylamide	*Rhodococcus sp.*	1985	Nitto Chemical Ltd, Japan
D-aspartic acid and L-alanine	*Pseudomonas dacunhae*	1988	Tanabe Seiyaku Co., Japan
L-carnitine	*Agrobacterium sp.*	1993	Lonza, Czech.Rep.
2-keto-L-gulonic acid	*Acetobacter sp.*	1999	BASF, Merck, Cerestar, Germany

2.2 From gastric juice to SweetzymeT – The history of enzymatic biotransformations

Enzymes were in use for thousands of years before their nature was gradually understood. No one really knows when the calf stomach was used as a catalyst for the first time in the manufacture of cheese.

As early as **1783**, Spallanzani showed that gastric juice secreted by cells could digest meat *in vitro*. In **1836**, Schwan called the active substance pepsin [26]. In **1876**, Kühne (figure 12) presented a paper to the Heidelberger Natur-Historischen und Medizinischen Verein, suggesting that such non-organized ferments should be called **e n z y m e s** [27]. At that time two terms were used: "organized ferment" such as cell-free yeast extract from Büchner, and "unorganized ferment" such as gastric juice secreted by cells. Today the terms "intracellular" and "extracellular" are used. Kühne also presented some interesting results from his experiments with trypsin. The word "enzyme" comes from Greek for "in yeast" or "leavened" [28].

Microorganisms synthesize numerous enzymes, each having its own function. **Intracellular** enzymes operate inside the cell in a protected and highly structured environment, while **extracellular** enzymes are secreted from the cell, thus working in the medium surrounding the microorganism.

The commercial usage of extracellular microbial enzymes started in the West around **1890**, thanks to the Japanese entrepreneur Takamine. He settled down in the United States and started an enzyme factory based on Japanese technology. The principal product was called takadiastase. This was a mixture of amylolytic and proteolytic enzymes prepared by cultivation of *Aspergillus oryzae*. In France, Boidin and Effront developed bacterial enzymes in **1913**. They found that the hay bacillus, *Bacillus subtilis*, produces an extremely heat-stable α-amylase when grown in still cultures on a liquid medium prepared by extraction of malt or grain [29].

In **1894**, Emil Fischer [30,31] observed in his studies of sugars that the enzyme called emulsin catalyzes the hydrolysis of β-methyl-D-glucoside, while the enzyme called maltase is active towards the α-methyl-D-glucoside as substrate (figure 13).

This led Fischer to suggest his famous "lock–and-key" theory of enzyme specificity, which he would describe in his own words as follows: "To use a picture, I would say that enzyme and the glucoside must fit into each other like a lock and key, in order to effect a chemical reaction on each other" [1].

In **1913**, Michaelis and Menten published a theoretical consideration of enzymatic catalysis. This consideration envisaged the formation of a specific enzyme-substrate complex which further decomposed and yielded the product with the release of the enzyme. This led to the development of the famous Michaelis-Menten equation to describe the typical saturation kinetics observed with purified enzymes and single substrate reactions [32].

By **1920**, about a dozen enzymes were known, none of which had been isolated [33]. Then, in **1926**, Sumner [34] crystallized urease from jack bean, *Canavalia ensiformis,* and announced that it was a simple protein.

Separat-Abdruck aus den Verhandlungen des Heidelb. Naturhist.-Med.
Vereins. N. S. I. 3. Verlag von Carl Winter's Universitätsbuchhandlung
in Heidelberg.

Ueber das Verhalten verschiedener organisirter
und sog. ungeformter Fermente.

Ueber das Trypsin (Enzym des Pankreas).

Von **W. Kühne.**

1876

Fig. 12 W. F. Kühne [27].

Ueber das Verhalten verschiedener organisirter und sog. ungeformter Fermente.

Sitzung am 4. Februar 1876.

Hr. W. Kühne berichtet über das Verhalten verschiedener organisirter und sog. ungeformter Fermente. Um Missverständnissen vorzubeugen und lästige Umschreibungen zu vermeiden schlägt Vortragender vor, die ungeformten oder nicht organisirten Fermente, deren Wirkung ohne Anwesenheit von Organismen und ausserhalb derselben erfolgen kann, als *Enzyme* zu bezeichnen. — Genauer untersucht wurde besonders das Eiweiss verdauende Enzym des Pankreas, für welches, da es zugleich Spaltung der Albuminkörper veranlasst, der Name *Trypsin* gewählt wurde. Das Trypsin vom Vortr. zuerst dargestellt und zwar frei von durch dasselbe noch verdaulichen und zersetzbaren Eiweissstoffen, verdaut nur in alkalischer, neutraler, oder sehr schwach sauer reagirender Lösung. Dasselbe wird durch nicht zu kleine Mengen Salicylsäure, welche das Enzym in bedeutenden Quantitäten löst, bei 40° C. gefällt, ohne dabei seine specifische Wirksamkeit zu verlieren. Wird die Fällung in Sodalösung von 1 pCt. gelöst, so verdaut sie höchst energisch unter Bildung von Pepton, Leucin, Tyrosin u. s. w. Nur übermässiger Zusatz von Salicylsäure bis zur Bildung eines dicken Krystallbreies vernichtet die enzymotischen Eigenschaften. Dies Verhalten war kaum zu erwarten, seit Kolbe und J. Müller die hemmende, selbst vernichtende Wirkung kleiner Mengen Salicylsäure auf einige Enzyme hervorgehoben hatten. Die Beobachtungen des Vortr., der ausser dem Trypsin noch das Pepsin eingehender untersuchte, stehen jedoch mit den Angaben von J. Müller, nach welchen Salicylsäure bei einem Gehalte der

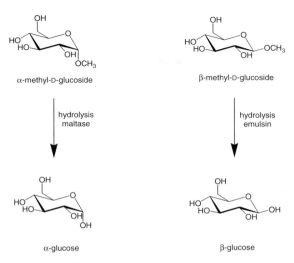

α-methyl-D-glucoside

β-methyl-D-glucoside

hydrolysis
maltase

hydrolysis
emulsin

α-glucose

β-glucose

Fig. 13 Emil Fischer' s substrates.

Northrop and his colleagues [26] soon supported Sumner's claim that an enzyme could be a simple protein. They isolated many proteolytic enzymes beginning with pepsin in **1930** by applying classical crystallization experiments. By the late **1940s** many enzymes were available in pure form and in sufficient quantity for investigation of their chemical structure. Currently, more than 3,000 enzymes have been catalogued [35]. The ENZYME data bank contains information related to the nomenclature of enzymes [36]. The current version contains 3,705 entries. It is available through the ExPASy WWW server (http://www.expasy.ch/). Several hundreds of enzymes can be obtained commercially [37].

In **1950**, there was still no evidence that a given protein had a unique amino acid sequence. Lysosyme was the first enzyme whose tertiary-structure (figure 14) was defined in **1966** with the help of X-ray crystallography [38].

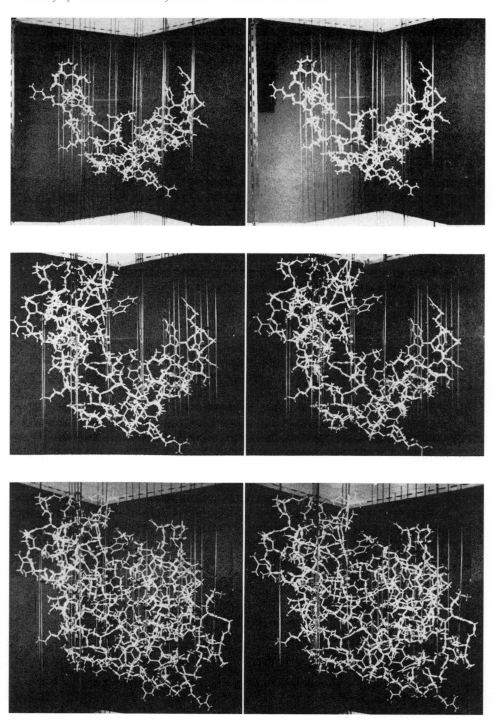

Fig. 14 Stereo photographs of models of part of the lysozyme molecule [38].

Further, ribonuclease A was one of the first enzymes made on a laboratory scale by organic chemistry methods. In **1969**, Gutte and Merrifield synthesized its whole sequence in 11,931 steps [39].

By **1970**, the complete molecular structures of several enzymes had been established and plausible reaction mechanisms could be discussed [26].

Hill (**1897**) was the first to show that the biocatalysis of hydrolytic enzymes is reversible [40].

Pottevin (**1906**) went further and demonstrated that crude pancreatic lipase could synthesize methyl oleate from methanol and oleic acid in a largely organic reaction mixture [41].

While the first benefit for the industry from the microbiological development had come early, the investigations with isolated enzymes hardly influenced the industry at that time. Consequently, industrial enzymatic biotransformations have a much shorter history than microbial biotransformations in the production of fine chemicals.

Invertase was probably the first immobilized enzyme to be used commercially for the production of Golden Syrup by Tate & Lyle during World War II, because sulfuric acid as the preferred reagent was unavailable at that time (figure 15) (42).

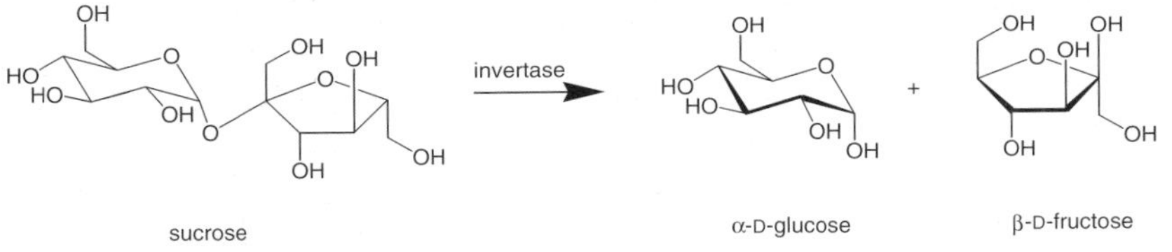

sucrose   α-D-glucose   β-D-fructose

Fig. 15 Inversion of sucrose by invertase.

Yeast cells were autolysed and the autolysate clarified by adjustment to pH 4.7, followed by filtration through a calcium sulphate bed and adsorption into bone char. A bone char layer containing invertase was incorporated into the bone char bed, which was already used for syrup decolorisation. The scale of operation was large, the bed of invertase-char being 60 cm deep in a 610 cm deep bed of char. The preparation was very stable since the limiting factor was microbial contamination or loss of decolorising power rather than the loss of enzymatic activity. The process was cost-effective but the product did not have the flavor quality of the acid-hydrolysed material. This is the reason why the immobilized enzyme was abandoned once the acid became available again [42].

Industrial processes for L-amino acid production based on the batch use of soluble aminoacylase were already in use in **1954**. However, like many batch processes with soluble enzymes, they had their disadvantages such as higher labor costs, complicated product separation, low yields, high enzyme costs and non-reusability of enzyme. During the **mid-1960s** the Tanabe Seiyaku Co. of Japan was trying to overcome these problems by using immobilized aminoacylases. In **1969**, they started the industrial production of L-methionine by aminoacylase immobilized on DEAE-Sephadex in a packed bed reactor (figure 16). This was the first full scale industrial use of an immobilized enzyme. The most important advantages are the relative simplicity and ease of control [44].

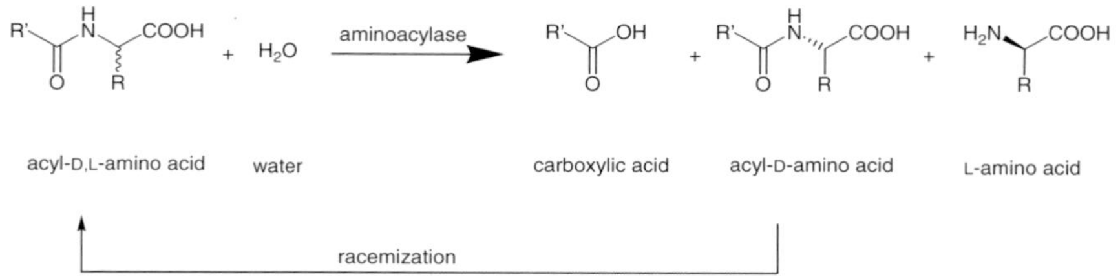

acyl-D,L-amino acid   water   carboxylic acid   acyl-D-amino acid   L-amino acid

racemization

Fig. 16 L-Amino acid production catalyzed by aminoacylase.

In a membrane reactor system developed at Degussa-Hüls AG in Germany in **1980** [45], native enzymes, either pure or of technical grade, are used in homogeneous solution for the large scale production of enantiomerically pure L-amino acids (figure 17).

Fig. 17 Enzyme membrane reactor (Degussa-Hüls AG, Germany).

A membrane reactor is particularly well suited for cofactor-dependent enzyme reactions, especially if the cofactor is regenerated by another enzyme reaction and retained by the membrane in modified form [46]. There are several advantages of carrying out biocatalysis in membrane reactors over heterogeneous enzymatic catalysis: there are no mass transfer limitations, enzyme deactivation can be compensated for by adding soluble enzyme and the reactors can be kept sterile more easily than immobilized enzyme systems. The product is mostly pyrogen free (major advantage for the production of pharmaceuticals), because the product stream passes through an ultrafiltration membrane. Scale-up of membrane reactors is simple because large units with increased surface area can be created by combining several modules.

The enzymatic isomerization of glucose to fructose (figure 18) represents the largest use of an immobilized enzyme in the manufacture of fine chemicals.

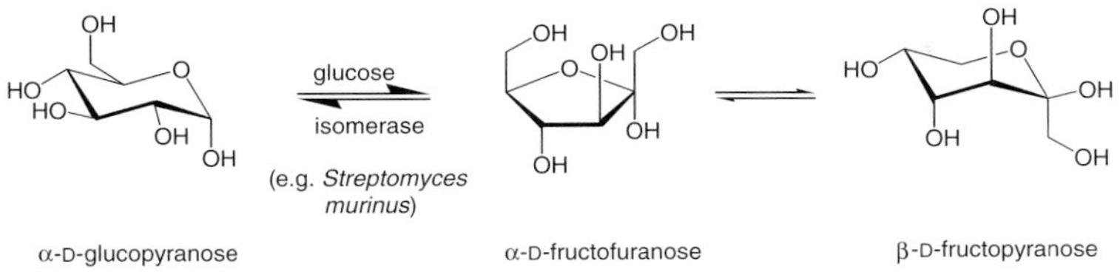

Fig. 18 Isomerization of glucose to fructose.

High-fructose corn syrup HFCS has grown to become a large-volume biotransformation product [47]. While sucrose is sweet, fructose is approximately 1.5 times sweeter and consequently high quality invert syrups (i.e. hydrolyzed sucrose) may be produced. Invert syrups contain glucose and fructose in a 1:1 ratio. However, the food industry needed a long time to become acquainted with the glucose isomerase potential to produce high quality fructose syrups from glucose. Again, the Japanese were the first to employ soluble glucose isomerase to produce high quality fructose syrups in **1966**. At the beginning of **1967**, Clinton Corn Processing Company, Iowa, USA, was the first company to manufacture enzymatically produced fructose corn syrup [47].The glucose-isomerase catalyzed reversible reaction gave a product containing about 42 % of fructose, 50 % of glucose and 8 % of other sugars. Due to various reasons, economic viability being the more important among them, the first commercial production of fructose syrups using glucose isomerase immobilized on a cellulose ion-exchange polymer in a packed bed reactor plant started only in **1974**. It was initiated by Clinton Corn Processing [44]. In **1976**, Kato was the first company in Japan to manufacture HFCS in a continuous process as opposed to a batch process. In **1984**, it became the first company to isolate crystalline fructose produced in this process by using an aqueous separation technique.

The glucose isomerase Sweetzyme T, produced by Novo, Denmark is used in the starch processing industry in the production of high fructose syrup. The key to its long life is immobilization. The enzyme is chemically bound to a carrier, making the particles too large to run out through the sieve at the bottom of the isomerization columns. Sweetzyme T is packed into columns where it is used to convert glucose into fructose. The record for the longest lifetime of a column is 687 days, held by a Japanese company called Kato Kagaku in Kohwa near Nagoya. The reaction conditions are pH 7.5 and T = 55 °C. Though enzyme activity is reduced at this temperature, its stability and productivity are considerably improved [48].

The engineers from Kato used to say: "The better the substrate you put in, the better the results you get out". Each column at Kato contains 1,800 kg of Sweetzyme T. The column needs to be changed when the flow rate decreases to about 10 % of the initial value. Sweetzyme T displays a linear decay curve under steady state operating conditions. With regard to productivity, the yield from the record-breaking column was 12,000 kg of fructose syrup (containing 42 % fructose) (dry substance)/ kg of Sweetzyme T. The normal column productivity was 8,000–10,000 kg / kg enzyme. The 687 days' record for Sweetzyme T is also a world record in the starch industry [48] (figure 19).

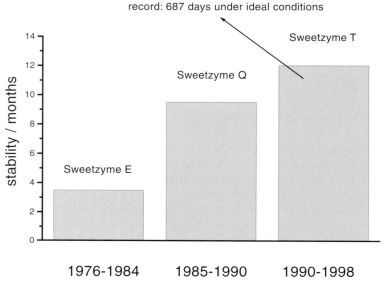

record: 687 days under ideal conditions

Fig. 19 Improved biocatalyst stability by biocatalyst engineering at Novo.

"Central del Latte" of Milan, Italy, was the first company which commercially hydrolyzed milk lactose with immobilized lactase using SNAMprogetti technology [49]. An industrial plant with a capacity of 10 tons per day is situated in Milan. The entrapped enzyme is lactase obtained from yeast and the reaction is performed batchwise at low temperature. Lactase hydrolyses lactose, a sugar with poor solubility properties and a relatively low degree of sweetness, to glucose and galactose (figure 20).

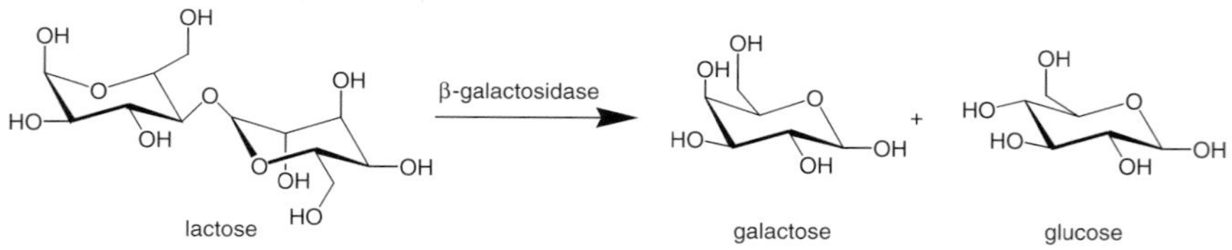

Fig. 20 β-Galactosidase catalyzed hydrolysis of lactose to galactose and glucose.

After the processed milk reaches the desired degree of hydrolysis of lactose, it is separated from the enzyme fibers, sterilized, and sent for packing and distribution. SNAMprogetti's process enables the manufacture of a high-quality dietary milk at low cost. This milk has a remarkable digestive tolerance, pleasant sweetness, unaltered organoleptic properties, and good shelf-life. It does not contain foreign matter. The industrial plant is shown in figure 21.

Fig. 21 Industrial plant for processing low – lactose milk [49].

Penicillin G, present in *Penicillum notatum* and discovered by Fleming in **1929**, revolutionized chemotherapy against pathogenic microorganisms. Today, β-lactam antibiotics such as penicillins and cephalosporins are very widely used. Thousands of semisynthetic β-lactam antibiotics are being synthesized to find more effective compounds. Most of these compounds are prepared from 6-aminopenicillanic acid (6-APA), 7-aminocephalosporanic acid (7-ACA) and 7-amino-desacetoxycephalosporanic acid (7-ADCA).

At present, 6-APA is mainly produced either by chemical deacylation or by enzymatic deacylation using penicillin amidase from penicillin G or V. This process, which exemplifies the best known usage of an immobilized enzyme in the pharmaceutical industry, is being used since around **1973** (figure 22). Several chemical steps are replaced by a single enzymatic reaction. Organic solvents, the use of low temperature (-40 °C) and the need for absolutely anhydrous conditions, which made the process difficult and expensive, were no longer necessary in the enzymatic process [50].

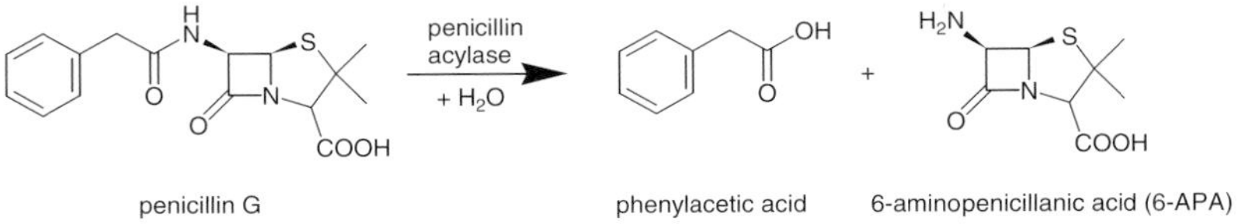

penicillin G   phenylacetic acid   6-aminopenicillanic acid (6-APA)

Fig. 22 Enzymatic synthesis of 6-aminopenicillanic acid (6-APA).

For many years enzymatic 7-ACA production was nothing but a dream. This changed in **1979**, when Toyo Jozo, Japan, in collaboration with Asahi Chemical Industry, also Japan, developed and succeeded in the industrial production of 7-ACA by a chemoenzymatic two-step process starting from cephalosporin C (figure 23):

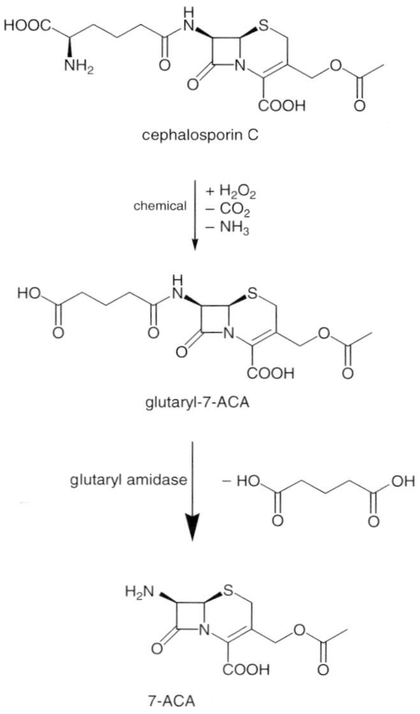

cephalosporin C

chemical | + H_2O_2 / – CO_2 / – NH_3

glutaryl-7-ACA

glutaryl amidase | – HO...OH

7-ACA

Fig. 23 Two-step process of 7-ACA production from cephalosporin C.

The chemical process requires highly purified cephalosporin C as raw material. A number of complicated reaction steps are carried out at –40 °C to –60 °C, and the reaction time is long. Furthermore, hazardous reagents, such as phosphorous pentachloride, nitrosyl chloride and pyridine are used in this process. The removal of such reagents causes significant problems. Therefore, the development of an enzymatic process was a dream for a long time. In the enzymatic process, liberated glutaric acid reduces the pH and inhibits the glutaryl-7-ACA amidase, the enzyme that catalyzes the deacylation of cephalosporin C. Because of this change in pH the reaction rate is decreased, requiring strict pH control during the reaction process. For these reasons, a recirculation bioreactor with immobilized glutaryl-7-ACA amidase and an automatic pH controller were designed for the 7-ACA production. The bioreactor for industrial 7-ACA production is shown in figures 24 and 25. The process has been in operation at Asahi Chemical Industry since **1973**. It is reported that about 90 tons of 7-ACA are thus produced annually [51].

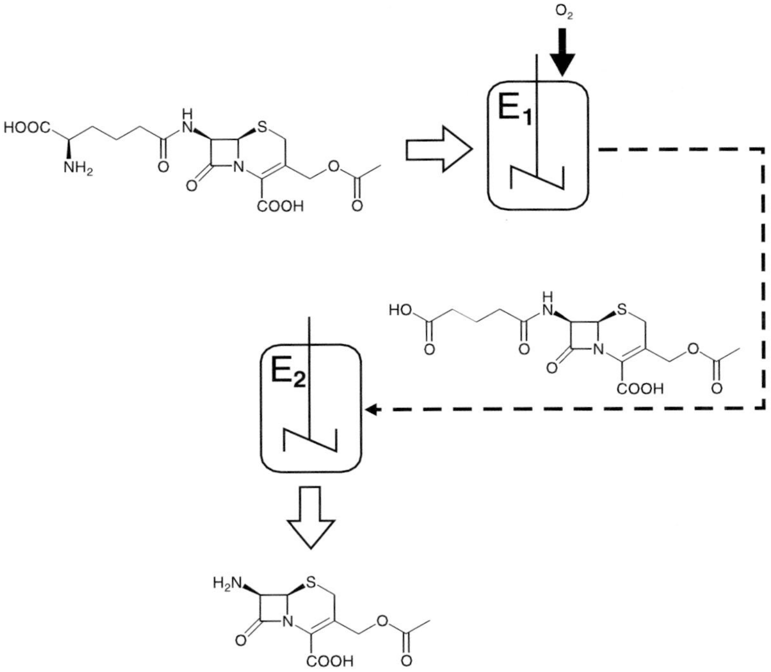

Fig. 24 Flow scheme for the production of 7-ACA. Production carried out at Asahi Chemical Industry. (E_1 = D-aminoacid oxidase; E_2 = glutaryl amidase).

Fig. 25 The bioreactor plant for 7-ACA production carried out at Asahi Chemical Industry (Reprinted from Ref. [51], p. 83 by courtesy of Marcel Dekker Inc.).

Four technological advances, having major impact on enzymatic biotransformations, were required for the acceptance of enzymes as 'alternative catalysts' in industry [52].

The first technological advance was the development of large-scale techniques for the release of enzymes from the interior of microorganisms [53]. Although the majority of industrial purification procedures are based on the same principles as those employed at laboratory scale, the factors under consideration while devising industrial scale purification regimes are somewhat different. When isolating enzymes on an industrial scale for commercial purposes, a prime consideration has to be the cost of production in relation to the value of the end product. Therefore, techniques used on a laboratory scale are not always suitable for large scale work [54]. Production and isolation of an intracellular microbial enzyme are quite expensive. The costs of the usage of water-soluble protein as catalyst for biotransformations can be justified only by its repeated use [55].

The second technological advance was the development of techniques for large-scale immobilization of enzymes. As mentioned earlier, the first enzyme immobilized in the laboratory was invertase, adsorbed onto charcoal in the year **1916** [56]. However, only after the development of immobilization techniques on a large scale occured in the **1960s**, many different industrial processes using immobilized biocatalysts have been established. The historical invertase column operating since **1968** on a laboratory scale is shown in figure 26.

Fig. 26 Historical invertase column [49].

It was shown that by increasing the concentration of sucrose, the efficiency of the fiber-entrapped invertase (which hydrolyses sucrose) can be increased. This occurred because the substrate, which is an inhibitor of the enzyme, could not reach high concentration levels inside the microcavities of the fibers owing to diffusion limitations [49].

Table II lists some industrial biotransformations performed by isolated enzymes.

Table II: Selected historical, industrial applications of isolated enzymes.

Product	Biocatalyst	Operating since	Company
L-amino acid	aminoacylase	1954, 1969	Tanabe Seiyaku Co. Ltd., Japan
6-aminopenicillanic acid	penicillin acylase	1973	SNAMProgetti and others[*]
low lactose milk	lactase	1977	Central del Latte, Milan, Italy (SNAMProgetti technology)
7-amino-cephalosporanic acid	D-amino acid oxidase	1979	Toyo Jozo and Asahi Chemical Industry, Japan

[*] Beecham, Squibb, Astra Lakenedal, Bayer, Gist-Brocades, Pfizer, Bristol Myers, Boehringer Mannheim, Biochemie, Novo, Hindustan Antibiotics

The first Enzyme Engineering Conference was held at Hennicker, New Hampshire, in **1971**. The term "immobilized enzymes" describing "enzymes physically confined at or localized in a certain region or space with retention of their catalytic activity and which can be used repeatedly and continuously" was adopted at this conference [57].

The third technological advance was the development of techniques for biocatalysis in organic media. The usage of very high proportions of organic solvents for increasing the solubility of reactants was examined in **1975** in the reaction with isolated cholesterol oxidase to produce cholestenone [58]. The enzymatic synthesis was believed to be incompatible with most organic syntheses carried in nonaqueous media. This changed after Klibanov [59] recognized in **1986** that most enzymes could function quite well in organic solvents. Since that time different processes involving an organic phase have been established in industry (Table III).

Table III: Industrial biotransformations involving poorly water-soluble reactants.

Process	Biocatalyst	Operating since	Company
fat interesterification	lipase	1979, 1983	Fuji Oil, Unilever
ester hydrolysis	lipase	1988	Sumitomo
transesterification	lipase	1990	Unilever
aspartame synthesis	thermolysin	1992	DSM
acylation	lipase	1996	BASF

The fourth and most recent technological advance is recombinant DNA technology. This technology is only now being widely used for biotransformations.

Generally, microorganisms isolated from nature produce the desired enzyme at levels which are too low to offer a cost-effective production process. Consequently, the modification of the organism would be highly desirable for process development. Currently, there are three principal approaches available for strain improvement. The first one, *direct evolution* [60], i.e. improvement by mutation and selection, has been successfully used in many industrial microbiological fields for many years. In **1978**, Clarke showed that evolution processes can be performed on a laboratory scale. Microevolution occuring in bacterial cultures grown in the chemostat gives rise to altered enzyme specificity, enabling microorganisms to degrade some unusual synthetic organic compounds. Successive mutational steps could be responsible for the evolution of new enzymatic specificities. The rate of production of existing enzymes and the expression of previously dormant genes are also typically affected by this event [61]. The second method is *hybridization*. It involves modification of the cellular genetic information by transference of DNA from another strain. The third method is *recombinant DNA technology*, whereby genetic information from one strain can be manipulated *in vitro* and then inserted into the same or another strain.

Recombinant DNA technology has dramatically changed enzyme production, because enzymes are synthesized in cells by the normal protein synthesis methods [62,63]. A 5–10 year period required for classical enzyme development can be reduced to 1–2 years. Protein engineering, in combination with recombinant expression systems, allows to plug in a new enzyme variant and to be very quick at manufacturing levels [64]. Novel microbial catalysts, together with recent advances in molecular biology, offer scientists an opportunity to rapidly evolve selected genes and considerably improve bacterial biocatalysts [65]. For example, a method for the rapid generation of thermostable enzyme variants has been developed [66,67]. This is done by introducing the gene coding for a given enzyme from a mesophilic organism into a thermophile. Variants that retain the enzymatic activity at higher growth temperatures of the thermophile are selected. This can be accomplished by constructing the artificial environment in which only the evolutionary adaptation of the enzyme can permit cell growth. This strategy can be readily extended to the general method of screening mutant enzymes. Another example is random mutation, developed as a method for highly efficient generation of mutant enzymes. The cloned gene coding for a given enzyme can be mutated either chemically or enzymatically *in vitro*. The mutant enzymes can be readily screened because mutant genes can be separated from intact genes. Various mutant enzymes have a change in properties, such as substrate specificity, thermal stability and coenzyme selectivity, have been isolated by this technique. These methods do not require predictive strategies, unlike, for example, site-directed mutagenesis. It is hoped that in course of time they will make enzymes excellent catalysts fulfilling all the requirements for industrial use. This research field may be referred to as **biocatalyst engineering** [23].

2.3 Advantages of biotransformations over classical chemistry

Enzymes are proteins, things of beauty and a joy forever [26]

Biocatalysis is a relatively green technology. Enzyme reactions can be carried out in water at ambient temperature and neutral pH, without the need for high pressure and extreme conditions, thereby saving process energy. Biocatalysis has proven to be a useful supplementary technology for the chemical industry, allowing, in some cases, reactions which are not easily conducted by classical organic chemistry or, in other cases, allowing reactions which can replace several chemical steps. Today, highly chemo-, regio- and stereoselective biotransformations can simplify manufacturing processes and make them even more economically attractive and environmentally acceptable [68].

Both new discoveries and incrementalism describe how the industrial enzyme business changed during **1996**. Enzymes have competed well with chemical methods for resolution but not with synthesis. Ibuprofen, phenylethylamine and acrylamide are commonly cited as compounds using enzyme-based chiral processes. There is also an unconfirmed suspicion that the fat substitute Olestra, because of some of its structural features, may require enzymatic steps for synthesis. The outlook for industrial enzymes is positive. The suppliers have extensive portfolios of promising new enzymes in their product pipelines. The range of customers considering the utilization of enzymes, as a replacement to conventional chemical methods, appears to be growing. New niche applications continue to be discovered in otherwise mature segments [69]. It appears that enzyme-based processes are gradually replacing conventional chemical-based methods. Finally, the latest literature on enzymology suggests that other biocatalysts will add to future sales, both in established and new markets. The enzyme "nitrogenase", converting dinitrogen to ammonia, a basic chemical compound, has been discovered recently [70]. Dream reactions of organic chemists might become true in the future, with biocatalysts where functional or chiral groups are introduced into molecules by utilizing H_2, O_2 or CO_2. Recently Aresta reported of a carboxylase enzyme that utilizes CO_2 in the synthesis of 4-hydroxybenzoic acid starting from phenyl-phosphate [71].

Although the production of D-amino acids is currently of great interest, there has been no known industrial manufacture of D-amino acids except for D-*p*-hydroxyphenylglycine and D-phenylglycine. Chemical methods are not suitable for large scale production of D-amino acids at the moment due to low yield and high cost. Most L-amino acids are efficiently manufactured by fermentation, but D-amino acids are hardly produced by fermentation, apart from a few exceptions, because it is difficult to obtain high optical purity and productivity. Enzymatic methods are most plausible for the industrial manufacture of D-amino acids with respect to optical purity and productivity. D-Amino acids such as D-*p*-hydroxy phenylglycine and D-phenylglycine are produced from D,L-hydantoins. From an industrial point of view, availability of cheap starting materials and the development of suitable biocatalysts are most important. The number of substrates that are available on an industrial scale is limited. Based on these criteria, synthetic intermediates of D,L-amino acids and L-amino acids produced by biotransformations would be the most important starting materials for the production of D-amino acids. The enzymatic production of D-amino acids is classified into three categories based on the starting materials [72]:

1. D,L-Amino acids (D-amino acylase)
2. Synthetic intermediates (D,L-hydantoin:D-hydantoinhydrolase; D,L-amino acid amides:D-amidase)
3. Prochiral substrates (α-keto acids, L-amino acids; D-transaminase and amino acid racemase)

The fed batch process [73] used in the production of L-DOPA, having a final product concentration of 110 g · L^{-1}, has many advantages over the classical chemical process, such as: a single reaction step, water as the only reaction by-product, no need for optical separation, shorter production cycle of three days, simple down-stream processing and process sustainability. L-DOPA is a metabolic precursor of dopamine, a very important drug in the treatment of Parkinsonism.

It is difficult to directly assess the true commercial value of biocatalysis, because the real value of the products made using the biocatalysts must be taken into account. Of course, its major advantage lies in stereoselective reactions. A good example of its technological power and commercial potential is the afore-mentioned stereoselective hydroxylation of steroids.

In comparison to fermentation processes fewer side-products are formed in enzymatic biotransformations, complex and expensive fermenters are not required, aeration, agitation and sterility need not necessarily be maintained and the substrate is not diverted into the formation of *de novo* cellular biomass [50]. Isolated biocatalysts are especially useful if the reaction they catalyze is about to be completed, if they are resistant to product inhibition, and if they are active in the presence of low concentrations of substrate (such as in detoxification reactions where pollutants are present in the waste stream). "One-pot" multi-enzyme reactions are much more feasible than a combined use of several chemical catalysts or reagents, especially as the latter often have to be used in reactors made of special resistant materials to tolerate extreme conditions, such as the use of concentrated acids under elevated temperatures and pressures [50].

References

1. Sheldon, R.A. (1993) Chirotechnology, pp.105, Marcel Dekker, New York.
2. Turner, M.K.(1998) Perspectives in biotransformations, in: Biotechnology, **Vol 8,** (Rehm, H.-J., Reed, G. eds.), Biotransfromations I (Kelly, D.R. ed.), pp 9, Wiley-VCH, Weinheim.
3. Mitchel, C.A. (1916) Vinegar: Its Manufacture and Examination, Griffin, London.
4. Mori, A. (1993) Vinegar production in a fluidized bed reactor with immobilized bacteria, in: Industrial Application of Immobilized Biocatalysts (Tanaka, A., Tosa, T., Kobayashi, T. eds.) pp. 291–313, Marcel Dekker, New York
5. Ebner, H., Sellmer, S., Follmann, H. (1996) Acetic acid, in: Biotechnology, (Rehm, H.J., Reed, G., Pühler, A., Stadler, P. eds.), **Vol 6:** Products of Primary Metabolism, (Roehr, M. ed.) pp.383, VCH, Weinheim
6. Pasteur, L. (1858) Mémoire sur la fermentation de l'acide tartrique, C. R. Acad. Sci. (Paris) **46**, 615–618
7. Pasteur, L. (1862) Suite a une précédente communication sur les mycodermes; Nouveau procéde industriel de fabrication du vinaigre, Compt. Rend. **55**, 28–32
8. Sebek, O.K. (1982) Notes on the historical development of microbial transformations, Microb. Transform. Bioact. Compd. **1**, 1–8
9. Buchner, E. (1897) Alkoholische Gärung ohne Hefezellen, Ber. Dtsch. Chem. Ges. **30 (**117), 1110–1113
10. Neuberg, C., Hirsch, J. (1921) Über ein Kohlenstoffketten knüpfendes Ferment (Carboligase), Biochem. Z. **115**, 282–310
11. Hildebrandt, G., Klavehn, W. (1930) Verfahren zur Herstellung von L-1-Phenyl-2-methylamino-propan-1-ol. Knoll A.-G. Chemische Fabriken in Ludwigshafen, Ger. Pat. DE 548 459
12. Kluyver, A.J., de Leeuw, F.J. (1924) *Acetobacter suboxydans,* een merkwaardige azijnbacterie, Tijdschr. Verg. Geneesk. **10** , 170
13. Reichstein, T., Grüssner, H. (1934) Eine ergiebige Synthese der L-Ascorbinsäure (Vitamin C) Helv. Chim. Acta **17**, 311–328
14. Peterson, D.H., Murray, H.C., Epstein, S.H., Reineke, L.M., Weintraub, A., Meister, P.D., Leigh, H.M. (1952) Microbiological oxygenation of steroids. I. Introduction of oxygen at carbon-11 of progesterone, J. Am. Chem. Soc. **74**, 5933–5936

15. Sarett, L.H. (1946) Partial synthesis of pregnene-4-triol-17(β),20(β),21-dione-3,11 and pregnene-4-diol-17(β),21-trione-3,11,20 monoacetate, J. Biol. Chem. **162**, 601–631

16. Sebek, O.K., Perlman D. (1979) Microbial transformation of steroids and sterols, in: Microbial Technology, 2nd ed. **Vol. 1**, pp 484–488, Academic Press, New York

17. Ensley, D.B., Ratzkin, J.B., Osslund, D.T., Simon, J.M., (1983) Expression of naphthalene oxidation genes in *Escherichia coli* results in the biosynthesis of indigo, Science **222**, 167–169

18. Wick, C.B. (1995) Genencor International takes a green route to blue dye, Gen. Eng. News, January 15 (**2**), 1,22

19. Anderson, S., Berman-Marks, C., Lazarus, R., Miller, J., Stafford, K., Seymour, J., Light, D., Rastetter, W., Estell, D., (1985) Production of 2-keto-L-gulonate, an intermediate in L-ascorbate synthesis, by genetically modified *Erwinia herbicola*, Science **230**, 144–149

20. McCoy, M., (1998) Chemical makers try biotech paths, Chem. Eng. **76** (25), 13–19

21. Motizuki, K. et al. (1966) Method for producing 2-keto-L-gulonic acid, US 3,234,105

22. Neidleman, S.L. (1980) Use of enzymes as catalysts for alkene oxide production, Hydrocarbon Proc. **59** (11), 135–138

23. Ikemi, M., (1994) Industrial chemicals: enzymatic transformation by recombinant microbes, Bioproc. Technol. **19**, 797–813

24. Nagasawa T., Yamada H., (1989) Microbial transformations of nitriles, TIBTECH **7**, 153–158.

25. Kobayashi, M., Nagasawa, T., Yamada, H. (1992) Enzymatic synthesis of acrylamide: a success story not yet over, TIBTECH, **10**, 402–408

26. Perham, R.N. (1976) The protein chemistry of enzymes, in "Enzymes: One Hundred Years" (Gutfreund, H. ed.) FEBS Lett. **62** Suppl. E20-E28

27. Kühne, W. (1876) Über das Verhalten verschiedener organisierter und sogenannter ungeformter Fermente. Über das Trypsin (Enzym des Pankreas), Verhandlungen des Heidelb. Naturhist.-Med. Vereins. N.S.I3, Verlag von Carl Winter's, Universitätsbuchhandlung in Heidelberg

28. Gutfreund, H. (1976) Wilhelm Friedrich Kühne; an appreciation. in: "Enzymes: One Hundred Years" (Gutfreund H. ed.) FEBS Lett. **62**, Suppl. E1-E12

29. Aunstrup, K. (1979) Production, isolation and economics of extracellular enzymes, in: Appl. Biochem. Bioeng., **Vol 2** Enzyme Technology, (Wingard, L.B., Katchalski-Katzir, E., Goldstein, L. eds.) pp. 27–69, Academic Press, New York

30. Fischer, E. (1894) Ber. Dtsch. Chem. Ges. **27**, 2895

31. Fischer, E. (1894) Ber. Dtsch. Chem. Ges. **27**, 3189

32. Michaelis, L., Menten, M.L. (1913) The kinetics of invertin action, Biochem. Z. **49**, 333–369

33. Schoffers, E., Golebiowski, A., Johnson C.R. (1996) Enantioselective synthesis through enzymatic asymmetrization, Tetrahedron **52**, 3769–3826

34. Sumner, J.B.(1926) The isolation and crystallization of the enzyme urease, J. Biol. Chem. **69**, 435–441

35. International Union Of Biochemistry and Molecular Biology (1992) Enzyme Nomenclature, Academic Press Inc., San Diego

36. Bairoch, A. (1999) The ENZYME data bank in 1999, Nucleic Acids Res. **27 (1), 310–311**

37. Data Index 4 (1996) Alphabetical listing of industrial enzymes and source, in: Industrial Enzymology, (Godfrey, T., West, S. eds.) 2ndEd. pp. 583–588, Stockton Press,

38. Phillips, D.C. (1967) The hen-egg-white lysozyme molecule, Proc. Natl. Acad. Sci. US. **57**, 484–495

39. Gutte, B., Merrifield, R.B. (1969) The total synthesis of an enzyme with ribonuclease A activity, J. Am. Chem. Soc. **91**, 501–502

40. Hill, A.C. (1897) Reversible zymohydrolysis, J. Chem. Soc. **73,** 634–658

41. Pottevin, H. (1906) Actions diastasiques réversibles. Formation et dédoublement des ethers-sels sous l'influence des diastases du pancréas. Ann. Inst. Pasteur **20**, 901–923

42. Cheetham, P.S.J. (1995) The applications of enzymes in industry, in: Handbook of Enzyme Biotechnology (Wiseman, A. ed.), pp 420, Ellis Harwood, London

43. Chaplin, M.F., Bucke, C. (1990) Enzyme Technology, pp. 190, Cambridge Univ. Press, Cambridge

44. Trevan, M.D., (1980) Immobilized Enzymes. An Introduction and Applications in Biotechnology, p.71, John Wiley, New York

45. Bommarius, A.S., Drauz, K., Groeger, U., Wandrey, C. (1992) Membrane bioreactors for the production of enantiomerically pure α-amino acids, in: Chirality in Industry (Collins, A.N., Sheldrake, G.N., Crosby, J. eds.), pp. 372–397, John Wiley, New York

46. Kragl, U., Vasic-Racki, D., Wandrey, C. (1993) Continuous processes with soluble enzymes, Ind. J. Chemistry, **32 B**, 103–117

47. Antrim, R.L., Colilla, W., Schnyder, B.J. (1979) Glucose isomerase production of high-fructose syrups, in: Appl. Biochem. Bioeng. ,**Vol 2** Enzyme Technology, (Wingard, L.B., Katchalski-Katzir, E., Goldstein, L. eds.) pp. 97–207, Academic Press, New York

48. (1998) 687 days is the record for Sweetzyme T, BioTimes, Novo Nordisk, 1/98.
49. Marconi, W., Morisi, F., (1979) Industrial application of fiber-entrapped enzymes, in: Appl. Bio-chem. Bioeng. **Vol 2** Enzyme Technology, (Wingard, L.B., Katchalski-Katzir, E., Goldstein, L. eds.) pp. 219–258, Academic Press, New York
50. Cheetham, P.S.J. (1995) The application of enzymes in industry, in: Handbook of Enzyme Bio-technology (Wiseman, A. ed.), pp. 493–498, Ellis Harwood, London
51. Matsumoto, K. (1993) Production of 6-APA, 7-ACA, and 7-ADCA by immobilized penicillin and cephalosporin amidases, in: Industrial Application of Immobilized Biocatalysts (Tanaka, A., Tosa, T., Kobayashi, T. eds.) pp. 67–88, Marcel Dekker, New York
52. Lilly, M.D.(1994) Advances in biotransformation processes, Chem. Eng. Sci. **49**, 151–159
53. Hetherington, P.J., Follows, M., Dunnill, P., Lilly, M.D. (1971) Release of protein from baker's yeast (*Saccharomyces cerevisiae*) by disruption in an industrial homogenizer, Trans. Inst. Chem. Eng. **49**, 142–148
54. Atkinson, T., Scawen, M.D., Hammond, P.M. (1987) Large scale industrial techniques of enzyme recovery, in: Biotechnology, (Rehm, H.J., Reed, G. eds), **Vol 7a** Enzyme Technology (Kennedy, J.F. ed.), pp. 279–323, VCH, Weinheim
55. Kennedy, J.F., Cabral, J.M.S. (1987) Enzyme immobilization, in: Biotechnology, (Rehm, H.J., Reed, G. eds), **Vol 7a** Enzyme Technology (Kennedy, J.F. ed.), pp. 347–404, VCH, Weinheim
56. Nelson, J.M., Griffin E.G. (1916) Adsorption of invertase, J. Am. Chem. Soc. **38**, 1109–1115
57. Powel, L.W. (1996) Immobilized enzymes, in: Industrial Enzymology, (Godfrey, T., West, S. eds.), 2nd Ed., p. 267, Stockton Press
58. Buckland, B.C., Dunnill, P., Lilly, M.D. (1975) The enzymatic transformation of water-insoluble reactants in nonaqueous solvents. Conversion of cholesterol to cholest-4-ene-3-one by a *Nocardia* sp., Biotechnol. Bioeng. **17**, 815–826
59. Klibanov, A.M. (1986) Enzymes that work in organic solvents, CHEMTECH **16**, 354–359.
60. Arnold, F.H., Morre, J.C. (1997) Optimizing industrial enzymes by directed evolution, in: New Enzymes for Organic Synthesis, Adv. Biochem. Eng. Biotechnol. **58**, pp. 2–14, Springer, Berlin
61. Borriss, R. (1987) Biotechnology of enzymes, in: Biotechnology **Vol 7a** (Rehm, H.J., Reed, G. eds.) Enzyme Technology (Kennedy, J.F. ed.) pp. 35–62, VCH, Weinheim
62. Gerhartz, W. (1990) Enzymes in industry, (Gerhartz, W. ed.) p. 11, VCH, Weinheim.
63. Clarke, P.H. (1976) Genes and enzymes, in: "Enzymes: One Hundred Years" (Gutfreund H. ed.) FEBS Lett. **62 Suppl.**, E37-E46
64. Hodgson, J. (1994) The changing bulk biocatalyst market, Bio/Technology **12** (August), 789–790
65. Wacket L.P. (1997) Bacterial biocatalysis: stealing a page from nature's book, Nature Biotechnol. **15** ,415–416
66. Matsumura, M., Aiba, S. (1985) Screening for thermostable mutant of kanamycin nucleotidyl-transferase by the use of a transformation system for a thermophile, *Bacillus stearothermophilus*, J. Biol. Chem. **260**, 15298–15303
67. Liao, H., McKenzie, T., Hageman, R. (1986) Isolation of a thermostable enzyme variant by clon-ing and selection in a thermophile, Proc. Natl. Acad.Sci. USA, **83**, 576–580
68. Petersen, M., Kiener, A. (1999) Biocatalysis. Preparation and functionalization of *N*-heterocycles. Green Chem. **1**, 99–106
69. Wrotnowski C. (1996) Unexpected niche applications for industrial enzymes drives market growth, Gen. Eng. News, February 1, 14–30
70. Rawls, R.L. (1998) Breaking up is hard to do, Chem. Eng. News **76** (25), 29–34
71. Aresta, M., Quaranta, E., Liberio, R., Dileo, C., Tommasi, I. (1998) Enzymatic synthesis of 4-OH-benzoic acid from phenol and CO_2: the first example of a biotechnological application of a carboxylase enzyme, Tetrahedron **54**, 8841–8846
72. Yagasaki, M.,Ozaki, A. (1998) Industrial biotransformations for the production of D-amino acids, J. Mol. Cat. B: Enzymatic **4**, 1–11
73. Enie, P., Nakazawa, H., Tsuchida, T., Namerikawa, T., Kumagai, H. (1996) Development of L-DOPA production by enzymic synthesis, Japan Bioindustry Letters, **13** (1), 2–4.

3 Enzyme Classification

CHRISTOPH HOH, MURILLO VILLELA FILHO

Institute of Biotechnology
Forschungszentrum Jülich GmbH
D-52425 Jülich, Germany

3.1 The Enzyme Nomenclature

In early times of biochemistry there were no guidelines for naming enzymes. The denomination of newly discovered enzymes was given arbitrarily by individual workers. This practice had proved to be inadequate. Occasionally two different enzymes had the same name while in other cases two different names were given to the same enzyme. Furthermore, there emerged denominations which provided no clue about the catalyzed reaction (e.g. catalases, or pH 5 enzyme).

With the great progress experienced by biochemistry in the 1950's, a large number of enzymes could be isolated and characterized. By this time it became evident that it was necessary to regulate the enzyme nomenclature. So, the International Union of Biochemistry and Molecular Biology (IUBMB), formerly International Union of Biochemistry (IUB), set up in consultation with the International Union for Pure and Applied Chemistry (IUPAC), an Enzyme Commission in charge of guiding the naming and establishing a systematic classification for enzymes. In 1961, the report of the commission was published. The proposed classification was used to name 712 enzymes. This work has been widely used as a guideline for enzyme nomenclature in scientific journals and textbooks ever since. It has been periodically updated, new entries have been included or old ones have been deleted, while some other enzymes have been reclassified. The sixth complete edition of the Enzyme Nomenclature (1992) contains 3196 enzymes [1]. Five supplements to the Enzyme Nomenclature with various additions and corrections have been published until today [2,3,4,5,6] signaling the constantly growing number of new enzyme entries. An update documentation of the classified enzymes is available on the ENZYME data bank server [7,8].

The Enzyme Nomenclature suggests two names for each enzyme, a **recommended name** convenient for every day use and a **systematic name** used to minimize ambiguity. Both names are based on the nature of the catalyzed reaction. The recommended name is often the former trivial name, sometimes after little change to prevent misinterpretation. The systematic name also includes the involved substrates. This taxonomy leads to the classification of enzymes into six main classes (Table 1).

Table 1: The main enzyme classes

Enzyme class	Catalyzed reaction
1. Oxidoreductases	oxidation-reduction reactions
2. Transferases	transfer of functional groups
3. Hydrolases	hydrolysis reactions
4. Lyases	group elimination (forming double bonds)
5. Isomerases	isomerization reactions
6. Ligases	bond formation coupled with a triphospate cleavage

As the systematic name may be very extensive and uncomfortable to use, the Enzyme Commission (EC) has also developed a numeric system based on the same criteria, which can be used together with the recommended name to specify the mentioned enzyme. According to this system, each enzyme is assigned a four-digit EC number (Table 2). The first digit denotes the main class that specifies the catalyzed reaction type. These are divided into subclasses, according to the nature of the substrate, the type of the transferred functional group or the nature of the specific bond involved in the catalyzed reaction. These subclasses are designated by the second digit. The third digit reflects a further division of the subclasses according to the substrate or co-substrate, giving origin to sub-sub-classes. In the fourth digit a serial number is used to complete the enzyme identification.

Table 2: Constitution of the four-digit EC number

EC number EC (i).(ii).(iii).(iv)	
(i)	the main class, denotes the type of catalyzed reaction
(ii)	sub-class, indicates the substrate type, the type of transferred functional group or the nature of one specific bond involved in the catalyzed reaction
(iii)	sub-subclass, expresses the nature of substrate or co-substrate
(iv)	an arbitrary serial number

As an example, aminoacylase (*N*-acyl-L-amino-acid amidohydrolase, according to the systematic nomenclature), an enzyme used in the industrial production of L-methionine, has the classification number EC 3.5.1.14 (see process on page 300). The first number (i = 3) indicates that this enzyme belongs to the class of hydrolases. The second number (ii = 5) expresses that a carbon-nitrogen bond is hydrolyzed and the third number (iii = 1) denotes that the substrate is a linear amide. The serial number (iv = 14) is needed for full classification of the enzyme.

As the biological source of an enzyme is not included in its classification, it is important to mention this together with the enzyme number for full identification. So the enzyme used in the production of "acrylamide" should be mentioned as "nitrilase (EC 4.2.1.84) from "*Rhodococcus rhodochrous*" (see process on page 362).

An important aspect concerning the application of the enzyme nomenclature is the direction how a catalyzed reaction is written for purposes of classification. To make the classification more transparent the direction should be the same for all enzymes of a given class, even if this direction has not been demonstrated for all enzymes of this class. Many examples for the use of this convention can be found in the class of oxidoreductases.

A further implication of this system is the impossibility of full classification of an enzyme if the catalyzed reaction is not clear. Complete classification of the enzymes only depends on the natural substrates. Non-natural substrates are not considered for the classification of the biocatalyst.

Finally, it is important to emphasize that the advantageous influence of the enzyme classification is not limited to biochemistry's enzyme nomenclature. It is also very beneficial for organic preparative chemists because it facilitates the choice of enzymes for synthetic applications. Since the classification of the enzymes is based on the catalyzed reactions it helps chemists to find an appropriate biocatalyst for a given synthetic task. An analogous nomenclature for chemical catalysts has not been set up until today.

The number of existing enzymes in nature is estimated to reach the 25,000 mark [9]. It is one essential part of biochemistry and related sciences to try to find and identify them. The scientist isolating and characterizing a new enzyme is free to report the discovery of that "new" biocatalyst to the Nomenclature Committee of the IUBMB and may form a new systematic name for this enzyme. An appropriate form to draw the attention of the editor of the Enzyme Nomenclature to enzymes and other catalytic entities missing from this list is available online [10].

3.2 The Enzyme Classes

The following part of this chapter aims at giving a compact overview of the six main enzyme classes and their subclasses. Since the industrial bioprocesses and biotransformations illustrated in the following chapters of the book are divided according to the involved enzymes and their classes, this short survey should provide the reader with the most important information on the enzyme classes.

The six main enzyme classes are resumed separately by giving a general reaction equation for every enzyme subclass according to the Enzyme Nomenclature. The reaction equations are picturized in a very general manner pointing out just the most important attributes of the catalyzed reactions. The authors would like to emphasize that no attempt has been made to provide a complete summary of the reactions catalyzed by the enzymes listed in the Enzyme Nomenclature. The reaction schemes have been elaborated to give reaction equations being as general and clear as possible and as detailed as necessary.

An important point that needs to be considered in this context concerns the enzymes classified as EC (i).99 or EC (i).(ii).99. These enzymes are either very substrate specific and therefore cannot be classified in already existing enzyme subclasses (or sub-subclasses) or a substrate of these enzymes has not been completely identified yet.

For instance, in the enzyme main class EC 5.(ii).(iii).(iv) (isomerases), the EC number 5.99 only describes "other isomerases" that cannot be classified within the other existing subclasses EC 5.1 to EC 5.5. It is important to point out that the enzymes classified with a 99-digit have not been considered in the reaction equations unless stated explicitly. The catalyzed reactions of these enzymes differ exceedingly from those of the other enzymes in the same main division.

The following short remarks on the generalized reaction schemes should help the reader to understand the illustrated enzyme catalyzed reactions:

1. Each main enzyme class is introduced by a short paragraph giving a general idea of the respective enzymes.
2. By generalizing nearly all catalyzed reactions of one enzyme subclass to only one or a few reaction equations, some details of the single reactions had to be neglected, e.g. specification of the cofactor, reaction conditions (pH, temperature), electric charge or stoichiometry. Correct protonation of the substrates and products depending on the pH value of each reaction mixture has not been taken into consideration as well. Also, the enzyme itself does not appear in the reaction schemes of this chapter.
3. If the catalyzed reaction leads to a defined equilibrium, only one direction of this reaction is considered according to its direction in the Enzyme Nomenclature. In consequence, no equilibrium arrows are used in any reaction scheme of this chapter.
4. Enzymes of a given subclass may show some frequently appearing common properties or some very worthwhile uniqueness. These qualities are taken into account by additional comments below the reaction schemes.

EC 1 Oxidoreductases

The enzymes of this first main division catalyze oxidoreduction reactions, which means that all these enzymes act on substrates through the transfer of electrons. In the majority of the cases the substrate that is oxidized is regarded as hydrogen donor. Various cofactors or coenzymes serve as acceptor molecules. The systematic name is based on *donor:acceptor oxidoreductase*.

Whenever possible the nomination as a *dehydrogenase* is recommended. Alternatively, the term *reductase* can be used. If molecular oxygen (O_2) is the acceptor, the enzymes may be named as *oxidases*.

EC 1.1 Acting on CH-OH group of donors

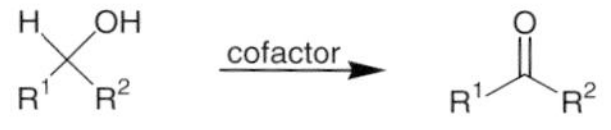

R^1 = hydrogen, organic residue
R^2 = hydrogen, organic residue, alcoxy residue

The sub-subclasses are defined by the type of cofactor.

EC 1.2 Acting on aldehyde or oxo group of donors

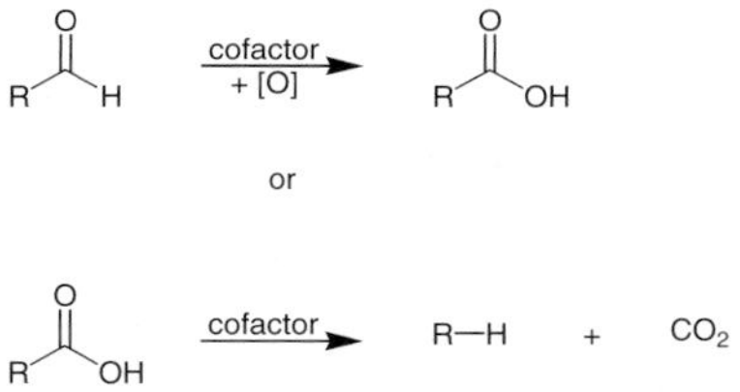

or

R = hydrogen, organic residue

Analogous with the first depicted reaction, the aldehyde can be oxidized to the respective thioester with coenzyme A (CoA). In the case of oxidation of carboxylic acids, the organic product is not necessarily bound to hydrogen as suggested in the figure. It can also be bound to the cofactor. The sub-subclasses are classified according to the cofactor.

EC 1.3 Acting on the CH-CH group of donors

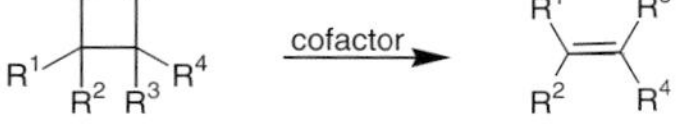

R^{1,2,3,4} = hydrogen, organic residue

In some cases the residues can also contain heteroatoms, e.g. dehydrogenation of *trans*-1,2-dihydroxycyclohexa-3,5-diene to 1,2-dihydroxybenzene (catechol). Further classification is based on the cofactor.

EC 1.4 Acting on the CH-NH$_2$ group of donors

R^{1,2} = hydrogen, organic residue

In most cases the imine formed is hydrolyzed to give an oxo-group and ammonia (deaminating). The division into sub-subclasses depends on the cofactor.

EC 1.5 Acting on the CH-NH group of donors

R^{1,2} = hydrogen, organic residue
R^3 = organic residue

In some cases the primary product of the enzymatic reaction may be hydrolyzed. Further classification is based on the cofactors.

EC 1.6 Acting on NAD(P)H

A = acceptor

Generally enzymes that use NAD(P)H as reducing agent are classified according to the substrate of the reverse reaction. Only enzymes which need some other redox carrier as acceptors to oxidize NAD(P)H are classified in this subclass. Further division depends on the redox carrier used.

EC 1.7 Acting on other nitrogen compounds as donors

$$N_{red}R_3 \xrightarrow{\text{cofactor}} N_{ox}R_3$$

R = hydrogen, organic residue, oxygen

The enzymes that catalyze the oxidation of ammonia to nitrite and the oxidation of nitrite to nitrate belong to this subclass. The subdivision is based on the cofactor.

EC 1.8 Acting on sulfur group of donors

$$\text{(S)}_{red} \xrightarrow{\text{cofactor}} \text{(S)}_{ox}$$

(S)$_{red}$ = sulfide, sulfite, thiosulfate, thiol, etc.

(S)$_{ox}$ = sulfite, sulfate, tetrathionate, disulfite, etc.

The substrates may be either organic or inorganic sulfur compounds. The nature of the cofactor defines the further classification.

EC 1.9 Acting on a heme group of donors

$$\boxed{\text{heme}}\!-\!Fe^{2+} \xrightarrow{\text{cofactor}} \boxed{\text{heme}}\!-\!Fe^{3+}$$

The sub-subclasses depend again on the cofactor.

EC 1.10 Acting on diphenols and related substances as donors

X_d = OH, NH$_2$
X_a = O, NH

The aromatic ring may be substituted; ascorbates are also substrates for this subclass. The primary product may undergo further reaction. The subdivision in four sub-subclasses depends on the cofactor.

EC 1 Oxidoreductases

EC 1.11 Acting on a peroxide as acceptor

$$H_2O_2 + D_{red} \longrightarrow H_2O + D_{ox}$$

D = donor

The single sub-subclass contains the peroxidases.

EC 1.12 Acting on hydrogen as donor

$$H_2 + A^+ \longrightarrow H^+ + A{-}H$$

Sub-subclass 1.12.1 contains enzymes using NAD^+ and $NADP^+$ as cofactors. Other hydrogenases are classified under 1.12.99. Enzymes using iron-sulfur compounds as cofactor are listed under 1.18.

EC 1.13 Acting on single donors with incorporation of molecular oxygen

$$A + O_2 \longrightarrow AO_{(2)}$$

If two oxygen atoms are incorporated, the enzyme belongs to the sub-subclass 1.13.11 and if only one atom of oxygen is used the enzyme is classified as 1.13.12. All other cases are classified under 1.13.99.

EC 1.14 Acting on paired donors with incorporation of molecular oxygen

$$A + O_2 \xrightarrow{\text{cofactor}} AO_{(2)}$$

The classification into sub-subclasses depends on whether both oxygen atoms or just one is bonded to the substrate. The difference to subclass 1.13 is the requirement of a cofactor.

EC 1.15 Acting on superoxide radicals as acceptor

$$O_2^{\bullet-} + O_2^{\bullet-} + H^+ \longrightarrow \tfrac{3}{2}O_2 + H_2O$$

The only enzyme classified under this subclass is superoxide dismutase.

38

EC 1.16 Oxidizing metal ions

$m \geq 0$
$n > m$

The two sub-subclasses are divided according to the cofactor.

EC 1.17 Acting on CH$_2$ groups

The origin of the oxidizing oxygen is either molecular oxygen or water.

EC 1.18 Acting on reduced ferredoxin as donor

ferredoxin$_{red}$ $\xrightarrow{\text{cofactor}}$ ferredoxin$_{ox}$

EC 1.19 With dinitrogen as acceptor

N_2 $\xrightarrow{\text{cofactor}}$ NH_3

The only enzyme classified under this subclass is nitrogenase.

EC 2 Transferases

The transferases are enzymes that transfer a chemical group from one compound (generally regarded as the donor) to another compound (generally regarded as the acceptor). Of all biological reactions, this class of biocatalysts is one of the most common [11]. To avoid any confusion, the following reaction schemes of the subclasses all show the same pattern: the donor molecule is always the first one among the substrates, the acceptor is the second one. If possible, some detailed information is given on the acceptor, but also a general denomination as A = acceptor has been chosen in three cases.

In general, the systematic names of these biocatalysts are formed according to the scheme *donor:acceptor* group*transferase*. In many cases, the donor is a cofactor (coenzyme) carrying the often activated chemical group to be transferred.

EC 2.1 Transferring one-carbon groups

A = acceptor
R = organic residue
Ⓒ = methyl-, hydroxymethyl-, formyl-, carboxyl-, carbamoyl- and amidino-groups

EC 2.2 Transferring aldehyde or ketone residues

R^1 = hydrogen or methyl residue
R^2 = methyl residue or polyol chain
R^3 = hydrogen or polyol chain

Three of the only four enzymes in this class depend on thiamin-diphosphate as a cofactor. The catalyzed reactions may be regarded as an aldol addition. Some enzymes also accept hydroxypyruvate as a donor to form CO_2 and the resulting addition product.

EC 2.3 Acyltransferases

X^1 = S, O, NH
X^2 = S, O, NH, CH$_2$
R^1 = hydrogen, alkyl-, aryl- or monophosphate residue
R^2 = hydrogen, alkyl- or aryl-residue
R^3 = alkyl-, aryl-, acyl- or monophosphate residue, aryl-NH

Transferred acyl-groups are often activated as coenzyme A (CoA) conjugates.

EC 2.4 Glycosyltransferases

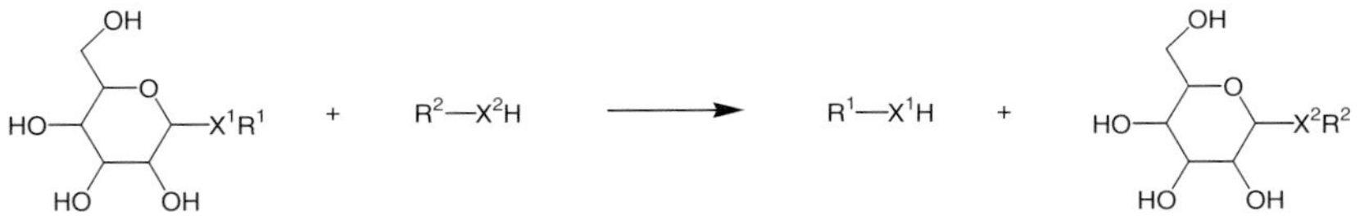

X^1 = O, PO$_4$$^{3-}$
X^2 = O, NH
R^1 = hydrogen, hexosyl, pentosyl, oligosaccharide, monophosphate
R^2 = hexosyl, pentosyl, oligosachharide, monophosphate, organic residue with OH- or NH$_2$-groups
X^1R^1 = nucleoside di- or monophosphates (e.g. UDP, ADP, GDP or CMP), purine

This enzyme class is subdivided into the hexosyl- (sub-subclass 2.4.1) and pentosyltransferases (sub-subclass 2.4.2). Although illustrating a hexosyl transfer in the figure, this general scheme is meant to describe both the enzyme sub-subclasses.

EC 2.5 Transferring alkyl or aryl groups, other than methyl groups

A = acceptor
X = OH, NH, SR, SO$_4$$^-$, mono-, di- or triphosphate
R = organic residue other than a methyl group

41

EC 2.6 Transferring nitrogenous groups

If NX = NH$_2$, then --- is a single bond.
If NX = NOH, then --- is a double bond.
R^1 = hydrogen, carboxy or methyl residue
R^2 = organic residue
R^3 = hydrogen, carboxy or hydroxymethyl residue
R^4 = organic residue

Pyridoxal-phosphate is the most frequently-appearing cofactor for these enzymes. For NX = NH$_2$ the substrates are often α-amino acids and 2-oxo acids.

EC 2.7 Transferring phosphorous-containing groups

or

X = OH, COOH, NH$_2$, PO$_4^{2-}$
R^1 = hydrogen, NDP, NMP, adenosine, monosaccharide residue, acyl residue, polyphosphate, histidine,
    syn-glycerol, organic residues carrying more functional groups
R^2 = hydrogen, monosaccharide residue, nucleosides, nucleotides, organic residues carrying more functional
    groups, proteins, polyphosphate
(P) = mono- or diphosphates

The enzymes transferring a phosphate residue from an ATP molecule to an acceptor are called kinases. The enzyme EC 2.7.2.2 (carbamate kinase) transfers a phosphate residue from an ATP molecule on CO$_2$ and NH$_3$ to form carbamoyl phosphate.

EC 2.8 Transferring sulfur-containing groups

R—(S) + A ⟶ R + A—(S)

A = acceptor, e.g. cyanide, phenols, alcohols, carboxylic acids, amino acids, amines, saccharides
R = sulfur atom, (phosphorous-) organic residue
(S) = sulfur atom, SO_3^{2-}, SH, CoA

EC 2.9 Transferring selenium-containing groups

The only enzyme classified under this subclass is L-seryl-tRNA (Sec) selenium transferase.

EC 3 Hydrolases

This third main class of enzymes plays the most important role in today's enzymatic industrial processes. It is estimated that approximately 80 % of all industrial enzymes are members of this enzyme class [12]. Hydrolases catalyze the hydrolytic cleavage of C-O, C-N, C-C and some other bonds, including P-O bonds in phosphates. The applications of these enzymes are very diverse: the most well-known examples are the hydrolysis of polysaccharides (see process on page 231), nitriles (see process on page 361 and 362), proteins or the esterification of fatty acids (see process on page 217). Most of these industrial enzymes are used in processing-type reactions to degrade proteins, carbohydrates and lipids in detergent formulations and in the food industry.

Interestingly, all hydrolytic enzymes could also be classified as transferases, since every hydrolysis reaction can be regarded as the transfer of a specific chemical group to a water molecule. But, because of the ubiquity and importance of water in natural processes, these biocatalysts are classified as hydrolases rather than as transferases.

The term *hydrolase* is included in every systematic name. The recommendation for the naming of these enzymes is the formation of a name which includes the name of the substrate and the suffix *–ase*. It is understood that the name of the substrate with this suffix means a hydrolytic enzyme.

EC 3.1 Acting on ester bonds

The nature of the substrate may differ largely, as shown in the three examples.

EC 3.1.1 Carboxylic ester hydrolase

R^1 = hydrogen, organic residue
R^2 = organic residue

EC 3.1.2 Thiolester hydrolase

R^1 = hydrogen, organic residue
R^2 = organic residue

EC 3.1.3 Phosphohydrolase ("phosphatase")

(P) = monophosphate
R = organic residue

EC 3.2 Glycosidases

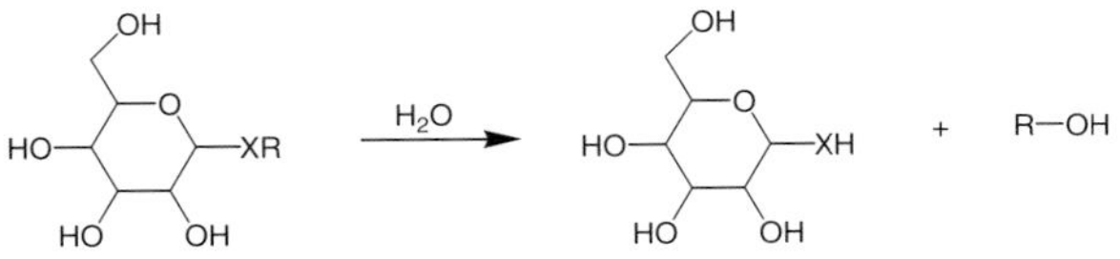

X = O, N or S
R = organic residue

The illustration shows the hydrolysis of a hexose derivative although pentose derivatives are also accepted as substrates.

EC 3.3 Acting on ether bonds

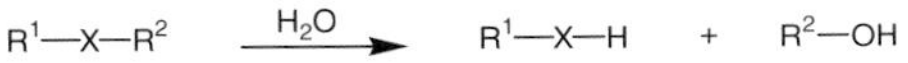

R^1—X—R^2 $\xrightarrow{H_2O}$ R^1—X—H + R^2—OH

X = O or S
$R^{1,2}$ = organic residue

EC 3.4 Acting on peptide bonds

$R^{1,2}$ = part of amino acids or proteins

EC 3.5 Acting on carbon-nitrogen bonds other than peptide bonds

R = organic residue

45

For some nitriles a similar reaction takes place. The enzyme involved is called nitrilase (EC 3.5.5).

$$R—C{\equiv}N \xrightarrow{\ H_2O\ } \underset{R}{\overset{O}{\|}}{C}—OH \quad + \quad NH_3$$

R = aromatic, heterocyclic and certain unsaturated aliphatic residues

EC 3.6 Acting on acid anhydrides

$$R—\overset{\overset{O}{\|}}{\underset{\underset{OH}{|}}{P}}—O—A \xrightarrow{\ H_2O\ } R—\overset{\overset{O}{\|}}{\underset{\underset{OH}{|}}{P}}—OH \quad + \quad HO—A$$

A = phosphate, organic phosphate, sulfate
R = organic residue, hydroxy group

EC 3.7 Acting on carbon-carbon bonds

$$R^1{\overset{O}{\|}}{C}—CH_2—{\overset{O}{\|}}{C}R^2 \xrightarrow{\ H_2O\ } R^1{\overset{O}{\|}}{C}—OH \quad + \quad CH_3{\overset{O}{\|}}{C}R^2$$

R^{1,2} = organic residue, hydroxy group

There is only one sub-subclass.

EC 3.8 Acting on halide bonds

$$R_3C—X \xrightarrow{\ H_2O\ } R_3C—OH \quad + \quad HX$$

X = halogen
R = hydrogen, organic residue, hydroxy group

EC 3.9 Acting on phosphorous-nitrogen bonds

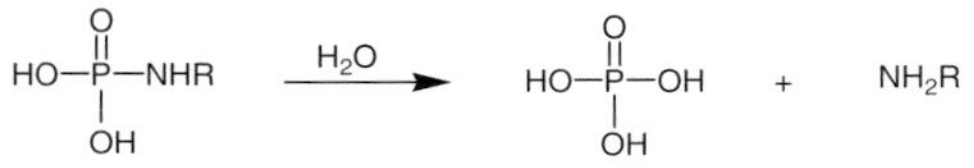

R = organic residue

The only enzyme classified under this subclass is phosphoamidase.

EC 3.10 Acting on sulfur-nitrogen bonds

$\circledS$ = sulfon group
R = organic residue

There is only one subdivision of this subclass.

EC 3.11 Acting on carbon-phosphorous bonds

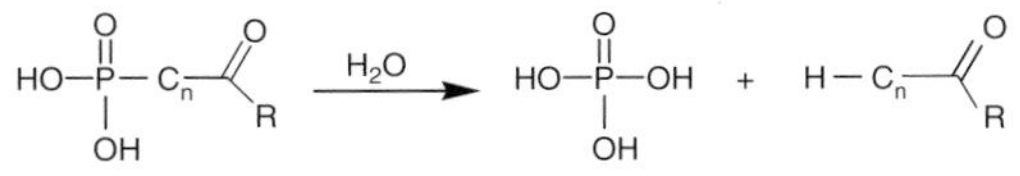

R = CH_3, OH
n = 0, 1

If n = 0, the product is an aldehyde.

EC 3.12 Acting on sulfur-sulfur bonds

(S_1) = sulfate

(S_2) = thiosulfate

The only enzyme classified under this subclass is trithionate hydrolase.

EC 4 Lyases

From the commercial perspective, these enzymes are an attractive group of catalysts as demonstrated by their use in many industrial processes (see chapter 5). The reactions catalyzed are the cleavage of C-C, C-O, C-N and some other bonds. It is important to mention that this bond cleavage is different from hydrolysis, often leaving unsaturated products with double bonds that may be subject to further reactions. In industrial processes these enzymes are most commonly used in the synthetic mode, meaning that the reverse reaction – addition of a molecule to an unsaturated substrate – is of interest. To shift equilibrium these reactions are conducted at very high substrate concentrations which results in very high conversions to the desired products. For instance, a specific type of lyase, the *phenylalanine ammonia lyase* (EC 4.1.99.2), catalyzes the formation of an asymmetric C-N bond yielding the L-amino acid dihydroxy-L-phenylalanine (L-DOPA). This amino acid is produced on a ton scale and with very high optical purities (see process on page 342).

Systematic denomination of these enzymes should follow the pattern *substrate group-lyase*. The hyphen should not be omitted to avoid any confusion, e.g. the term *hydro-lyase* should be used instead of *hydrolyase*, which looks quite the same as a *hydrolase*.

In the recommended names, terms like *decarboxylase*, *aldolase* or *dehydratase* (describing the elimination of CO_2, an aldehyde or water) are used. If the reverse reaction is much more important, or the only one known, the term *synthase* may be used.

EC 4.1 Carbon-carbon lyases

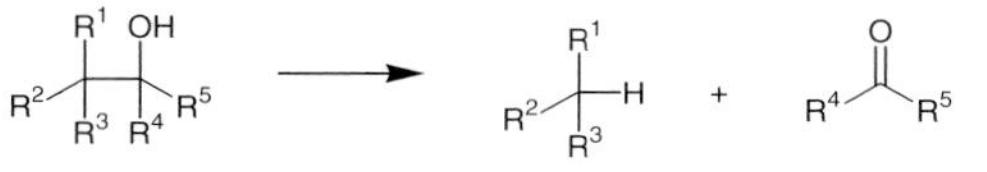

$R^{1,2,3,4,5}$ = hydrogen, organic residue

If the substrate is a carboxylic acid, one of the products will be carbon dioxide. If the substrate is an aldehyde, carbon monoxide may be a product.

EC 4.2 Carbon-oxygen lyases

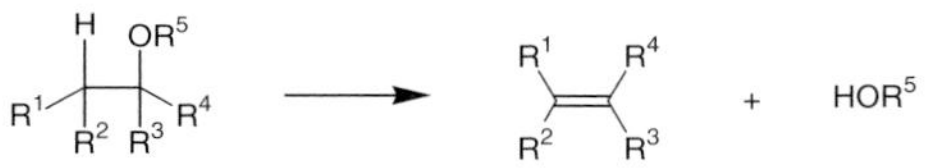

$R^{1,2,3,4,5}$ = hydrogen, organic residue

A further addition of water to the product may lead to an oxo acid. This is the case for some amino acids, where ammonia is then eliminated.

EC 4.3 Carbon-nitrogen lyases

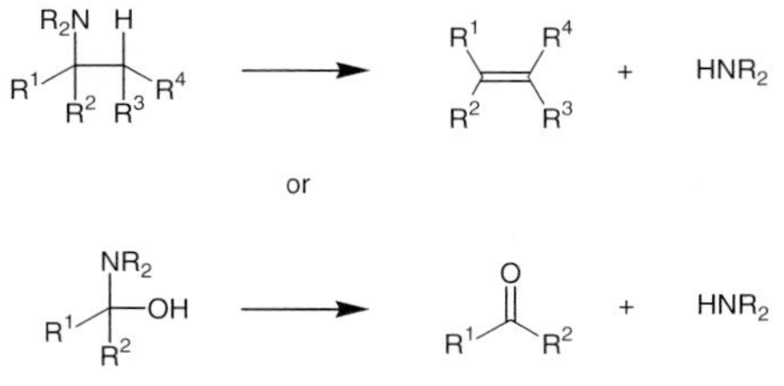

or

R = organic residue

The resulting double bond may change its position in order to deliver a more stable product, for instance in the case of keto-enol tautomerism. The product may also undergo a further reaction.

49

EC 4.4 Carbon-sulfur lyases

Ⓢ = SH, (di)substituted sulfide, sulfur-oxide, SeH
R = organic residue

According to the Enzyme Nomenclature the carbon-selenium lyase also be-longs to this subclass. Similar to other lyases, further reactions may occur on the product. In the case of disubstituted sulfides, there is no hydrogen bonded to the sulfur in the product.

EC 4.5 Carbon-halide lyases

X = halogen
R = organic residue

The primary product may also undergo further reaction. In the case of dihalo-substituted methane the sequential reaction will lead to the aldehyde. Amino compounds may react under elimination of ammonia to oxo compounds. If thio-glycolate is a cofactor, a sulfur-carbon bond will replace the halogen-carbon one.

EC 4.6 Phosphorous-oxygen lyase

or

Ⓟ = monophosphate
R = organic residue

With the exception of EC 4.6.1.4 all enzymes of this subclass lead to cyclic products.

EC 5 Isomerases

This enzyme class only represents a small number of enzymes, but nevertheless one of them plays a major role in todays industry. This enzyme, known as *glucose isomerase* (EC 5.3.1.5), catalyzes the conversion of D-glucose to D-fructose which is necessary in the production of high-fructose corn syrup (HFCS) (see process on page 387). This syrup is a substitute for sucrose and is used by the food and beverage industries as a natural sweetener.

In general, the isomerases catalyze geometric or structural changes within one single molecule. Depending on the type of isomerism, these enzymes may be called as *epimerases*, *racemases*, cis-trans-*isomerases*, *tautomerases* or *mutases*.

EC 5.1 Racemases and epimerases

$X = NH_2, NHR, NR_2, OH, CH_3, COOH$
$R^{1,2}$ = organic residue

EC 5.2 *cis-trans*-Isomerases

$X = C$ or N
$R^{1,2,3,4}$ = organic residue

If $X = N$ the substrate is an oxime. In this case R^4 represents the single electron pair.

EC 5.3 Intramolecular oxidoreductases

General scheme for the subclasses 5.3.1–5.3.4

$$R^1—X_{Ox.}—Z_{Red.}—R^2 \longrightarrow R^1—X_{Red.}—Z_{Ox.}—R^2$$

$R^{1,2}$ = organic residue

General scheme for the subclass 5.3.99

$$R^1—X_{Ox.}(Y)_n—Z_{Red.}—R^2 \longrightarrow R^1—X_{Red.}(Y)_n—Z_{Ox.}—R^2$$

n = 0,1,2
$R^{1,2}$ = organic residue

For these enzymes the centers of oxidation and reduction in the substrate need not to be adjacent.

To avoid misunderstandings the sub-subclasses 5.3.1–5.3.4 are presented separately.

EC 5.3.1 Interconverting aldoses and ketoses

$R^{1,2}$ = hydrogen, organic residue

If R^1 is a hydrogen atom, then R^2 is any organic residue and *vice versa.*

EC 5.3.2 Interconverting keto-enol-groups

$R^{1,2}$ = hydrogen, organic residue

EC 5.3.3 Transposing C=C bonds

$R^{1,2,3,4,5}$ = organic residue

EC 5.3.4 Transposing S-S bonds

The cysteine residues are parts of proteins.

EC 5.4 Intramolecular transferases (mutases)

This enzyme sub-class can be divided into two groups.

The enzymes belonging to 5.4.1 and 5.4.2 catalyze the transfer of a functional group from one oxygen atom to another oxygen atom of the same molecule.

TG = transferred groups are acyl or orthophosphate groups
$R^{1,2}$ = organic residue
n = 0 or 4

The enzymes classified under 5.4.3 catalyze the transfer of a whole amino group from one carbon atom of a molecule to a neighboring atom of the same molecule.

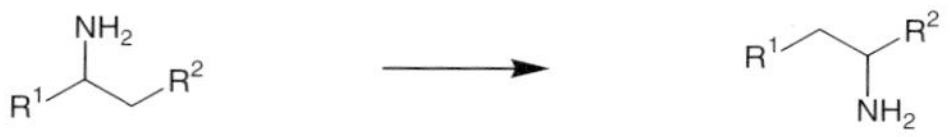

$R^{1,2}$ = organic residue

53

EC 5.5 Intramolecular lyases

X = O, CH$_2$
R^{1,2} = organic residue

EC 6 Ligases

In contrast to all other five enzyme classes this last main division in the Enzyme Nomenclature is the only one where no member is used for the production of any fine chemical in an industrial process. Nevertheless these biocatalysts play a major role in genetic engineering and genetic diagnostics, since specific enzymes in this class called DNA ligases catalyze the formation of C-O bonds in DNA synthesis. This reaction is essential in genetic engineering sciences, allowing connection of two DNA strings into a single one.

To generalize, ligases are enzymes catalyzing a bond formation between two molecules. This reaction is always coupled with the hydrolysis of a pyrophosphate bond in ATP or a similar triphosphate. The bonds formed are, e.g., C-O, C-S, and C-N bonds.

The systematic names should be formed on the system *X:Y ligase*.

EC 6.1 Forming carbon-oxygen bonds

Ⓟ = diphosphate
R = organic residue

The tRNA-hydroxy group is the 2′- or 3′-hydroxy group of the 3′-terminal nucleoside.

54

EC 6.2 Forming carbon-sulfur bonds

P = diphosphate
R = organic residue
NTP = nucleotide triphosphate (ATP, GTP)
NMP = nucleotide monophosphate (AMP, GMP)

The thiol group is the terminal group of the coenzyme A molecule.

EC 6.3 Forming carbon-nitrogen bonds

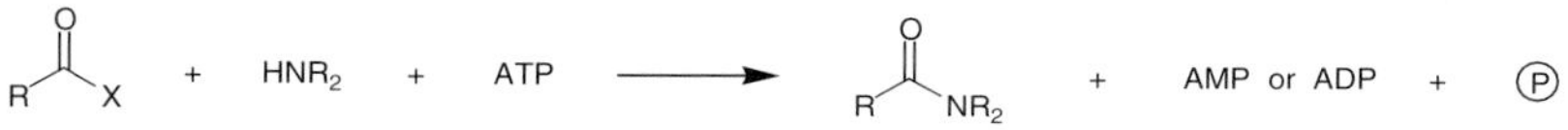

P = monophosphate, diphosphate
X = OH, H, COOH
R = hydrogen, organic residue

CO_2 is the substrate for the enzyme EC 6.3.3.3. There are exceptions to this reaction pattern, like ligase EC 6.3.4.1, that catalyzes the following reaction:

P = diphosphate

EC 6.4 Forming carbon-carbon bonds

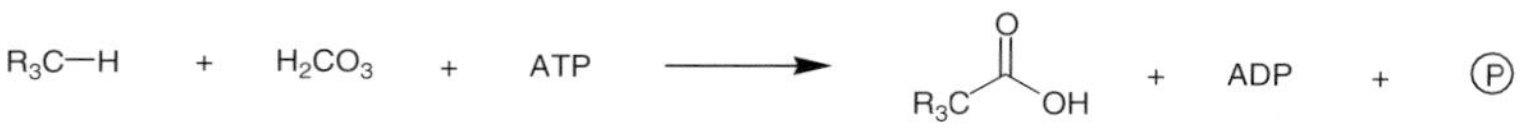

P = monophosphate
R = hydrogen, organic residue

EC 6.5 Forming phosphoric ester bonds

$\fbox{(nucleotide)$_n$}$—OH + $\fbox{(nucleotide)$_m$}$—Ⓟ + ATP ⟶ $\fbox{(nucleotide)$_n$}$—Ⓟ—$\fbox{(nucleotide)$_m$}$

+ AMP + Ⓟ

Ⓟ = monophosphate
R = hydrogen, organic rest

These enzymes are repair enzymes for DNA. The enzyme EC 6.5.1.2 uses NAD$^+$ as cofactor.

References

[1] International Union Of Biochemistry and Molecular Biology (1992) Enzyme Nomenclature, Academic Press Inc., San Diego

[2] Supplement 1: Nomenclature Committee of the International Union of Biochemistry and Molecular Biology (NC-IUBMB) (1994), Eur. J. Biochem. **223**, 1–5

[3] Supplement 2: Nomenclature Committee of the International Union of Biochemistry and Molecular Biology (NC-IUBMB) (1995), Eur. J. Biochem. **232**, 1–6

[4] Supplement 3: Nomenclature Committee of the International Union of Biochemistry and Molecular Biology (NC-IUBMB) (1996), Eur. J. Biochem. **237**, 1–5

[5] Supplement 4: Nomenclature Committee of the International Union of Biochemistry and Molecular Biology (NC-IUBMB) (1997), Eur. J. Biochem. **250**, 1–6

[6] Supplement 5: Nomenclature Committee of the International Union of Biochemistry and Molecular Biology (NC-IUBMB) (1999), Eur. J. Biochem. **264**, 610–650

[7] Appel, R.D., Bairoch, A., Hochstrasser, D.F. (1994) A new generation of information tools for biologists: the example of the ExPASy WWW server, Trends Biochem. Sci. **19**, 258–260

[8] Bairoch, A. (1999) The ENZYME data bank in 1999, Nucleic Acids Res. **27**,310–311 (available through http://www.expasy.ch/enzyme/)

[9] Kindel, S. (1981) Technology **1**, 62

[10] http://www.expasy.ch/enzyme/enz_new_form.html

[11] Ager, D.J. (1999) Handbook of Chiral Chemicals, Marcel Dekker Inc., New York/Basel

[12] Wrotnowski, C. (1997) Unexpected niche applications for industrial enzymes drives market growth, Gen. Eng. News **Feb.1**, 14 + 30

4 Basics of Bioreaction Engineering

Andreas Liese

Institute of Biotechnology
Forschungszentrum Jülich GmbH
D-52425 Jülich, Germany

Karsten Seelbach

Corporate Process Technology
Degussa-Hüls AG
D-45764 Marl, Germany

Nagaraj N. Rao

Rane Rao Reshamia Laboratories Pvt. Ltd
Turbhe Naka, Navi Mumbai – 400 705, India

The prerequisite for a process development is rational design. The starting point is the availability of the reactants as well as of the catalyst (figure 1). This is a very important point that has not only a practical impact but also an economical one. The next step should be the characterization of the reaction system by the

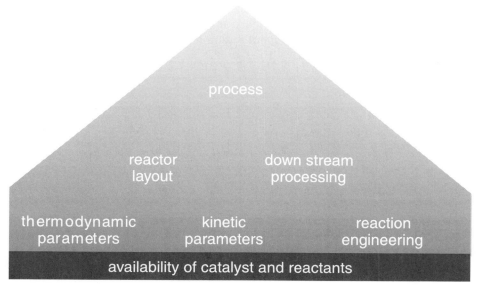

Fig. 1 Objective – rational process design.

kinetic, thermodynamic and reaction engineering parameters. These together then determine the type of reactor to be chosen and how the down stream processing looks like, thereby forming the total process.

In this chapter some fundamental aspects of reaction engineering, kinetics and enzymatic synthesis are described that are needed for the understanding of the data given in chapter 5.

We will start with definitions of key reaction engineering terms that are used throughout the book. These are followed by an introduction to enzyme biosynthesis and a brief overview of general characteristics of the different enzyme classes. Further topics discussed are the fundamental types of reactors and their mode of operation.

4.1 Definitions

4.1.1 Conversion

The conversion is the number of converted molecules per number of starting molecules:

$$X_s = \frac{n_{s0} - n_s}{n_{s0}} \tag{1}$$

X_s conversion of substrate s

n_{s0} amount of substrate s at the start of the reaction (mol)

n_s amount of substrate s at the end of the reaction (mol)

The conversion has to be maximized firstly to avoid recycling of unconverted reaction solution and secondly to minimize reactor volumes. On the other hand, high conversions can result in long reaction times or high amounts of catalyst being employed. Unwanted subsequent reactions of the product will then result in the formation of by-products.

4.1.2 Yield

The yield is the number of synthesized molecules per number of starting molecules:

$$\eta_p = \frac{n_p - n_{p0}}{n_{s0}} \cdot \frac{|\nu_s|}{|\nu_p|} \tag{2}$$

η_p yield of product p

n_{p0} amount of product p at the start of the reaction (mol)

n_p amount of product p at the end of the reaction (mol)

ν_s stoichiometric factor for substrate s

ν_p stoichiometric factor for product p

In combination with the conversion or the selectivity it describes how many product molecules are synthesized in relation to the starting amount of substrate molecules. The described yield is the analytical one. Often the isolated yield is given instead, which describes the synthesized amount of product after down stream processing. The latter does not help in understanding single reaction steps and developing correct kinetic models. If an entire process is considered, the overall yield can be calculated by multiplication of all single yields.

4.1.3 Selectivity

The selectivity is the number of synthesized product molecules per number of converted molecules:

$$\sigma_p = \frac{n_p - n_{p0}}{n_{s0} - n_s} \cdot \frac{|\nu_s|}{|\nu_p|} \tag{3}$$

σ_p selectivity to component p

n_{s0} amount of substrate s at the start of the reaction (mol)

n_s amount of substrate s at the end of the reaction (mol)

n_{p0} amount of product p at the start of the reaction (mol)

n_p amount of product p at the end of the reaction (mol)

ν_s stoichiometric factor for substrate s

ν_p stoichiometric factor for product p

The selectivity describes the synthesized product molecules in relation to the substrate molecules converted. Selectivity has to be as close to '1' as possible to avoid waste of educt. It belongs to the most important economical factors.

If only a very short reaction course is looked at, the selectivity leads to the differential form. This is interesting for gaining information on the synthesis of by-products at every step of conversion. It is decisive for estimating whether a premature stop of the reaction is efficient with regard to the overall yield of the reaction.

The combination of conversion, yield and selectivity leads to the equation:

$$\eta = \sigma \cdot X \tag{4}$$

4.1.4 Enantiomeric excess

The enantiomeric excess (*ee*) is the difference in the number of both enantiomers per sum of the enantiomers:

$$ee_R = \frac{n_R - n_S}{n_R + n_S} \tag{5}$$

ee_R enantiomeric excess of (*R*)-enantiomer

n_R amount of (*R*)-enantiomer (mol)

n_S amount of (*S*)-enantiomer (mol)

The enantiomeric excess describes the enantiomeric purity of an optically active molecule. Small differences in the constellation sequence of the binding partners of one central atom lead to big differences in chemical behavior, in biological pathways and recognition. Since not only different organoleptic properties for both the enantiomers can be found, but also contrary pharmacological effects, it is most important to find syntheses with clear enantiomeric selectivities. Since biological catalysts (enzymes) have improved performance as a result of evolution, they often fulfill this task in the best way. Many pharmaceuticals as well as herbicides and fungicides used to be sold as racemic mixtures, if the unwanted enantiomer did not have a deleterious effect on the organism. But in the last decades, attempt has often been made to switch over to one enantiomer. These products are called racemic switches. Less of the unwanted enantiomer is produced and plant capacity is increased. Organisms are exposed to lesser quantities of chemicals. The environmental benefits are also quite significant.

4.1.5 Turnover number

The turnover number (*tn*) is the number of synthesized molecules per number of used catalyst molecules:

$$tn = \frac{n_p}{n_{cat}} \cdot \frac{1}{|\nu_p|} \tag{6}$$

tn   turnover number

n_p amount of product *p* at the end of the reaction (mol)

ν_p stoichiometric factor for the product *p*

n_{cat} amount of catalyst (mol)

The turnover number is a measure of the efficiency of a catalyst. Especially when using expensive catalysts, the *tn* should be as high as possible to reduce the cost of the product. It is very important to name defined reaction parameters in combination with the *tn* to make this value comparable. Instead of the *tn*, the deactivation rate or half life may be given.

The turnover number can also be given for cofactors / coenzymes.

4.1.6 Turnover frequency

The turnover frequency (*tof*) is defined as the number of converted molecules per unit of time:

$$tof = \frac{\partial n_s}{\partial t \cdot n_{catalyst}} \tag{7}$$

tof   turnover frequency (s^{-1})

∂n_s differential amount of converted substrate (mol, µmol)

∂t differential time for conversion (s)

$n_{catalyst}$ mole of catalyst (mol, µmol)

The turnover frequency expresses the enzyme activity. It is noteworthy that chemical catalysts are very slow in comparison to enzymes. An example is the epoxidation catalyst Mn-Salen with a *tof* of 3 h^{-1} and the enzymatic counterpart chloroperoxidase (CPO) with a *tof* of 4,500 h^{-1}. But usually the great advantage in activity is lost when educt/product solubility, stability and molecular mass of the enzyme are also considered (here molecular masses are: 635 $g \cdot mol^{-1}$ and 42,000 $g \cdot mol^{-1}$, respectively) [1, 2].

4.1.7 Enzyme activity

The enzyme activity is defined as the reaction rate per unit weight of catalyst (protein):

$$V = \frac{\partial n_s}{\partial t \cdot m_{catalyst}} \tag{8}$$

V maximum activity of enzyme at defined conditions ($katal \cdot kg^{-1}$, $U \cdot mg^{-1}$)

∂n_s differential amount of converted substrate (mol, µmol)

∂t differential time for conversion (s, min)

$m_{catalyst}$ mass of catalyst (kg, mg)

The SI-unit for the enzyme activity is katal ($kat = mol \cdot s^{-1}$, 1 kat = $6 \cdot 10^7$ U), which results in very low values so that often µ-, n- or pkat are used. More practical is the Unit (1 U = 1 $µmol \cdot min^{-1}$). It is very important that the activity be named with the used substrate and with all necessary reaction conditions, like temperature, buffer salts, pH-value etc. The concentration of a protein solution can be determined using indirect photometric tests (e. g. Bradford [3]) or analytical methods (e. g. electrophoresis). Only when the enzyme contains a photometrically active component, e. g. a heme protein, the concentration can be easily determined by direct photometric absorption. The activity can also be given per reaction volume ($U \cdot mL^{-1}$), if the enzyme concentration cannot be determined due to missing molecular mass or the inability to analyze the mass of solubilized enzyme.

4.1.8 Deactivation rate

The deactivation rate is defined as the loss of catalyst activity per unit of time:

$$V_1 = V_0 \cdot e^{-k_{deact} \cdot (t_1 - t_0)} \tag{9}$$

k_{deact} deactivation rate $(\text{min}^{-1}, \text{h}^{-1}, \text{d}^{-1})$

V_0 enzyme activity at the start of the measurement $(\text{U} \cdot \text{mg}^{-1})$

V_1 enzyme activity at the end of the measurement $(\text{U} \cdot \text{mg}^{-1})$

t_0 start time of the measurement (min, h, d)

t_1 end time of the measurement (min, h, d)

The deactivation rate expresses the stability of a catalyst.

4.1.9 Half life

The half life is defined as the time in which the activity is halved:

$$V_1 = V_0 \cdot e^{-k_{deact} \cdot (t_1 - t_0)} \tag{10}$$

$$V_2 = V_0 \cdot e^{-k_{deact} \cdot (t_2 - t_0)} \tag{11}$$

$$V_1 = \tfrac{1}{2} \cdot V_2 \tag{12}$$

$$\Rightarrow \quad t_{1/2} = \frac{\ln(2)}{k_{deact}} \tag{13}$$

$t_{1/2}$ half life of catalyst (min, h, d)

V_x enzyme activity at time t_x $(\text{U} \cdot \text{mg}^{-1})$

t_x time of measurement (min, h, d)

k_{deact} deactivation rate $(\text{min}^{-1}, \text{h}^{-1}, \text{d}^{-1})$

The half life expresses the stability of a catalyst. The activity usually shows a typical exponential decay. Therefore the half life can be calculated and it gives an extent of the catalyst deactivation independent of considered time differences.

4.1.10 Catalyst consumption

The biocatalyst consumption (*bc*) is defined as the mass or activity of catalyst consumed per mass of synthesized product:

$$bc = \frac{m_{catalyst}}{m_{product}} \tag{14}$$

bc biocatalyst consumption ($g \cdot kg^{-1}$ or $U \cdot kg^{-1}$)

$m_{catalyst}$ mass or activity of catalyst used for synthesized mass of product (g or U)

$m_{product}$ mass of synthesized product (g)

If an expensive catalyst is used, the biocatalyst consumption should be as low as possible to decrease the biocatalyst consumption cost in production. Often, pharmaceutical products are valued so high that in discontinuous reactions the catalyst can be discarded without recycling. Since the catalyst stability can change with conversion due to deactivating by-products, it is interesting to look at the differential catalyst consumption to find the optimal conversion for the end of the reaction and separating the reaction solution from the catalyst.

4.1.11 Residence time

The residence time (τ) is defined as the quotient of reactor volume and feed rate:

$$\tau = \frac{V_R}{F} \tag{15}$$

τ residence time or reaction time (h)

V_R reactor volume (L)

F feed rate ($L \cdot h^{-1}$)

The residence time describes the average time of a molecule in the reactor. Since the residence times of different molecules are not the same, usually the average residence time is used. Diffusion effects and non-ideal stirring in a continuously operated stirred tank reactor (CSTR) or back mixing in plug flow reactors results in a broad distribution of single residence times. For a detailed simulation of the process this distribution has to be taken into account. For example, one educt molecule could leave the reactor directly after it was fed into the reactor or it could stay in the reaction system forever. Therefore, the selectivity can be strongly influenced by a broad distribution.

4.1.12 Space-time yield

The space-time yield (*STY*) is the mass of product synthesized per reactor volume and time. It is also named as the **volumetric productivity**:

$$STY = \frac{m_p}{\tau \cdot V_R} \tag{16}$$

STY space-time yield ($g \cdot L^{-1} \cdot d^{-1}$)

m_p mass of synthesized product (g)

τ residence time or reaction time (d)

V_R reactor volume (L)

The space-time yield expresses the productivity of a reactor system. The *STY* should be as high as possible to decrease investment costs of a plant. A low *STY* means low product concentrations or bad reaction rates. Low concentrations lead to more complex down stream processing, while low reaction rates necessitate larger reactor volumes. The reactor volume for heterogeneously catalyzed reactions is usually the volume of the catalyst itself.

4.2 Biosynthesis and immobilization of biocatalysts

This chapter tries to give a brief introduction to the biosynthesis of enzymes, which are the biological catalysts. For a more detailed introduction the reader is referred to textbooks [4–6]. Additionally the immobilization of biocatalysts is discussed that is often used on an industrial scale to reduce the catalyst costs and to increase the stability.

4.2.1 Types of biocatalysts

The biocatalyst is always described as a whole cell or an enzyme. In the first case we face a mini-reactor with all necessary cofactors and sequences of enzymes concentrated in one cell. In the second case the main catalytic unit is isolated and purified. In both cases optimization is possible. Furthermore, multi-step biosynthetic pathways can be changed to prevent degradation of the desired product or produce precursors normally not prioritized in the usual pathway. All these changes for the whole cell lead to an optimized mini-plant. The optimization of the main catalyst is comparable with catalyst development inside a reaction system. In this chapter the composition and biological synthesis of enzymes as well as the different genetic terms (e.g. mutation, cloning, etc.) that are used in chapter 5 are briefly explained.

Whole cells can be bacteria, fungi, plant cells or animal cells. They are subdivided into the two groups of prokaryotic cells and eukaryotic cells.

4.2.1.1 Prokaryotic cells

Prokaryotic cells are the "lowest microorganisms" and do not possess a true nucleus. The nuclear material is contained in the cytoplasm of the cell. They reproduce by cell division. They are relatively small in size (0.2 to 10 μm) and exist as single cells or as mycelia. When designing bioreactors, an adequate supply of nutrients as well as oxygen into the bioreactor must be assured, since the cells, e.g. bacteria, grow rapidly. Parameters such as pH, oxygen feed rate and temperature in the bioreactor must be optimized. Perhaps the most widely used prokaryotic microorganism in industrial biotransformations is *Escherichia coli,* which is a native to the human intestinal flora.

4.2.1.2 Eukaryotic cells

Eukaryotic cells are higher microorganisms and have a true nucleus bounded by a nuclear membrane. They reproduce by an indirect cell division method called mitosis, in which the two daughter nuclei normally receive identical compliments of the number of chromosomes characteristic of the somatic cells of the species. They are larger in size (5–30 μm) and have a complex structure. When eukaryotic cells are used as biocatalysts, high or low mechanical stress must be avoided by using large stirrers at slow speed and by eliminating dead zones in the fermenter. *Saccharomyces cerevisiae* and *Zymomonas mobilis* represent the most important eukaryotic cells used in industrial biotransformations.

4.2.2 Enzyme structure and biosynthesis

An enzyme is an accumulation of one or more polypeptide chains in the form of a protein. It is unique in being capable of accelerating or producing by catalytic action a transformation in a substrate for which it is often specific.

The three-dimensional structure of an enzyme is determined at different levels [7]:

- Primary structure: sequence of connected amino acids of a protein chain.
- Secondary structure: hydrogen bonds from the type of R–N-H – O=C-R are responsible for the formation of the secondary structure, the α-helix or the β-sheet, of one protein chain.
- Tertiary structure: hydrogen and disulfide bonds, as well as ionic and hydrophobic forces lead to the tertiary structure, the folded protein chain.
- Quaternary structure: If several protein chains are combined in the form of subunits, the quaternary structure is formed. Not covalent bonds, but molecular interactions occuring in the secondary, tertiary and quaternary structures, are responsible for the formation of the well-functioning catalytic system.

The amino acid sequence of a protein is determined by the nucleic acids, which are the non-protein constituents of nucleoproteins present in the cell nucleus. The nucleic acids are complex organic acids of high molecular weight consisting of chains of alternate units of phosphate and a pentose sugar with a purine and a pyrimidine base attached to the sugar. The DNA (desoxyribonucleic acid) consists of four bases, namely adenine, thymine, cytosine and guanine, whereas the RNA (ribonucleic acid) contains uracil instead of thymine (figure 2). In RNA, the sugar is ribose instead of 2-desoxyribose. Figure 3 shows the matching bases in a short section of a DNA strand.

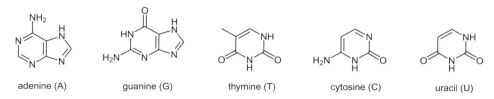

adenine (A)   guanine (G)   thymine (T)   cytosine (C)   uracil (U)

Fig. 2 Bases of nucleic acids.

Two chains are combined to form a helical structure, but only thymine–adenine (T-A) and cytosine-guanine (C-G) can be coupled. The phosphoric ester can be bound on the 3′- or 5′-hydroxy group of the sugar component (figure 3):

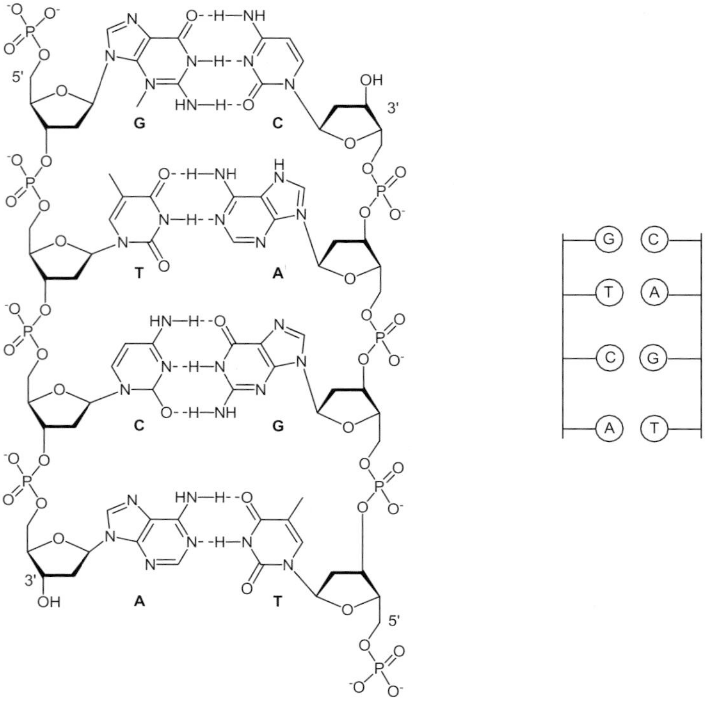

Fig. 3 Matching bases in a short section of a DNA strand.

The process of transcribing DNA, transferring mRNA out of the nucleus and reading mRNA to build a protein, e.g. an enzyme, is called **gene expression**. The expression of a gene that is encoding a protein can be divided into the two main steps of **transcription** and **translation**. In prokaryotic cells that do not contain a nucleus both steps of transcription and translation take place at the same time and at the same place, in the cytosol.

Transcription

In transcription, a double-stranded DNA serves as a template to synthesize a messenger RNA (mRNA) with the help of a RNA-polymerase. Starting from the linear genetic determinants on a DNA strand, the flow of cellular information descends to specify the structure of the proteins through the transcription into mRNA molecules having complimentary base sequences to the parent DNA strands. In short, the information of a gene is transferred to a mRNA by a RNA-polymerase and can thus leave the nucleus. The enzyme polymerase needs a start sequence, the promoter, to begin with the building of the mRNA. The transcription ends when a termination sequence is found (figure 4).

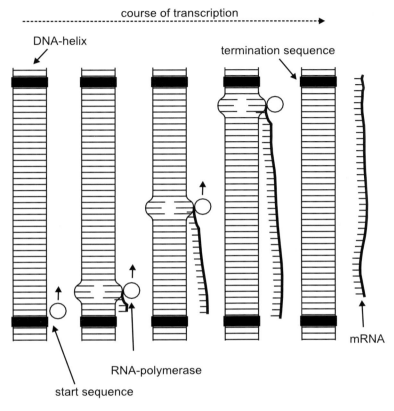

Fig. 4 Transcription of DNA information on mRNA.

Translation

In translation, the mRNA serves as a template to form the protein molecule. The information that is contained in the mRNA can be translated into a protein building sequence by the ribosomes. The ribosome needs a binding sequence where it binds to the mRNA. It moves along the mRNA until it finds a start sequence (AUG). Here the protein biosynthesis begins. Each amino acid incorporated into the protein is defined by the combination of three nucleotides known as the codon. Of the 64 possible codons, 61 are used to code 20 amino acids and the remaining three are used as stop signals for translation. The translation table of the codons depends on the organism. A transfer-RNA (tRNA), which looks like a clover leaf due to intramolecular bonds, is used as a carrier for one special amino acid at one end. The tRNA binds to the mRNA and the corresponding amino acid is connected to the end of the existing protein chain. This procedure is repeated until a stop sequence (stop codon) is found (UAA, UAG or UGA). Figure 5 shows the translation of the mRNA to a protein:

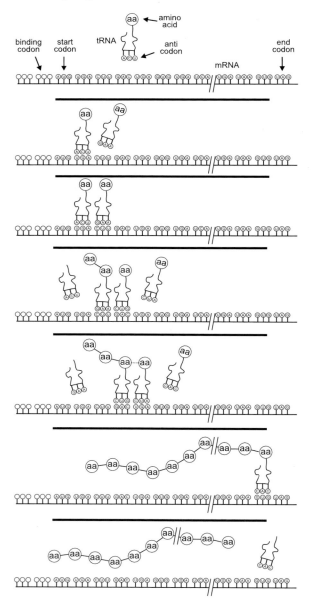

Fig. 5 Synthesis of protein by translation of mRNA.

In some cases the protein is modified post-translationally (e.g. for insulin production).

4.2.3 Cloning of biocatalysts

The aim of cloning is to improve the expression of a specific gene, which is responsible for the synthesis of the desired enzyme, a protein. In principle, it should be sufficient to cut out the gene fragment and transfer it into a cell. In this cell the protein will be synthesized but the aim is to produce a high quantity of protein, which is not possible with only one cell. Therefore the cell has to be divided and multiplied. But during this process only the cell typical DNA will be multiplied for each cell during cell division. This replication procedure is catalyzed by the DNA-polymerase which needs a special start sequence, the promoter (see above). This sequence usually does not exist for DNA-fragments. Therefore a cloning vector has to be found which contains the interesting gene and an origin of replication. Vectors are DNA molecules which serve as a recipient or carrier for foreign DNA. They carry an origin of DNA replication and genetic markers which allow them to be detected in host cells. The most commonly used vectors are the plasmids, which are self-replicating rings of DNA and are not contained in the main set of chromosomes of a cell. Figure 6 shows the difference in cell replication using the single DNA-fragment and the modified plasmid:

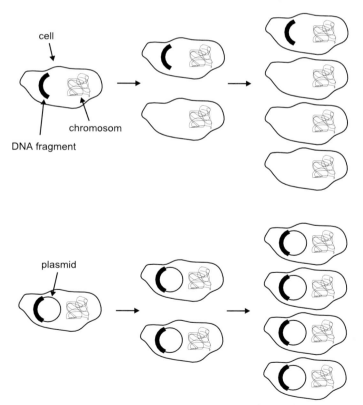

Fig. 6 The use of plasmids for the replication of the interesting DNA-fragment.

The desired fragment can be inserted into the plasmid ring by use of restriction enzymes (restriction endonucleases) which cut a nucleotide sequence to sticky or blunt ends [8]. Thus one chain is longer than the other or both are of the same length (figure 7). Blunt ends have to be converted to sticky ends in order to insert them into a plasmid ring by the use of a ligase:

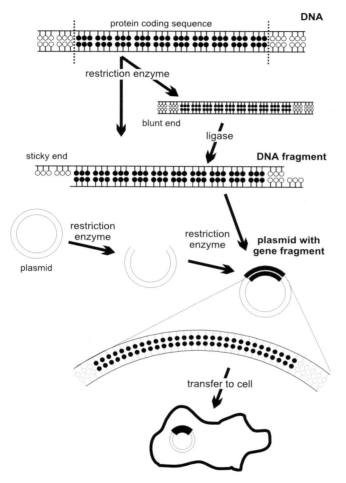

Fig. 7 The insertion of a DNA-fragment into a plasmid vector.

For genes from bacteria and viruses (prokaryotic organisms) this procedure works in the above-described manner. But, if a more complex gene from an eukaryotic organism is to be expressed, the protein cannot be translated since some codons do not contain any coding informations (**introns**), as opposed to the informative **exons**. These introns have to be removed prior to translation (**splicing**) (figure 8):

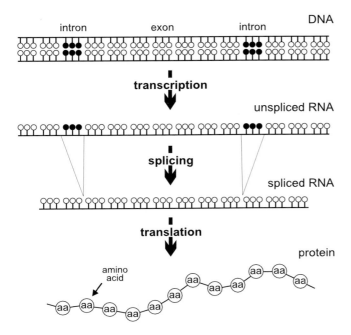

Fig. 8 Splicing of RNA before translation is possible.

Only few organisms employed for cloning can be used for splicing. For example, yeast can be cloned, but cloning is very inefficient and is therefore seldom used. A better method can be the isolation of the mRNA from the protein-producing organism. The mRNA can be converted to the DNA segment named cDNA by use of reverse transcriptase. cDNA (complimentary DNA) is a double-stranded DNA copy of the mRNA and serves as a template for protein biosynthesis. The blunt ends can be converted into sticky ends by a ligase and inserted to the plasmid by the restriction enzymes. A further method, which can only be applied if the amino acid sequence and thus the nucleic base sequence is known, is to synthesize the RNA step by step by chemical methods. Figure 9 summarizes the different approaches:

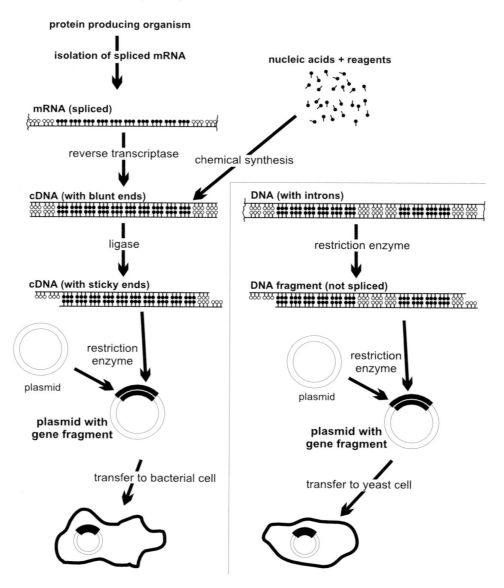

Fig. 9 Approaches to solve the problem of removing uninformative genetic material.

There are several factors to be considered while optimizing cloning and gene expression:

- The plasmid should have only one position for each restriction enzyme to prevent mismatching of the sequence in a plasmid.
- Further, the promoter should be strong so that the expression of the gene sequence is favored for production of high concentrations of the protein.
- The recombinant organism should be genetically stable.

4.2.4 Screening and mutation of biocatalysts

Screening for interesting enzymes can be directed by understanding biological pathways. If the biological reactions are known for several conversions of structural elements in the intermediates, the isolation of the corresponding catalyst can be started. If a catalyst can be isolated, purified and produced in small amounts for preliminary investigations of substrate spectrum and kinetics, the next step would be the sequencing of the amino acids and determination of the crystal structure to understand the mechanism of the biotransformation.

The enzymology of DNA, including the discovery and purification of restriciton enzymes and of DNA polymerases, have given the biotechnologist a new tool in the recent past for optimizing the protein, namely, **genetic engineering** or **protein engineering.** A cloned gene can now be overexpressed to manufacture relatively large quantities of the desired enzyme. Selective mutation of the gene – and, consequently, the amino acid sequence – leads to a modified enzyme with altered kinetic parameters or specificity.

Mutation can be carried out by several methods:

- physical (irradiation with high energy radiation)
- chemical (treatment with reactive molecules to change nucleic bases, insertion of analog bases, which results in mismatched bases pairing or fragments, which can be inserted during insertion to lead to a whole shift in the sequence of one chain)
- biochemical (enzymes to cut DNA-chains and to change selected nucleic bases)

However, the frequency of desired mutations and the possibility of combining desired properties in a recombination process are rather low due to the random nature of mutation and selection. Strain improvement depends, therefore, to a very large extent on proper and efficient selection methods which can detect and isolate one mutant among several thousands of cells.

Genetic engineering techniques also allow the production of enzymes of higher organisms by microorganisms by placement of the corresponding gene into the latter. Several examples are shown in chapter 5 where mutation improves dramatically K_M-values, activity and stability. The main objective of mutation and selection is to achieve higher overall productivity when using the biocatalyst, thus making biotransformations economically feasible.

4.2.5 Optimization of reaction conditions

Optimization of the reaction conditions is a very important point with regard to catalyst consumption, product specific catalyst costs and productivity.

In industrial biotransformations often the immobilization on a support is chosen to enhance the stability as well as to simplify the biocatalyst recovery. An alternative approach for an easy biocatalyst recovery is the use of a filtration unit [9,10,11]. For the separation of suspended whole cells or solubilized enzymes, membranes are often applied as filters. In membrane filtration, the specific pore size and charge of a membrane are used to separate different compounds by their physical size. Figure 10 shows the classification of different filtration types in the order of the pore sizes or cut off values.

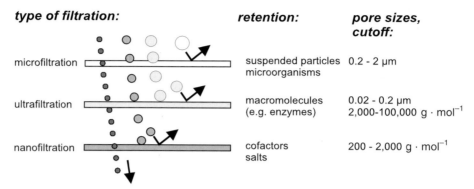

type of filtration:	**retention:**	**pore sizes, cutoff:**
microfiltration	suspended particles microorganisms	0.2 - 2 µm
ultrafiltration	macromolecules (e.g. enzymes)	0.02 - 0.2 µm 2,000-100,000 g $\cdot$ mol^{-1}
nanofiltration	cofactors salts	200 - 2,000 g $\cdot$ mol^{-1}

Fig. 10 Classification of different filtration types.

Microfiltration membranes are applied for the separation of whole cells (for examples see the processes on pages 137, 253, 369). Due to the lower molecular weight and physical size of enzymes, ultrafiltration membranes have to be used to retain them [12,13] (for examples see the processes on pages 103, 113, 125).

4.2.5.1 Immobilization of biocatalysts

If a good biocatalyst is found for a specific reaction, one possibility for further improvement of its properties is immobilization. The best way in which the biocatalyst can be immobilized has to be found by experiment. This is dependent on the reaction, the stability of the biocatalyst, the possibilities for the immobilization of the biocatalyst and the activity of the immobilized biocatalyst. No straightforward plan for testing immobilization is known, but at least the biocatalyst structure should be considered.

The main advantages of immobilization are:

- easy separation of biocatalyst,
- lower down stream processing costs,
- possibility of biocatalyst recycling,
- better stability, especially towards organic solvents and heat,
- use of fixed bed reactors and,
- easier realization for continuous production.

The main disadvantages of immobilization are:

- loss of absolute activity due to immobilization process,
- lower activity of immobilized biocatalyst compared to non-immobilized biocatalyst as used in processes with membrane filtration,
- additional costs for carrier or immobilization matrix and immobilization procedure,

- carrier or matrix cannot be recycled and
- diffusion limitations lowering reaction rates.

In spite of these disadvantages, immobilization has become an indispensable part of industrial biotransformations. The most common methods for the immobilization are entrapment in matrices, cross-linking and covalent binding (figure 11):

Entrapment

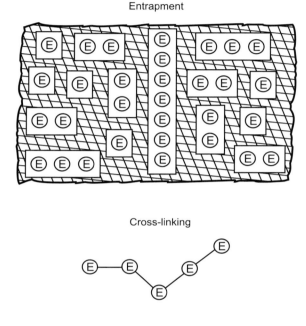

Cross-linking

Carrier binding

covalent   adsorption

Fig. 11 Common immobilization methods.

Entrapment

The biocatalyst can be entrapped in natural or synthetic gel matrices. A very simple method is the entrapment in sodium alginate, a natural polysaccharide. The water soluble alginate is mixed with the biocatalyst solution and dropped into a calcium chloride solution in which water-insoluble alginate beads are formed (figure 12):

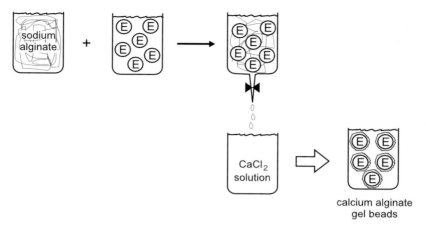

Fig. 12 Alginate gel formation.

Another naturally-occuring polysaccharide widely used for immobilization is κ-carrageenan. In a manner similar to the alginate method, a mixture of κ-carrageenan in saline and biocatalyst solution (or suspension) is dropped into a solution of a gelling reagent like potassium chloride. Ammonium, calcium and aluminum cations also serve as good gelling reagents. The gel can be hardened by glutaraldehyde, hexamethylenediamine or other cross-linking reagents, often enhancing biocatalyst stability (figure 13):

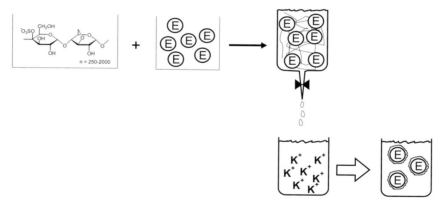

Fig. 13 κ-Carrageenan gel formation.

An often-used synthetic immobilization method employs polyacrylamide gel. The biocatalyst, a monomer (acrylamide) and a cross-linker (e.g. N,N'-methylenebisacrylamide) are mixed and polymerized by starting the reaction with an initiator (e.g. potassium persulfate) in the presence of a stimulator (e.g. 3-dimethylamino propionitrile) (figure 14):

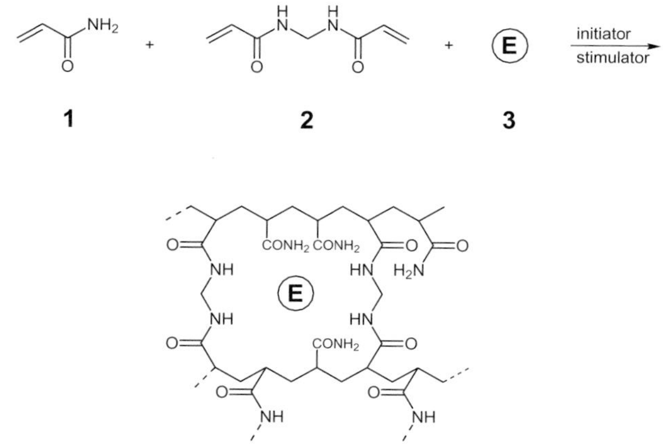

1 = acrylamide
2 = N,N'-methylenebisacrylamide
3 = enzyme
4 = polyacrylamide gel

Fig. 14 Polyacrylamide gel formation.

Cross-linking

The most popular cross-linking reagent is glutaraldehyde, although other bi- or multifunctional reagents can be used instead. The single biocatalyst (protein molecule or cell) is cross-linked to insoluble macromolecules or cell pellets (figure 15):

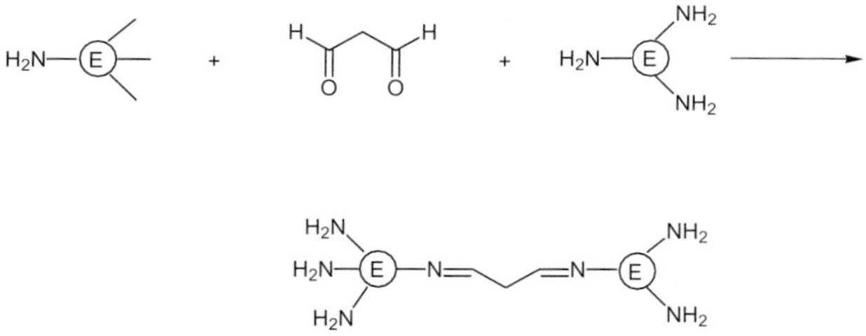

Fig. 15 Cross-linking with glutaraldehyde.

Covalent binding

The biocatalyst, in this case usually an isolated enzyme, can be attached to a carrier by a reaction of amino- or acid-groups of the proteins. Generally, amino acid residues, which are not involved in the active site or in the substrate-binding site of the enzyme, can be used for covalent binding with carriers. Usually the carrier is a polymer (polysaccharide, polysiloxane, polyacrylamide etc.) bearing hydroxy groups or amino groups on its surface. To combine the enzyme with these groups, different activation methods can be applied. These activated carriers are commercially available. Examples are spacers with epoxy groups, which are activated by cyanogen bromide or other activating groups such as acid azide, leading to a spontaneous reaction with the amino group of the biocatalyst (figure 16):

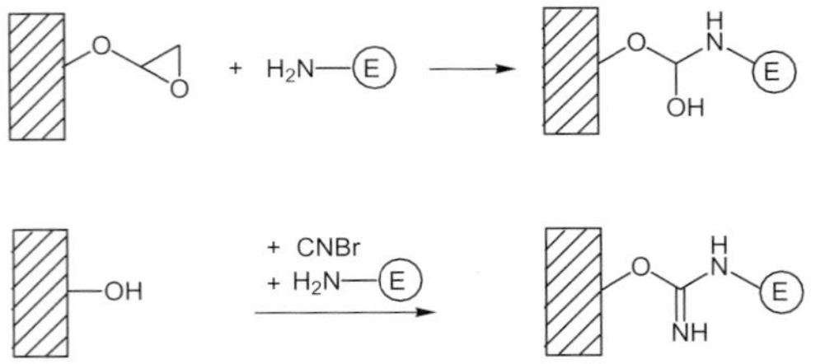

Fig. 16 Two examples for carrier-coupling using the amino group of an enzyme.

If the acid group is to be coupled to a carrier, the enzyme has to be activated too, e.g. by reaction with a diimide. The activated carrier and enzyme can now be coupled (figure 17):

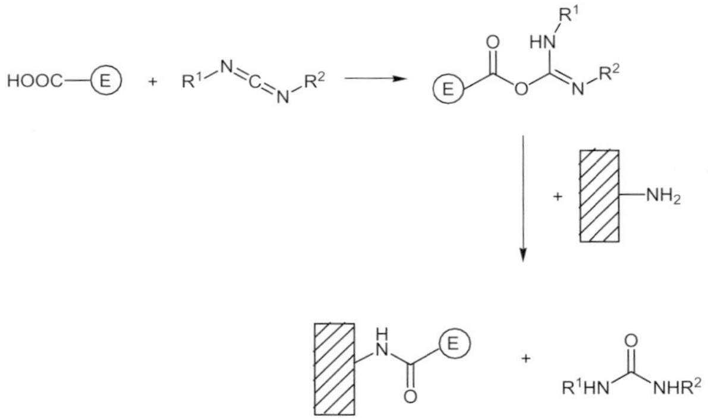

Fig. 17 Example for carrier-coupling using the acid group of an enzyme.

4.2.5.2 Reaction parameters

The reaction parameters have to be optimized for the reaction with respect to high space-time yields and high stability of the biocatalyst, meaning low production costs.

Parameters that can be varied are:

- pH,
- temperature,
- solvents,

- buffer salts,
- cofactors,
- immobilization methods,
- substrate and product concentrations,
- addition of antioxidants or stabilizers,
- reactor material or coating and
- physical treatment (stirring, pumping, gas-liquid phases, etc.).

During optimization of the space-time yield it is necessary to consider catalyst costs. Especially the combination of stability and activity has to be considered. Sometimes it is desirable to work at very low temperatures with low reaction rates, which have to be compensated by a high amount of biocatalyst if the turnover number can nevertheless be increased. In other cases the turnover number will have a lower priority since as much product as possible is to be synthesized.

No pragmatic rule exists for the best strategy to optimize reaction conditions. Empirical methods based on statistical methods have good chances of being successful (e.g. genetic algorithms).

The most important improvements can be made in finding and constructing an optimal catalyst.

4.3 Characteristics of the different enzyme classes

For selecting reaction conditions and deciding on the reaction layout it is also important to consider the special properties and limitations that are specific for the different enzyme classes. In the following paragraphs a brief overview is given. References pointing to the related industrial biotransformations in chapter 5 are inserted for illustration.

EC 1: Oxidoreductases

Oxidoreductases are all cofactor dependent. The reducing or oxidizing equivalent is either supplied or taken by the cofactor. The most commonly needed cofactors are $NADH/NAD^+$, $NADPH/NADP^+$, $FADH/FAD^+$, ATP/ADP and PQQ [14–16]. Since some of them like NADH or NADPH are quite expensive, an effective cofactor regeneration system is required to design a cost-effective process. In the literature and in industry mainly three applied approaches can be found to solve this problem. If working with isolated enzymes, either a second enzyme can be used (in the case of NADH the best approach is to use a formate dehydrogenase that utilizes formate and produces CO_2 (see page 103 and 125)) or the cofactor can be regenerated by applying a second substrate (figure 18).

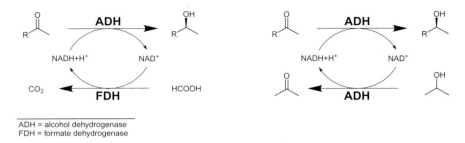

ADH = alcohol dehydrogenase
FDH = formate dehydrogenase

Fig. 18 Different approaches to cofactor recycling.

Another approach is the application of whole cells with glucose for example as a C-source (see page 121). Through this approach the multicatalyst-system of the whole cell itself is used for the regeneration. There are also electrochemical regeneration methods known which have not yet made it to an industrial process [17].

EC 2: Transferases

In nature, transferases play a far larger role than in industrial biotransformations. Here only a few are known. This may be due to the fact that often equilibrium reactions impede quantitative yields (see page 175), coupling reactions arise and the group-transferring substrates are very expensive or their corresponding products are not easily recycled. Nonetheless, these reactions could gain importance in future, should the latter mentioned problems be solved in a chemoenzymatic synthesis. The very high regio- and stereoselectivities in transferase-catalyzed reactions are the main reasons for their increasing utility in synthesis (see page 179). This property leads to the *one enzyme – one linkage* principle.

EC 3: Hydrolases

If hydrolases are used for the kinetic resolution of a racemate, the maximal yield is limited to 50% by the enzyme itself (figure 19a). There are different possible ways to increase the yield up to 100%. If the target compound is the preferentially formed product enantiomer the dynamic resolution process can be applied (figure 19b) [18].

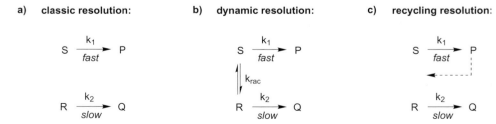

Fig. 19 Different methods of kinetic resolution.

The strategy is a fast isomerization of the substrate enantiomers ($k_{rac} \gg k_1$) [19]. In the case of the *in situ* racemization of amino acid esters the racemization can readily be induced by the addition of pyridoxal-5-phosphate [20]. Alternatively, racemases can be applied (see page 314) [21]. A dynamic resolution is also indicated if the substrate is chirally labile [22]. If the target compound is the more slowly converted substrate enantiomer (figure 19c), one possibility is the recycling of the product under chemical conditions leading to racemization (see page 300). This is only applicable if $k_2 \ll k_1$.

There are also methods available employing chemical stereo-inversion for the production of chiral alcohols, e.g. the synthetic pyrethroid insecticide (*S*)-4-hydroxy-3-methyl-2-prop-2-ynyl-cyclopent-2-enone (see page 208). The lipase-catalyzed kinetic resolution is carried out by hydrolyzing the acylated (*S*)-enantiomer. Subsequently, the cleaved alcohol is sulfonated in the presence of the acylated compound with methanesulfonyl chloride (figure 20) [23].

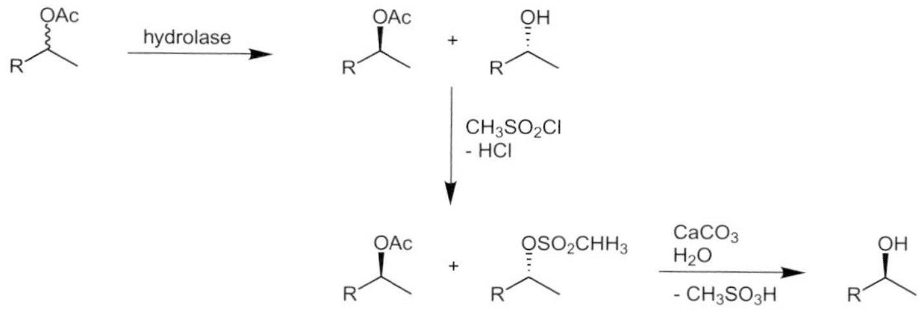

Fig. 20 Resolution coupled to chemical stereo-inversion.

Of key importance is the fact that the hydrolysis of the sulfonated enantiomer in the presence of small amounts of calcium carbonate takes place under inversion of the chiral center in contrast to the hydrolysis of the acylated enantiomer, which is carried out under retention. By this means, an enantiomeric excess of 99.2% and a very high yield are achieved for the (*R*)-alcohol.

EC 4: Lyases

Lyases are of growing significance for industrial biotransformations, since the predominant bond-breaking (lyase) reactions taking place in nature can be reversed (bond formation, synthetase) under non-natural conditions (i.e. high reactant concentrations). Whenever the use of highly concentrated reactants becomes feasible (2.5 M aspartic acid, see page 334) or a product can be withdrawn from the reaction equilibrium (for instance, in a successive reaction) these reactions can be made to run quantitatively. Often a chiral center is generated during bond formation (see page 344). Even the simultaneous formation of two chiral centers is possible. The synthetase reaction leads to the construction of new bonds and is therefore of great importance in synthesis. Especially as a result of the technical evolution of enzymes their corresponding substrate spectra are currently being expanded.

EC 5: Isomerases

Racemases that can be classified as isomerases are of particular significance in kinetic racemic resolution, whenever it is possible to carry out the racemization under similar conditions as those present in a racemic resolution reaction. In this way, kinetic racemic resolutions can lead to yields of up to 100 %. This is the case in the production of D-amino acids and, since recent times, also in the production of L-amino acids (see page 314). Suitable racemases for other amino acid derivatives in technical applications are hardly known. Due to the above mentioned reasons, racemases will probably become the working field of choice for the technical evolution of enzymes. The most renowned enzyme of this group is certainly glucose isomerase (see page 387). In this case the isomerisation leads to an increase in value without the addition of further substrate. In this way, isomerases make it possible to use cheaper substrates (e.g. *N*-acetylglucosamine instead of acetylmannosamine in the preparation of neuraminic acid (see page 385)). Since the isomerization with epimerases does not necessarily give yield of 50 %, it is essential to examine if and how the problem of an undesirable state of equilibrium can be solved by appending a successive reaction.

EC 6: Ligases

For ligases in a narrower sense there are no known industrial biotransformations being carried out at a kg scale. Nonetheless, ligases play a significant role in nature, for instance in ribosomal peptide synthesis, in repairing DNA fragments and in genetic engineering (DNA-ligases).

Additionally, there are some issues that arise for the majority of all biotransformations. These are:

- Low solubility of reactants or products,
- Limited stability of biocatalysts.

With regard to the first point there are different solutions possible. One of the easiest is working in an aqueous-organic two-phase system. However, due to the limited stability of some enzymes in the presence of an interface or organic solvents this is not always possible. A more biocompatible approach is the addition of complexing agents like dimethylated cyclodextrins or adsorbing materials like XAD-7 resins used by Eli Lilly (see page 110). A solution based on reaction engineering is the membrane-stabilized interface as used by Sepracor, USA, in the case of the kinetic resolution of esters (see page 202) or the continuous extraction of reaction products as applied by the Research Center Juelich (see page 103).

There is a wide variety of methods available to increase the catalyst stability. These include the addition of antioxidants (e.g. dithiothreitol), the immobilization on supports, crosslinking of enzyme crystals, separation from deactivating reagents, variation of reaction conditions and optimization of the biocatalyst by the methods of genetic engineering. In an industrial environment it is more often the time that is limiting to find the proper method of catalyst stabilization.

4.4 Kinetics

In this chapter the fundamentals of enzyme kinetics will be discussed in brief. For a detailed description of enzyme kinetics and discussion of the different kinetic models please refer to the following publications: [24–28].

The determination of the kinetic parameters can be carried out in two different ways: either by measurement of the initial reaction rate at different reaction conditions or by batch experiments. In both cases a kinetic model needs to exist describing the reaction rate as a function of the concentrations of the different reaction components. The two methods differ in the number of variable components. In the case of the initial reaction rate determination, only the concentration of one compound is altered, whereby all others are constant. On the contrary, in the case of batch reactions, the time course of all concentrations of all (!) components is measured. Therefore, all mass balances (see equations 23 and 24) are needed for the determination of the kinetic parameters that form a system of coupled differential equations. The values of the kinetic parameters are determined by fitting the kinetic equations to the measured data by non-linear regression (figure 21). In the case of batch experiments this is supplemented by numerical integration of the reaction rate equations. An appropriate test of the kinetic model and the kinetic parameters is the simulation of the time-courses of batch reactor experiments with different starting concentrations of substrate. These are then compared to the actual batch experiments.

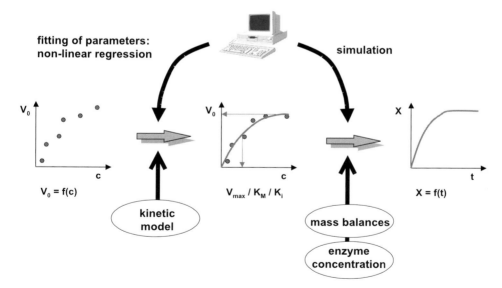

Fig. 21 Determination of kinetic parameters.

The fundamental description of enzyme kinetics dates back to Michaelis and Menten [24]. In 1913, they postulated in their theory on enzyme catalysis the existence of an enzyme-substrate (ES) complex that is formed in a reversible reaction out of substrate (S) and enzyme (E).

$$\text{E} + \text{S} \underset{k_{-1}}{\overset{k_1}{\rightleftharpoons}} \text{ES} \overset{k_2}{\longrightarrow} \text{E} + \text{P}$$

The rate limiting step is the dissociation of the ES complex ($k_{-1} \gg k_2$). The reaction rate is proportional to the 'rapid, preceding equilibrium'. As a consequence of the latter assumptions the following reaction rate equation is derived (equation 17):

$$v = \frac{V_{max} \cdot [S]}{K + [S]} \quad with: \quad K = K_S = \frac{k_{-1}}{k_1} \tag{17,18}$$

v reaction rate ($U \cdot mg^{-1}$)
V_{max} maximum reaction rate ($U \cdot mg^{-1}$)
K dissociation constant of ES complex (mM)

k_x reaction rate constant of reaction step x (min^{-1})
$[S]$ substrate concentration (mM)

Here K is identical to the dissociation constant K_S of the ES complex. Briggs and Haldane extended this theory in 1925 [29]. They substituted the assumption of the 'rapid equilibrium' by a 'steady state assumption'. This means that after starting the reaction a nearly steady state level of the ES complex is established in a very short time. The concentration of the ES complex is constant in time (d[ES]/dt = 0). In this assumption the constant K has to be enlarged by k_2, resulting in the Michaelis-Menten constant K_M.

$$K = K_M = \frac{k_{-1} + k_2}{k_1} \tag{19}$$

K_M Michaelis-Menten constant (mM)

The Michaelis-Menten constant does not describe any more the dissociation but is rather a kinetic constant. It denotes the special substrate concentration where half of the maximal activity is reached. Since the Michaelis-Menten constant approaches the dissociation constant K_S of the ES complex, it is valuable for estimating individual reaction kinetics. K_M values usually range from 10 mM to 0.01 mM. A low K_M value implies a high affinity between enzyme and substrate.

The function $v = f(S)$ is shown in figure 22:

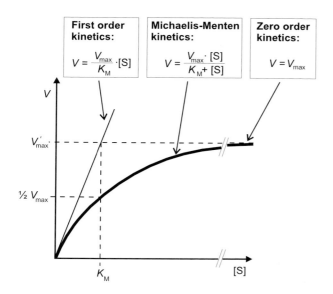

Fig. 22 Typical Michaelis-Menten-curve.

Here, two borderline cases have to be discussed. If the substrate concentration $[S]$ is far below the K_M-value a linear first-order kinetics results. The active sites of the enzyme are almost all free and the substrate concentration is rate limiting. If the substrate concentration is so high that all active sites are saturated, zero-order kinetics results.

By the above given Michaelis-Menten equation one-substrate reactions can be described. If a two-substrate reaction is to be addressed, two reaction rate terms are connected by multiplication. For a simple two-substrate reaction of A + B the double substrate kinetics for the forward reaction are:

$$v_{forward} = \frac{V_{max}^{forward} \cdot [A] \cdot [B]}{(K_{MA} + [A]) \cdot (K_{MB} + [B])} \tag{20}$$

A corresponding equation for the reverse reaction can be set up as well. The resulting total reaction rate equals the difference of forward and reverse reactions.

$$v = \frac{d[P]}{dt} = v_{forward} - v_{reverse} \tag{21}$$

The easiest way to describe a double substrate reaction is the already described kinetics (equations 20 and 21) derived from the single substrate Michaelis-Menten kinetics (equation 17). The disadvantages of this approach are:

- No information about the mechanism is included.
- Forward and reverse reactions are addressed as two totally independent reactions. No information about the equilibrium is included.

But opposite to these disadvantages there are also significant advantages of the Michaelis-Menten kinetics:

- Over broad ranges real reactors can be described with this simple type of kinetics.
- Kinetic parameters are independent of the definition of reaction direction.
- All parameters possess a graphical meaning.

A mechanistically correct description of the total reaction is only possible with a more complex model, e.g. ordered bi-bi, random bi-bi, ping-pong, etc [26]. In these models all equilibria leading to the formation of transition states are individually described. The single kinetic parameters do not have any more a descriptive meaning. The advantage of these mechanistic models is the exact description of the individual equilibria.

4.5 Basic reactor types and their mode of operation

While designing and selecting reactors for biotransformations, certain characteristic features of the biocatalysis have to be considered.

- Materials are processed in each active microbial cell, so that the main function of the bioreactor should be to provide and maintain the optimal conditions for the cells to perform the biotransformation.
- The performance of the biocatalysts depends on concentration levels and physical needs (such as salts and proper temperature, respectively). Microorganisms can adapt the structure and activity of their enzymes to the process condition, unlike isolated enzymes.
- The microbial mass can increase significantly as the biotransformation progresses, leading to a change in rheological behaviour. Also, metabolic products of cells may influence the performance of the biocatalyst.
- Microorganisms are often sensitive to strong shear stress.
- Bioreactors generally have to function under sterile conditions to avoid microbial contamination, so they must be designed for easy sterilization.
- In the case of both enzymes and whole cells as biocatalysts, the substrate and/ or the product may inhibit or deactivate the biocatalyst, necessitating special reactor layouts.
- Biotransformations with enzymes are usually carried out in a single (aqueous) or in two (aqueous/organic) phases, whereas the whole cells generally catalyze in gas-liquid-solid systems. Here, the liquid phase is usually aqueous.
- Foam formation is undesirable but often unavoidable in the case of whole cell biotransformations, where most processes are aerobic. Due consideration must be given to this aspect while designing or selecting a bioreactor.

Here only the three basic types of reactors are presented. All others are variations or deductions therefrom:

- Stirred-tank reactors (STRs) or batch reactors
- Continuously operated stirred-tank reactors (CSTRs)
- Plug-flow reactors (PFRs)

In contrast to the stirred tank reactor that is operated batchwise, the latter ones are operated continuously. By knowing the main characteristics of these fundamental reactors and some of their variations, it is possible to choose the appropriate reactor for a specific application [30]. This is especially important when dealing with a kinetically or thermodynamically limited system. In the following only the basic terms are explained. For further reading, the reader is referred to textbooks [31–40].

The *stirred tank reactor* is operated in a non-stationary way (figure 23). Assuming ideal mixing, the concentration is the same in every volume element as a function of time. With advancing conversion the substrate concentration decreases and the product concentration increases.

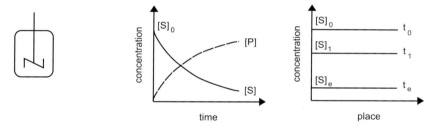

Fig. 23 Concentration-time and concentration-place profile for a stirred tank reactor.

This reactor type is widely used on an industrial scale. A variation is operation as repetitive batch or fed batch. *Repetitive batch* means that the catalyst is separated after complete conversion by filtration or even decantation. New substrate solution is added and the reaction is started again. *Fed batch* means that one reaction compound, in most cases the substrate, is fed to the reactor during the conversion.

The *continuously operated stirred tank reactor* works under product outflow conditions, meaning that the concentrations in every volume element are the same as those at the outlet of the reactor (figure 24).

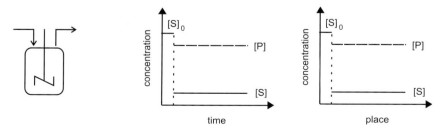

Fig. 24 Concentration-time and concentration-place profile for a continuously operated stirred tank reactor.

If the steady state is reached, the concentrations are independent of time and place. The conversion is controlled by the catalyst concentration and the residence time τ:

$$\tau = \frac{V}{F} \qquad (22)$$

τ residence time (h)

V   total reactor volume (L)

F   substrate feed rate $(L \cdot h^{-1})$

One very common application of the CSTR is the *cascade of n CSTRs* (figure 25). With increasing number n of reaction vessels the cascade is approximating the plug flow reactor. The product concentration increases stepwise from vessel to vessel.

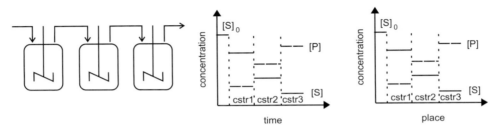

Fig. 25 Concentration-time and concentration-place profile for a cascade of continuously operated stirred tank reactors.

In the *plug flow reactor* the product concentration increases slowly over the length of the reactor (figure 26). Therefore, the average reaction rate is faster than in the continuously operated stirred tank reactor. In each single volume element in the reactor the concentration is constant in the steady state. In other words, the dimension of time is exchanged with the dimension of place in comparison to the stirred tank reactor.

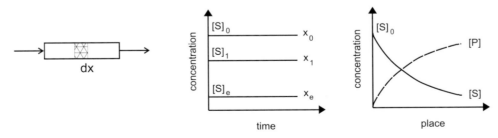

Fig. 26 Concentration-time and concentration-place profile for a plug-flow reactor.

If a reaction is limited by a substrate surplus or product inhibition the choice of the right reactor is important to yield a high reaction rate:

- Any reaction exhibiting *substrate surplus inhibition* should not be carried out in a batch reactor setup, since this results in longer reaction times. The high substrate concentration at startup lowers the reaction velocity. Here, a continuously operated stirred tank reactor is preferred. By establishing a high conversion in the steady state a low substrate concentration is achieved. Also the use of a fed batch results in a small substrate concentration.
- If *product inhibition* occurs either a stirred tank reactor, a plug flow reactor or a cascade of *n* continuously operated stirred tank reactors should be chosen. In all these reactors the product concentration increases over time. Alternatively a differential reactor with integrated product separation can be used.

4.5.1 Mass balances

The performance of the different reactor types concerning one reaction can be simulated mathematically. This is also the verification of the kinetic model of the reaction, since it should describe the course of the concentration for each compound with only a small error. The main part of the simulation model is the coupled system of differential equations of the first order, which are the mass balances of all reactants and products. The change in the concentration of one compound in time and in a volume element (= 'accumulation') is the sum of convection, reaction and diffusion.

$$\text{accumulation} = \text{convection} + \text{reaction} + \text{diffusion} \tag{23}$$

The convection term describes the change in the concentration of one compound in the reactor as the difference of the influx into the reactor and the efflux. The reaction term describes, by use of the kinetic model, the change of the concentration of one compound as a result of the reaction. The reaction velocity v is the sum of the individual reaction velocities describing consumption of a substrate or formation of a product. Diffusion is only given in the case where no ideal mixing is stated.

Depending on the reactor type chosen, the mass balance can be simplified, stating ideal mixing:

4.5.1.1 Stirred tank reactor:

The mass balance of each compound is defined by the reaction rate only, since no fluid enters or leaves the reactor. At a defined time the concentrations are the same in every volume element (diffusion = 0). There is no influx or efflux of substrate or products to a single volume element in time (convection = 0).

The mass balance is simplified to:

$$-\frac{d[S]}{dt} = v \tag{24}$$

The time t that is necessary to reach a desired conversion X can be determined by integrating the reciprocal rate equation from zero to the desired conversion X.

$$dt = -\frac{d[S]}{v} = \frac{[S]_0 \cdot dX}{v} \quad \Rightarrow \quad t = [S]_0 \cdot \int_0^X \frac{1}{v} \cdot dX \qquad (25)$$

4.5.1.2 Plug-flow reactor:

The change of reaction rate within a unit volume passing the reactor length is equivalent to a change corresponding to the residence time τ within the reactor. Diffusion is neglected in an ideal plug-flow reactor (diffusion = 0) and all the concentrations will not change with time in the steady state (accumulation = 0).

Just by exchanging t for τ equation 25 can be also used for the plug flow reactor to determine the residence time τ necessary to reach a desired conversion X.

$$\tau = [S]_0 \cdot \int_0^X \frac{1}{v} \cdot dX \qquad (26)$$

4.5.1.3 Continuously operated stirred tank reactor:

The concentration of substrate S within the reactor is effected by convection as well as by reaction. There is no diffusion between different volume elements (diffusion = 0) and in the steady state the concentrations will not change with time (accumulation = 0).

The mass balance is simplified to:

$$0 = -\frac{d[S]}{dt} = \frac{[S]_0 - [S]}{\tau} + v_S \qquad (27)$$

The residence time τ, that is necessary to reach a desired conversion, can be determined by equation 28:

$$\tau = [S]_0 \cdot \frac{1}{v} \cdot dX \qquad (28)$$

References

[1]   Deurzen, M.P.J. (1996) Selective oxidations catalyzed by chloroperoxidase, Thesis Delft University of Technology
[2]   Seelbach, K. (1997) Chloroperoxidase – Ein industrieller Katalysator? Regio- und enantioselective Oxidationen, Dissertation Mathematisch-Naturwissenschaftlichen Fakultät der Rheinischen Friedrich-Wilhelms-Universität Bonn
[3]   Bradford, M.M. (1976) A rapid and sensitive method for the quantitation of protein utilizing the principle of protein-dye binding, Anal. Biochem. 72, 248–254
[4]   Voet, D., Voet, J. (1995) Biochemistry, John Wiley & Sons, New York
[5]   Lehninger, A.L. (2000) Principles of Biochemistry, Worth Publishers, Inc., New York
[6]   Stryer, L. (1995) Biochemistry, W.H. Freeman and Co., San Francisco
[7]   Christen, H.R., Vögtle, F. (1990) Organische Chemie II, Von den Grundlagen zur Forschung, Otto Salle Verlag GmbH, Frankfurt
[8]   Primrose, S.W. (1987) Modern Biotechnology, Blackwell Scientific Publications, Oxford

[9]   Kragl, U. (1996) Immobilized enzymes and membrane reactors, in: Industrial Enzymology: The Application of Enzymes in Industry (Godfrey, T. and West, S., eds.) pp. 275–283, Macmillan Press LTD, London

[10] Noble, R.D., Stern, S.A. (1995) Membrane Separations Technology. Principles and Applications, Elsevier, Amsterdam

[11] Mulder, M. (1996) Basic Principles of Membrane Technology, Kluwer Academic Publishers, Dordrecht

[12] Wandrey, C., Wichmann, R., Bückmann, A.F., Kula, M.-R. (1980) Immobilization of biocatalysts using ultrafiltration techniques, in: Enzyme Engineering 5 (Weetall, H.H., Royer, G.P., eds.) pp. 453–456, Plenum Press, New York

[13] Flaschel, E., Wandrey, C., Kula, M.-R. (1983) Ultrafiltration for the separation of biocatalysts, in: Downstream Processing Advances in Biochemical Engineering / Biotechnology 26 (Fiechter, A., ed.) pp. 73–142, Springer Verlag, Berlin, Heidelberg, New York

[14] Chenault, H.K., Whitesides, G.M. (1987) Regeneration of nicotinamide cofactors for use in organic synthesis, Appl. Biochem. Biotechnol. 14, 147–197

[15] Fang, J.-M., Lin, C.-H. (1995) Enzymes in organic synthesis: oxidoreductions, J. Chem. Soc. Perkin Trans. 1, 967–978

[16] Wong, C.-H., Whitesides, G.M. (1994) Enzymes in Synthetic Organic Chemistry, Elsevier Science Ltd., Oxford

[17] Ruppert, R., Herrmann, S., Steckhan, E. (1988) Very efficient reduction of NAD(P)+ with formate catalysed by cationic rhodium complexes, J. Chem. Soc., Chem. Com. 1150–1151.

[18] Faber, K. (1997) Biotransformations of non-natural compounds: state of the art and future development, Pure & Appl. Chem. 69(8), 1613–1632

[19] Stecher, H., Faber, K. (1997) Biocatalytic deracemization techniques: Dynamic resolutions and stereoinversions, Synthesis 1–16

[20] Chen, S.-T., Huang, W.-H. (1994) Resolution of amino acids in a mixture of 2-methyl-2-propanol/water (19:1) catalyzed by alcalase via in situ racemization of one antipode mediated by pyridoxal 5-phosphate, J. Org. Chem. 59, 7580

[21] Kurihara, T. (1995) Isomerizations, in: Enzyme Catalysis in Organic Synthesis (Drauz, K., Waldmann, H., eds.) VCH Verlagsgesellschaft, Weinheim

[22] Kitamura, M., Tokunaga, M. (1993) Mathematical treatment of kinetic resolution of chirally labile substrates, Tetrahedron 49(9), 1853–1860

[23] Hirohara, H., Nishizawa, M. (1998) Biochemical synthesis of several chemical inseticide intermediates and mechanism of action of relevant enzymes, Biosci. Biotechnol. Biochem. 62, 1–9

[24] Michaelis, L., Menten, M.L. (1913) Die Kinetik der Invertinwirkung, Biochem. Z. 49, 333–369

[25] Bisswanger, H. (1979) Theorie und Methoden der Enzymkinetik, Verlag Chemie, Weinheim

[26] Segel, I.H. (1975) Enzyme Kinetics, John Wiley & Sons, Inc, New York

[27] Cornish-Bowden, A. (1981) Fundamentals of Enzyme Kinetics, Butterworth & Co Ltd., London

[28] Biselli, M., Kragl, U., Wandrey, C. (1995) Reaction engineering for enzyme-catalyzed biotransformations, in: Enzyme Catalysis in Organic Synthesis (Drauz, K., Waldmann, H., eds.) pp. 89–155, VCH Verlagsgesellschaft mbH, Weinheim

[29] Briggs, G.E., Haldane, J.B.S (1925) Biochem. J. 19, 338–339

[30] Kragl, U., Liese, A. (1999) Biotransformations, engineering aspects, in: The Encyclopedia of Bioprocess Technology: Fermentation, Biocatalysis & Bioseparation (Flickinger, M.C., Drew, S.W., eds.) pp. 454–464, John Wiley & Sons, New York

[31] Bailey, J.E., Ollis, D. F. (1986) Biochemical Engineering Fundamentals, McGraw-Hill, New York

[32] Baerns, M., Hofmann, H., Renken, A. (1992) Chemische Reaktionstechnik, Georg Thieme Verlag, Stuttgart

[33] Ertl, G., Knözinger, H., Weitkamp, J. (1997) Handbook of Heterogeneous Catalysis, Wiley-VCH, Weinheim

[34] Fitzer, E., Fritz, W., Emig, G. (1995) Technische Chemie, Springer Verlag, Berlin

[35] Fogler, S.H. (1998) Elements of Chemical Reaction Engineering, Prentice-Hall PTR

[36] Froment, G.F., Bischoff, K.B. (1990) Chemical reactor analysis and design, John Wiley & Sons, New York

[37] Jakubith, M. (1998) Grundoperationen und chemische Reaktionstechnik, Wiley-VCH, Weinheim

[38] Levenspiel, O. (1999) Chemical Reaction Engineering, Wiley-VCH, New York

[39] Richardson, J.F., Peacock, D.G. (1994) Chemical Engineering, Vol. 3, Chemical & Biochemical Reactors & Process Control, Pergamon Press, Oxford

[40] Westerterp, K.R., van Swaaij, W.P.M., Beenackers, A.A.C.M. (1984) Chemical Reactor Design and Operation, Wiley, New York

5 Processes

ANDREAS LIESE

Institute of Biotechnology
Forschungszentrum Jülich GmbH
D-52425 Jülich, Germany

KARSTEN SEELBACH

Corporate Process Technology
Degussa-Hüls AG
D-45764 Marl, Germany

ARNE BUCHHOLZ, JÜRGEN HABERLAND

Institute of Biotechnology
Forschungszentrum Jülich GmbH
D-52425 Jülich, Germany

In this chapter you will find industrial biotransformations sorted in the order of the enzyme classes (EC). One type of biotransformation is often carried out by several companies leading to identical or the same class of products. Here only one exemplary process is named. Only in cases where the reaction conditions differ fundamentally, resulting in a totally different process layout, these are listed separately (e.g. L-aspartic acid).

It is difficult to judge which processes are applied on an industrial scale. But even if not all of the following processes are used in the ton scale, they are at least performed by industrial companies to produce compounds for research or clinical trials in kg scale.

If you know of any new biotransformation carried out on an industrial scale, or you notice that we missed any important one, we would be pleased if you could supply us with the appropriate information. For your convenience you will find a form at the end of this book.

On the next pages you will find a process example with all necessary explanations for an easy understanding of all used parameters and symbols in the flow sheets.

By reading the example you will also see the maximum number of parameters we have tried to find for each process.

X.X.X.X = enzyme nomenclature number

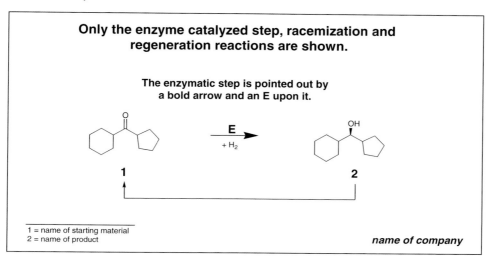

Only the enzyme catalyzed step, racemization and regeneration reactions are shown.

The enzymatic step is pointed out by a bold arrow and an E upon it.

E

+ H$_2$

1

2

1 = name of starting material
2 = name of product

name of company

Fig. X.X.X.X – 1

1) Reaction conditions

[N]:	molar concentration, mass concentration [molar mass] of component **N**
pH:	pH of reaction solution
T:	reaction temperature in °C
medium:	type of reaction medium: in most cases aqueous, but can also be several phases in combination with organic solvents
reaction type:	suggestion of enzyme nomenclature for the type of enzymatic catalyzed reaction
catalyst:	application of catalyst: solubilized / immobilized enzyme / whole cells
enzyme:	systematic name (alternative names)
strain:	name of strain
CAS (enzyme):	[CAS-number of enzyme]

2) Remarks

- Since in the chemical drawing on top of the page only the enzymatic step is shown, prior or subsequent steps, which might be part of the industrial process can be found here.

- Since it is often difficult to gain knowledge of the true industrial process conditions, those published in the past for the same reaction system are given.

- Beside the already mentioned topics you will find additional information regarding the discussed biotransformation, e.g. substrate spectrum, enzyme improvement, immobilization methods, and all other important information which does not fit to another category.

- If an established synthesis is replaced by a biotransformation, the classical, chemical synthesis can be found here as well.

3) Flow scheme

- The flow schemes are reduced to their fundamental steps. A list and explanation of the symbols is given in the next figures:

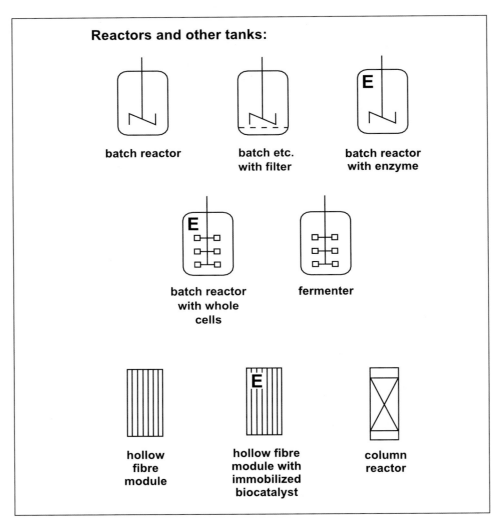

Fig. X.X.X.X – 2

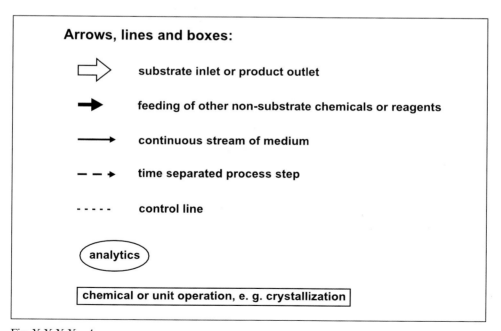

Separation and purification untis:

distillation

falling film evaporator

extraction

aqueous | organic

column for adsorption or ion-exchange

filtration

pervaporation

perv.

Fig. X.X.X.X – 3

Arrows, lines and boxes:

substrate inlet or product outlet

feeding of other non-substrate chemicals or reagents

continuous stream of medium

time separated process step

control line

analytics

chemical or unit operation, e. g. crystallization

Fig. X.X.X.X – 4

4) Process parameters

conversion:	molar conversion in %
yield:	molar yield in %
selectivity	molar selectivitiy in %
ee:	enantiomeric excess in %
chemical purity:	purity of component in %
reactor type:	fed or repetitive batch, CSTR, plug flow reactor
reactor volume:	reactor volume in L
capacity:	mass of product per year in $t \cdot a^{-1}$
residence time:	time for one batch reaction or residence time in continuous operated reactor in hours
space-time-yield:	mass of product per time and reactor volume in $kg \cdot L^{-1} \cdot d^{-1}$
down stream processing:	purification of raw material after reaction, e. g. crystallization, filtration, distillation
enzyme activity:	in U(nits = $\mu mol \cdot min^{-1}$) per mass of protein (mg) or volume of reaction solution (L)
enzyme consumption:	amount of consumed enzyme per mass of product
enzyme supplier:	company, country
start-up date:	start of production
closing date:	end of production
production site:	company, country
company:	company, country

5) Product application

- The application of the product as intermediate or the end-product are given here.

6) Literature

- Cited literature you will find here. Often a personal communication or direct information of the company provided us neccessary information.

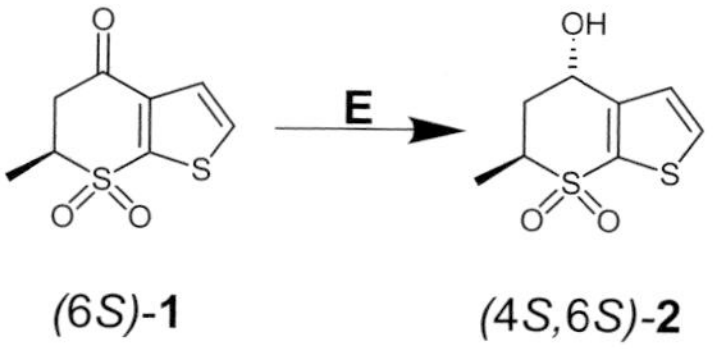

Alcohol dehydrogenase
Neurospora crassa

(6S)-**1**   (4S,6S)-**2**

1 = 5,6-dihydro-6-methyl-4H-thieno[2,3b]thiopyran-4-one-7,7-dioxide
2 = 5,6-dihydro-4-hydroxy-6-methyl-4H-thieno[2,3b]thiopyran-7,7-dioxide

Zeneca Life Science Molecules

Fig. 1.1.1.1 – 1

1) Reaction conditions

pH:	3.8–4.3
T:	33 °C
medium:	aqueous
reaction type:	redox reaction
catalyst:	suspended whole cells
enzyme:	alcohol NAD$^+$ oxidoreductase (alcohol dehydrogenase)
strain:	*Neurospora crassa*
CAS (enzyme):	[9031–72–5]

2) Remarks

- In comparison to the biotransformation the inversion of the *cis* alcohol in the chemical synthesis of trusopt™ (see product application) is not quantitative:

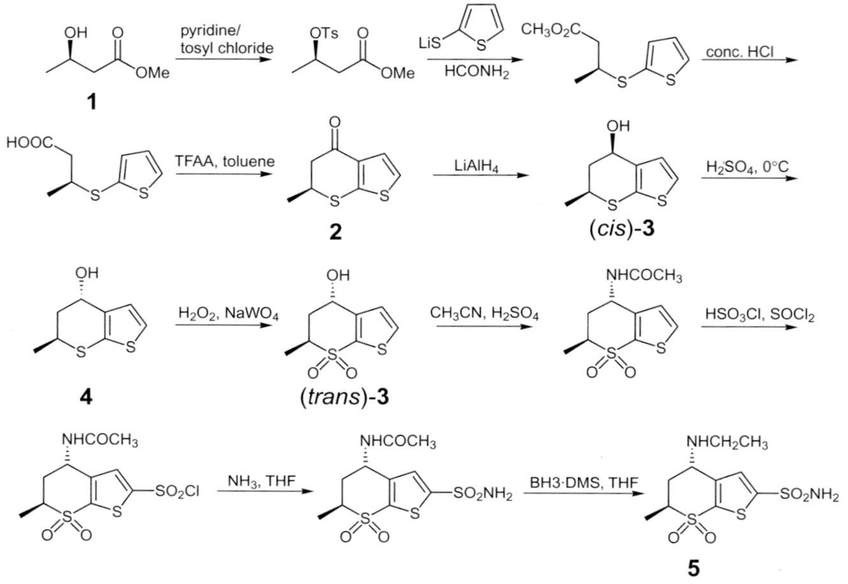

```
1 = (R)-3-hydroxy-methyl butyrate
2 = ketosulfide
3 = isomer of sulfide alcohol
4 = (4S)-alcohol
5 = trusopt
```

Fig. 1.1.1.1 – 2

- The biological route overcomes the problem of incomplete epimerization:

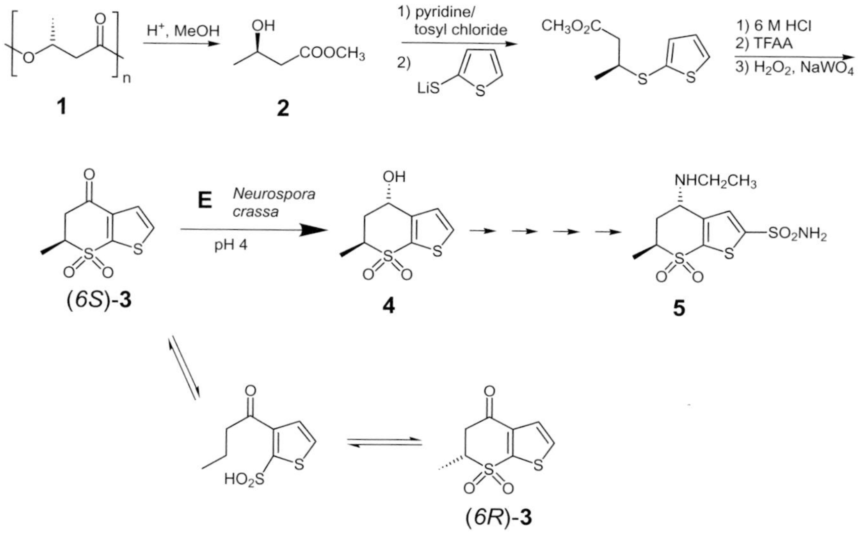

1 = biologically derived homopolymer
2 = (*R*)-3-hydroxy-methyl butyrate
3 = methyl ketosulfone
4 = (4*S*,6*S*)-hydroxy-sulfone
5 = MK-0507

Fig. 1.1.1.1 – 3

- The (*R*)-3-hydroxy-butyrate, which is responsible for the stereochemistry of the methyl group in the sulfone ring, can be produced by depolymerization of natural plastics.
- These plastics, e.g. biopol from Zeneca, are natural polymers produced by some microorganisms as storage compounds.
- The main problem of the chemoenzymatic synthesis is the epimerization of the (6*S*)-methyl ketosulfone in aqueous media above pH 5 (ring opening not possible for the reduced species).
- The reaction is carried out below pH 5 and to prevent the accumulation of the sulfone the addition is controlled by conversion to the product.
- The NADPH-specific enzyme could be purified.

3) Flow scheme

Not published.

4) Process parameters

yield:	> 85 %
ee:	> 98 %
chemical purity:	> 99 %
reactor type:	fed batch
capacity:	multi t
down stream processing:	crystallization
start-up date:	1994
company:	Zeneca Life Science Molecules, U.K.

5) Product application

- Intermediate in the synthesis of the carbonic anhydrase inhibitor trusopt (see remarks).

- Trusopt (invented and marketed by Merck & Co) is a novel, topically active treatment for glaucoma.

- Glaucoma is a disease of the eye characterized by increased intraocular pressure, which results in defects in the field of vision. If left untreated, the disease can cause irreversible damage to the optic nerve, eventually leading to blindness.

6) Literature

- Blacker, A.J., Holt, R.A. (1997) Development of a multistage chemical and biological process to an optically active intermediate for an anti-glaucoma drug, in: Chirality In Industry II (Collins, A.N., Sheldrake, G.N. and Crosby, J., eds.) pp. 246–261, John Wiley & Sons, New York

- Blacklock, T.J., Sohar, P., Butcher, J.W., Lamanec, T., Grabowski, E.J.J. (1993) An enantioselective synthesis of the topically-active carbonic anhydrase inhibitor MK-0507: 5,6-dihydro-(*S*)-4-(ethylamino)-(*S*)-6-methyl-4H-thieno[2,3-*β*]thiopyran-2-sulfonamide 7,7-dioxide hydrochloride, J. Org. Chem. **58**, 1672–1679

- Holt, R. A. (1996) Microbial asymmetric reduction in the synthesis of a drug intermediate, Chimica Oggi **9**, 17–20

- Holt, R. A. and Rigby, S. R. (1996) Process for microbial reduction producing 4(*S*)-hydroxy-6(*S*)methyl-thienopyran derivatives, Zeneca Limited, US 5580764

- Zaks, A., Dodds, D.R., (1997) Application of biocatalysis and biotransformations to the synthesis of pharmaceuticals, DDT **2**, 513–531

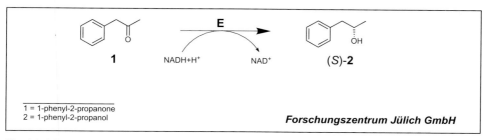

1 = 1-phenyl-2-propanone
2 = 1-phenyl-2-propanol

Forschungszentrum Jülich GmbH

Fig. 1.1.1.1 – 1

1) Reaction conditions

[**1**]:	0.015 M, 1.95 g·L^{-1} [134,18 g·mol^{-1}]
pH:	6.7
T:	25 °C
medium:	aqueous
reaction type:	redox reaction
catalyst:	solubilized enzyme
enzyme:	alcohol-NAD$^+$ oxidoreductase (alcohol dehydrogenase)
strain:	*Rhodococcus erythropolis*
CAS (enzyme):	[9031–72–5]

2) Remarks

- The cofactor regeneration is carried out with a formate dehydrogenase from *Candida boidinii* (FDH = formate dehydrogenase, EC 1.2.1.2) utilizing formate that is oxidized to CO_2:

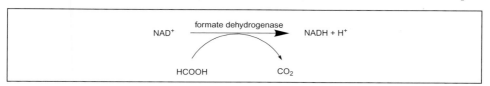

Fig. 1.1.1.1 – 2

- This reactor concept is especially attractive for starting materials of low solubility. The starting materials are directly titrated into the aqueous phase. The process consists of three loops: I: aqueous loop with a hydrophilic ultra-filtration membrane retaining the enzymes; II: permeated aqueous reaction solution products, starting materials and cofactors are passed through the tube phase of the extraction module; III: organic solvent phase, containing extracted products and starting materials.

- The charged cofactors (NAD$^+$/NADH) remain in the aqueous loops I and II. Therefore only deactivated cofactor needs to be replaced resulting in an economically high total turnover number (= ttn).

- The extraction module consists of microporous, hydrophobic hollow-fiber membranes. The organic extraction solvent is recycled by continuous distillation. The product remains in the bottom of the distillation column.

- Using this method very good space-time yields are obtainable in spite of the low substrate solubilities:

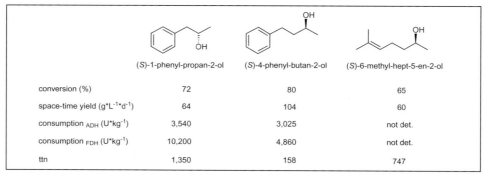

	(S)-1-phenyl-propan-2-ol	(S)-4-phenyl-butan-2-ol	(S)-6-methyl-hept-5-en-2-ol
conversion (%)	72	80	65
space-time yield (g*L^{-1}*d^{-1})	64	104	60
consumption $_{ADH}$ (U*kg^{-1})	3,540	3,025	not det.
consumption $_{FDH}$ (U*kg^{-1})	10,200	4,860	not det.
ttn	1,350	158	747

Fig. 1.1.1.1 – 3

3) Flow scheme

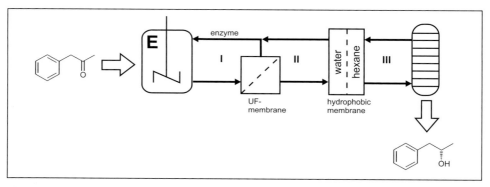

Fig. 1.1.1.1 – 4

4) Process parameters

conversion:	72 %
yield:	72 %
selectivity:	100 %
ee:	>99 %
reactor type:	CSTR (enzyme bimembrane reactor)
reactor volume:	0.05 L
residence time:	0.33 h
space-time-yield:	63.5 g·L^{-1}·d^{-1}
down stream processing:	distillation
enzyme activity:	0.95 U·mL^{-1}
enzyme consumption:	3.5 U·g^{-1}
enzyme supplier:	Institute of Enzyme Technology, University of Düsseldorf, Germany
start-up date:	1996
production site:	Jülich, Germany
company:	Forschungszentrum Jülich GmbH, Germany

5) Product application

- The trivial name for (*S*)-6-methyl-5-hepten-2-ol is (*S*)-(+)-sulcatol, a pheromone from the scolytid beetle *Gnathotrichus sulcatus / Gnathotrichus retusus.*

- (*S*)-1-Phenyl-2-propanol is used as an intermediate for the synthesis of amphetamines (sympathomimetics):

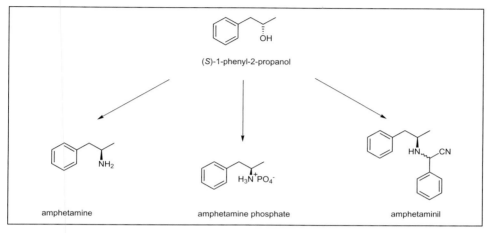

Fig. 1.1.1.1 – 5

- (*S*)-4-Phenyl-2-butanol is used as a precursor for anti-hypertensive agents and spasmolytics or anti-epileptics:

Fig. 1.1.1.1 – 6

6) Literature

- Bracher, F., Litz, T. (1994) Building blocks for the preparation of enantiomerically pure drugs containing a phenylalkylamine moiety, Arch. Pharm. **327**, 591–593

- Johnston, B., Slessor, K. (1979) Facile synthesis of the enantiomers of sulcatol, Can. J. Chem. **57**, 233–235

- Kragl, U., Kruse, W., Hummel, W., Wandrey, C. (1996) Enzyme engineering aspects of biocatalysis: Cofactor regeneration as example, Biotechnol. Bioeng. **52**, 309–319

- Kruse, W., Hummel, W., Kragl, U. (1996) Alcohol-dehydrogenase-catalyzed production of chiral hydrophobic alcohols. A new approach leading to a nearly waste-free process, Recl. Trav. Chim. Pays-Bas **115**, 239–234

- Kruse, W., Kragl, U., Wandrey, C. (1996) Verfahren zur kontinuierlichen enzymkatalysierten Gewinnung hydrophober Produkte, Forschungszentrum Jülich GmbH, DE 4436149 A1

- Kruse, W., Kragl, U., Wandrey, C. (1998) Process for the continuous enzymatic extraction of hydrophobic products and device suitable therefore, Forschungszentrum Jülich GmbH, Germany, US 5,795,750

- Liang, S.; Paquette, L.A. (1990) Biocatalytis-based synthesis of optically pure (C6)-functionalized 1-(*tert*-butyldimethyl-silyloxy)-2-methyl-(*E*)-heptenes; Tetrahedron Asym. **1**, 445–452

- Mori, K. (1975) Synthesis of optically active forms of sulcatol – The aggregation pheromone in the scolytid beetle, *Gnathotrichus sulcatus*, Tetrahedron **31**, 3011–3012

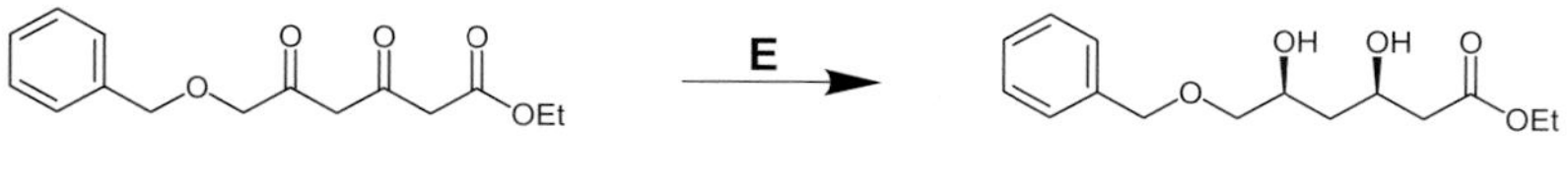

1 $(3R,5S)$-**2**

1 = 6-benzyloxy-3,5-dioxo-hexanoic acid ethyl ester
2 = 6-benzyloxy-3,5-dihydroxy-hexanoic acid ethyl ester

Bristol-Myers Squibb

Fig. 1.1.1.1 – 1

1) Reaction conditions

[**1**]:	0.036 M, 10 g · L^{-1} [278.30 g · mol^{-1}]
pH:	5.9
T:	33 °C
medium:	aqueous
reaction type:	redox reaction
catalyst:	solubilized enzyme (cell extract)
enzyme:	alcohol-NAD$^+$ oxidoreductase (alcohol dehydrogenase)
strain:	*Acinetobacter calcoaceticus*
CAS (enzyme):	[9031–72–5]

2) Remarks

- The bioreduction can be carried out with whole cells as well as with cell extract.

- To faciliate the cofactor regeneration of NADH, glucose dehydrogenase, glucose and NAD$^+$ is added to the reaction medium.

- The educt is prepared by the following method:

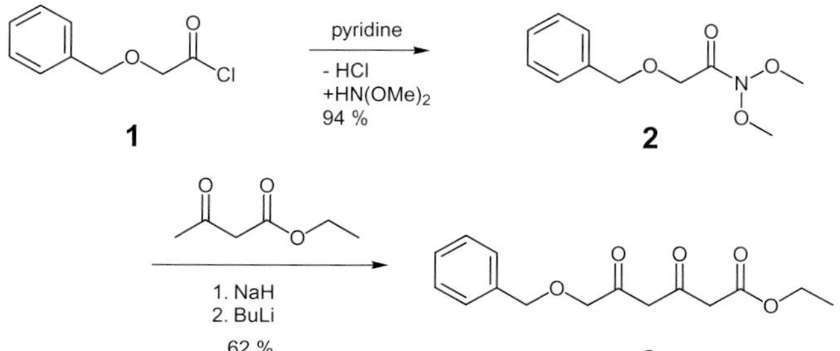

pyridine
- HCl
+HN(OMe)$_2$
94 %

1. NaH
2. BuLi
62 %

1 = benzyloxy-acetyl chloride
2 = 2-benzyloxy-*N,N*-dimethoxy-acetamide
3 = 6-benzyloxy-3,5-dioxo-hexanoic acid ethyl ester

Fig. 1.1.1.1 – 2

- This biotransformation is an alternative to the chemical synthesis via the chlorohydrin and selective hydrolysis of the acyloxy group. By chemical synthesis an overall yield of 41 % after final fractional distillation is achieved:

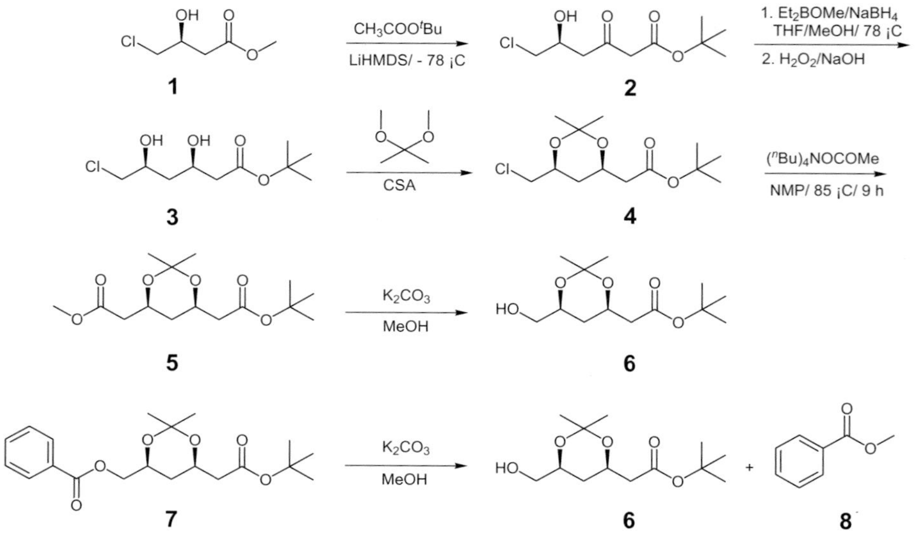

1 = 4-chloro-3-hydroxy-butyric acid methyl ester
2 = 6-chloro-5-hydroxy-3-oxo-hexanoic acid *tert*-butyl ester
3 = 6-chloro-3,5-dihydroxy-hexanoic acid *tert*-butyl ester
4 = (6-chloromethyl-2,2-dimethyl-[1,3]dioxan-4-yl)-acetic acid *tert*-butyl ester
5 = (6-*tert*-Butoxycarbonylmethyl-2,2-dimethyl-[1,3]dioxan-4-yl)-acetic acid methyl ester
6 = (6-hydroxymethyl-2,2-dimethyl-[1,3]dioxan-4-yl)-acetic acid *tert*-butyl ester
7 = (6-benzyl-2,2-dimethyl-[1,3]dioxan-4-yl)-acetic acid *tert*-butyl ester
8 = benzoic acid methyl ester

Fig. 1.1.1.1 – 3

3) Flow scheme

Not published.

4) Process parameters

yield:	92 %
ee:	99 %
reactor type:	batch
down stream processing:	centrifugation and extraction with methylene chloride
company:	Bristol-Myers Squibb, USA

5) Product application

- 6-Benzyloxy-(3R,5S)-dihydroxy-hexanoic acid ethyl ester is a key chiral intermediate for anticholesterol drugs that act by inhibition of hydroxy methyl glutaryl coenzyme A (HMG-CoA) reductase. The synthesis is shown in the following scheme:

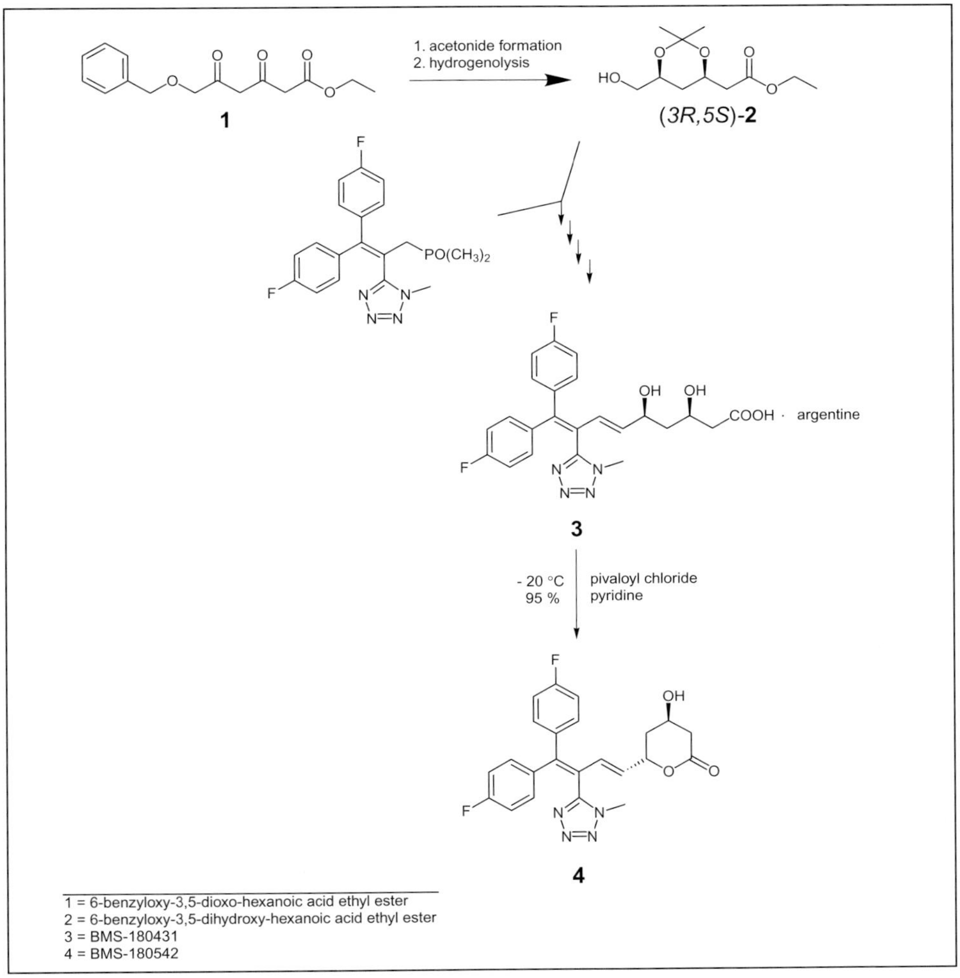

1. acetonide formation
2. hydrogenolysis

1

(*3R,5S*)-**2**

PO(CH$_3$)$_2$

3 · argentine

- 20 °C | pivaloyl chloride
95 % | pyridine

4

1 = 6-benzyloxy-3,5-dioxo-hexanoic acid ethyl ester
2 = 6-benzyloxy-3,5-dihydroxy-hexanoic acid ethyl ester
3 = BMS-180431
4 = BMS-180542

Fig. 1.1.1.1 – 4

6) Literature

- Patel, R.N., McNamee, C.G., Banerjee, A., Szarka, L.J. (1993) Stereoselective microbial or enzymatic reduction of 3,5-dioxo esters to 3-hydroxy-5-oxo, 3-oxo-5-hydroxy, and 3,5-dihydroxy esters, E.R. Squibb & Sons, Inc., EP 0569998A2

- Patel, R.N., Banerjee, A., McNamee, C.G., Brzozowski, D., Hanson, R.L., Szarka, L.J. (1993) Enzyme Microb. Technol. **15**, 1014–1021

- Sit, S.Y., Parker, R.A., Motoc, I., Han, W., Balasubramanian, N. (1990) Synthesis, biological profile, and quantitative structure-activity relationship of a series of novel 3-hydroxy-3-methylglutaryl coenzyme A reductase inhibitors, J.Med.Chem. **33** (11), 2982–2999

- Thottathil, J.K. (1998) Chiral synthesis of BMS-180542 and SQ33600; Tetrazole and phosphonic acid based HMG-CoA reductase inhibitors, Chiral USA 98, Chiral Technology – The way ahead, San Francisco, CA, USA, 18–19 May 1998

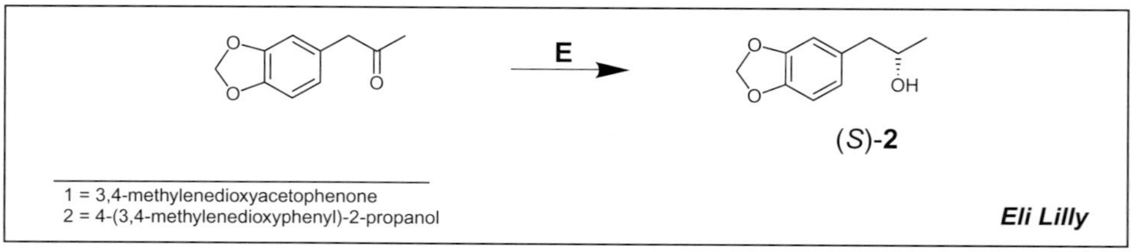

(S)-2

1 = 3,4-methylenedioxyacetophenone
2 = 4-(3,4-methylenedioxyphenyl)-2-propanol

Eli Lilly

Fig. 1.1.1.1 – 1

1) Reaction conditions

[**1**]:	< 0.011 M, 2 g·L^{-1} [178.18 g·mol^{-1}]
pH:	7.0
T:	33–35 °C
medium:	aqueous
reaction type:	reduction of keto group
catalyst:	suspended whole cells
enzyme:	alcohol-NAD$^+$ oxidoreductase (alcohol dehydrogenase)
strain:	*Zygosaccharomyces rouxii*
CAS (enzyme):	[9031–72–5]

2) Remarks

- Since the toxic limit of the substrate is 6 g·L^{-1} the substrate is adsorbed on XAD-7 resin (80 g·L^{-1} resin, resulting in a reaction concentration of 40 g·L^{-1} reaction volume). Using this method the volumetric productivity can be held at a high level.

- The advantage of the XAD-7 resin in comparison to organic solvents of the same polarity (log P < 2) is its non-toxic and non-denaturating character.

- The XAD-7 resin is reused for three times without any loss in performance.

- As reactor a Rosenmund agitated filter-dryer is used. A packed bed reactor would clog and an expanded bed reactor is not applicable due to the low density of XAD-7. The resin would be blown out of the top of it. The special agitator used has a hydraulic design that enables a filtration without clogging of the filter.

- The desorption of educt is limited to the equilibrium concentration of approximately 2 g·L^{-1} in the aqueous phase.

- To separate the yeast cells from the product that is adsorbed on the resin a 150 μm filter screen is used. In contrast to usual filtrations the resin (~ 500 μm) is retained and the yeast cells (~ 5 μm) stay in the filtrate. The product is liberated by washing the resin with acetone.

3) Flow scheme

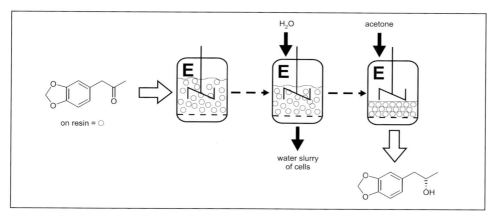

Fig. 1.1.1.1 – 2

4) Process parameters

yield:	96 %
ee:	> 99.9 %
chemical purity:	95 %
reactor type:	batch
reactor volume:	300 L
capacity:	kg scale
space-time-yield:	$75\ g \cdot L^{-1} \cdot d^{-1}$
down stream processing:	filtration, resin extraction
company:	Eli Lilly and Company, USA

5) Product application

- The product is converted to LY 300164, an orally active benzodiazepine:

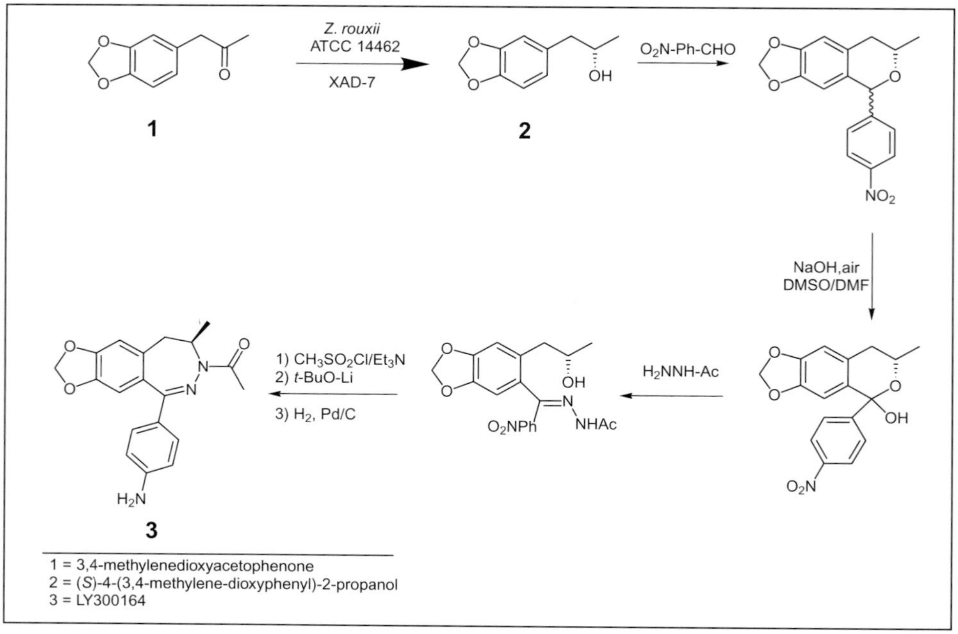

Fig. 1.1.1.1 – 3

- The product is tested for efficancy in treating amylotropic lateral sclerosis.

6) Literature

- Anderson, B.A., Hansen, M.M., Harkness, A.R., Henry, C.L., Vicenzi, J.T., Zmijewski, M.J. (1995) Application of a practical biocatalytic reduction to an enantioselective synthesis of the 5H-2,3-benzodiazepine LY300164, J. Am. Chem. Soc. **117**, 12358–12359

- Vicenzi, J.T., Zmijewski, M.J., Reinhard, M.R., Landen, B.E., Muth, W.L., Marler, P.G. (1997) Large-scale stereoselective enzymatic ketone reduction with *in-situ* product removal via polymeric adsorbent resins, Enzyme Microb. Technol. **20**, 494–499

- Zaks, A., Dodds, D.R. (1997) Application of biocatalysis and biotransformations to the synthesis of pharmaceuticals, Drug Discovery Today **2**, 12, 513–530

- Zmijewski, M.J., Vicenzi, J., Landen, B.E., Muth, W., Marler, P., Anderson, B. (1997) Enantioselective reduction of 3,4-methylene-dioxyphenyl acetone using *Candida famata* and *Zygosaccharomyces rouxii*, Appl. Microbiol. Biotechnol. **47**, 162–166

Lactate dehydrogenase
Staphylococcus epidermidis

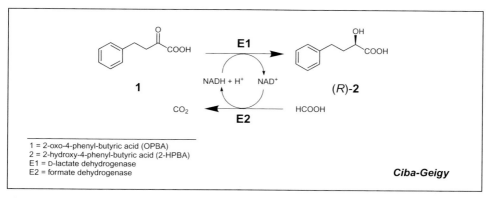

1 = 2-oxo-4-phenyl-butyric acid (OPBA)
2 = 2-hydroxy-4-phenyl-butyric acid (2-HPBA)
E1 = D-lactate dehydrogenase
E2 = formate dehydrogenase

Ciba-Geigy

Fig. 1.1.1.28 – 1

1) Reaction conditions

[**1**]:	0.2 M, 35.6 g $\cdot$ L^{-1} [178.18 g $\cdot$ mol^{-1}]
pH:	8.0
T:	30 °C
medium:	aqueous
reaction type:	redox reaction
catalyst:	solubilized enzyme
enzyme:	(*R*)-lactate-NAD oxidoreductase
strain:	*Staphylococcus epidermidis*
CAS (enzyme):	[9028–36–8]

2) Remarks

- Cofactor regeneration is carried out by formate dehydrogenase from *Candida boidinii* utilizing formate and producing CO_2. By the cofactor regeneration no by-product is formed that needs to be separated. In the steady state of the continuously operated stirred tank reactor a total turnover number of 900 is achieved.

- NAD$^+$ is added as cofactor with a concentration of 0.2 mM and ammonium formate with a concentration of 0.35 M.

- The optimal working pH would have been 6.5, but a pH of 8.0 is chosen due to better solubility of the hydroxy acid.

- The production is carried out in a continuously operated stirred tank reactor equipped with an ultrafiltration membrane (cutoff 10,000 Da) to retain the enzymes.

- For stabilization of the enzymes 0.15 % mercaptoethanol and 1 mM EDTA are added to the substrate solution.

- At a conversion of 90 % the pH is shifted from 6.2 (substrate solution) to 8.0. The reason is the different pK_a-values of 2-oxo-4-phenylbutyric acid (pK_a = 2.3) and (*R*)-2-hydroxy-4-phenylbutyric acid (pK_a = 3.76).

- The data given relate to the above process with isolated enzymes.

113

Alternatively immobilized whole cells of *Proteus vulgaris* can be used in a fixed bed reactor:

- By using small units of biomass in form of a packed bed of immobilized cells, high amounts of the hydroxy acid can be produced. The space-time-yield is high and the costs for down-stream processing are low (conversion > 99.5 %; ee = 99.8 %; space-time yield = 180 $g \cdot L^{-1} \cdot d^{-1}$; enzyme consumption = 159 $U \cdot kg^{-1}$).

- Per liter of fermentation broth 1.45 g cell mass (80 wt.-% water) can be isolated.

- The emulsion of cells, chitosan and solid silica acid is dropped into a polyphosphate solution resulting in ionotropic gel-beads.

- 1,000 g cell emulsion yields 400 g (= 800 mL) beads. The beads contain 0.25 g (wet cells) $\cdot g^{-1}$ (beads).

- Here electron mediators (= V) are used instead of NADH or NADPH:

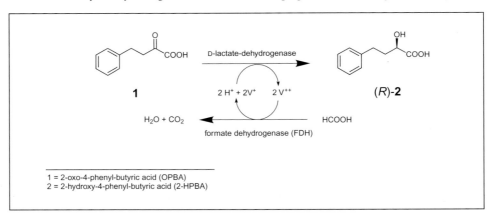

benzyl viologen   carbamoyl methyl viologen

Fig. 1.1.1.28 – 2

- The enzymes in the cells are capable of reducing and oxidizing the mediators. In this case the carbamoylmethylviologen is used. The following figure shows the regeneration cycle:

1 = 2-oxo-4-phenyl-butyric acid (OPBA)
2 = 2-hydroxy-4-phenyl-butyric acid (2-HPBA)

Fig. 1.1.1.28 – 3

- To avoid degassing of CO_2 the reaction is carried out under a pressure of 3 bar.

3) Flow scheme

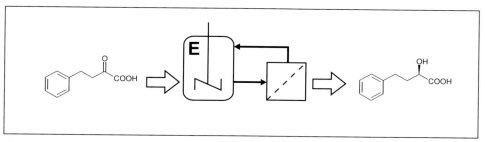

Fig. 1.1.1.28 – 4

4) Process parameters

conversion:	91 %
ee:	99.9 %
reactor type:	continuously operated stirred tank reactor (enzyme membrane reactor)
reactor volume:	0.2 L
residence time:	4.6 h (1 h)
space-time-yield:	165 g · L^{-1} · d^{-1} (410 g · L^{-1} · d^{-1})
down stream processing:	extraction, crystallization
enzyme consumption:	D-LDH: 150 U · kg^{-1} (= 1.5 % · d^{-1}); FDH = 150 U · kg^{-1}
enzyme supplier:	Sigma, Germany and Boehringer, Germany
company:	Ciba-Geigy, Switzerland

5) Product application

- The product is a precursor for different ACE-inhibtors (ACE = angiotensin converting enzyme).

- All ACE-inhibtors have the (*S*)-homophenylalanine moiety in common (see page 241).

6) Literature

- Fauquex, P. F., Sedelmeier, G. (1989) Biokatalysatoren und Verfahren zu ihrer Herstellung, Ciba-Geigy AG, EP 0371408

- Schmidt, E., Ghisalba, O., Gygax, D., Sedelmeier, G. (1992) Optimization of a process for the production of (*R*)-2-hydroxy-4-phenylbutyric acid – an intermediate for inhibitors of angiotensin converting enzyme, J. Biotechnol. **24**, 315–327

- Schmidt, E., Blaser, H.U., Fauquex, P.F., Sedelmeier, G., Spindler, F. (1992) Comparison of chemical and biochemical reduction methods for the synthesis of (*R*)-2-hydroxy-4-phenyl-butyric acid, in: Microbial Reagents in Organic Synthesis (S. Servi, ed.), pp. 377–388, Kluwer Academic Publishers, Netherlands.

115

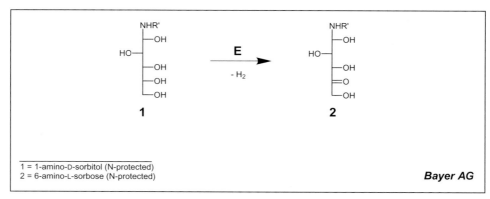

1 = 1-amino-D-sorbitol (N-protected)
2 = 6-amino-L-sorbose (N-protected)

Bayer AG

Fig. 1.1.99.21 – 1

1) Reaction conditions

[**1**]:	1 M (molecular weight depends on protecting group)
pH:	5.0
T:	32 °C
medium:	aqueous
reaction type:	oxidation
catalyst:	suspended whole cells
enzyme:	D-sorbitol:(acceptor)1-oxidoreductase (D-sorbitol dehydrogenase)
strain:	*Gluconobacter oxydans*
CAS (enzyme):	[9028–22–2]

2) Remarks

- The published synthesis of 1-desoxynojirimycin and its derivatives requires multiple steps and a laborious protecting group chemistry.

- To prevent undesired follow up reactions of 6-amino-D-sorbose in water the amino group has to be protected by, e.g., a benzyloxycarbonyl group. The protection of 1-amino-D-sorbitol is carried out in an aqueous medium at pH 8–10 with benzyloxycarbonyl chloride.

- The cells of *Gluconobacter oxydans* are produced by fermentation on sorbitol and used for the bioconversion step that is carried out in water without added nutrients.

- The cells are not immobilized; the very high specific substrate conversion rate would lead to severe limitations in immobilization beads.

- 6-Amino-L-sorbose (*N*-protected) is used as an intermediate in the miglitol production. 1-Desoxynojirimycin is produced by chemical intramolecular reductive amination of 6-amino-L-sorbose.

D-Sorbitol dehydrogenase
Gluconobacter oxydans

EC 1.1.99.21

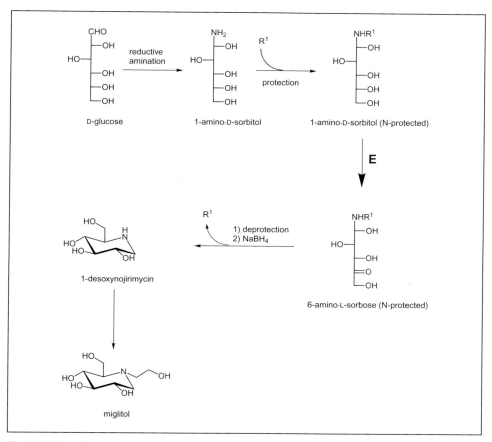

Fig. 1.1.99.21 – 2

117

D-Sorbitol dehydrogenase
Gluconobacter oxydans

EC 1.1.99.21

3) Flow scheme

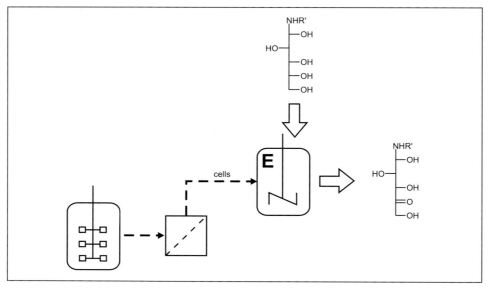

Fig. 1.1.99.21 – 3

4) Process parameters

yield:	90 %
reactor type:	batch
reactor volume:	> 10,000 L
company:	Bayer AG, Germany

5) Product application

- 6-Amino-L-sorbose is used as an intermediate for oral α-glucosidase-inhibitors.

- Derivatives of 1-desoxynojirimycin are pharmaceuticals for the treatment of carbohydrate metabolism disorders (e.g. diabetes mellitus).

6) Literature

- Kinast, G., Schedel, M. (1978) Verfahren zur Herstellung von 6-Amino-6-desoxy-L-sorbose, Bayer AG, DE 2834122 A1

- Kinast, G., Schedel, M. (1978) Herstellung von *N*-substituierten Derivaten des 1-Desoxynojirimycins, Bayer AG, DE 2853573 A1

- Kinast, G., Schedel, M. (1981) Vierstufige 1-Desoxynojirimycin-Synthese mit einer Biotransformation als zentralem Reaktionsschritt, Angew. Chem. **93**, 799–800

- Kinast, G., Schedel, M., Koebernick, W. (1981) Verfahren zur Herstellung von *N*-substituierten Derivaten des 1-Desoxynojirimycins, Bayer AG, EP 0049858 A2

118

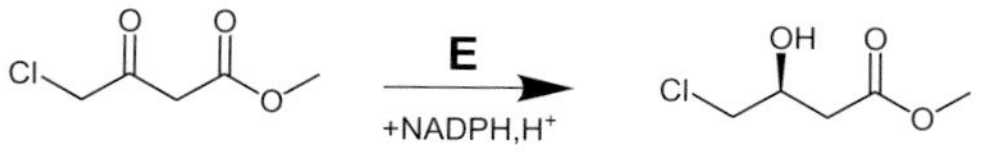

Dehydrogenase
Geotrichum candidum

EC 1.1.X.X

1 = 4-chloro-3-oxo-butanoic acid methyl ester
2 = 4-chloro-3-hydroxy-butanoic acid methyl ester

Bristol-Myers Squibb

Fig. 1.1.X.X – 1

1) Reaction conditions

[**1**]:	0.066 M, 10 g·L⁻¹ [150.56 g·mol⁻¹]
pH:	6.8
T:	28 °C
medium:	aqueous
reaction type:	reduction of keto group
catalyst:	suspended whole cells
enzyme:	dehydrogenase
strain:	*Geotrichum candidum* SC 5469

2) Remarks

- In initial experiments an enantiomeric excess of 96.9 % was reached. After heat-treatment (at 50 °C for 30 min) of the cells prior to use the ee increased to 99 %.

- The oxidoreductase was also isolated and purified 100-fold. It is NADP-dependant and the regeneration of cofactor in case of the isolated enzyme is carried out with glucose dehydrogenase. The isolated enzyme was immobilized on Eupergit C.

3) Flow scheme

Not published.

4) Process parameters

yield:	95 %
ee:	99 %
reactor type:	batch
reactor volume:	750 L
capacity:	multi kg
down stream processing:	filtration and extraction
enzyme supplier:	Bristol-Myers Squibb, USA
company:	Bristol-Myers Squibb, USA

119

5) Product application

- In general chiral β-hydroxy esters are versatile building blocks.

- (S)-4-Chloro-3-hydroxybutanoic acid methyl ester is used as chiral starting material in the synthesis of a cholesterol antagonist that inhibits the hydroxymethyl glutaryl CoA (HMG-CoA) reductase:

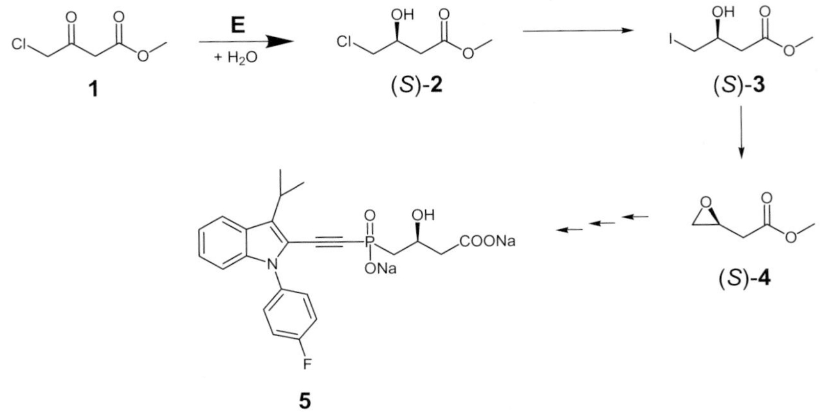

1 = 4-chloro-3-oxo-butanoic acid methyl ester
2 = 4-chloro-3-hydroxy-butanoic acid methyl ester
3 = 3-hydroxy-4-iodo-butanoic methyl ester
4 = oxiranyl-acetic acid methyl ester
5 = cholesterol antagonist

Fig. 1.1.X.X – 2

6) Literature

- Patel, R.N., McNamee, C.G., Banerjee, A., Howell, J.M., Robinson, R.S., Szarka, L.J. (1992) Stereoselective reduction of β-keto esters by *Geotrichium candidum*, Enzyme Microb. Technol. **14**, 731–738

- Patel, R.N. (1997) Stereoselective biotransformations in synthesis of some pharmaceutical intermediates, Adv. Appl. Microbiol. **43**, 91–140

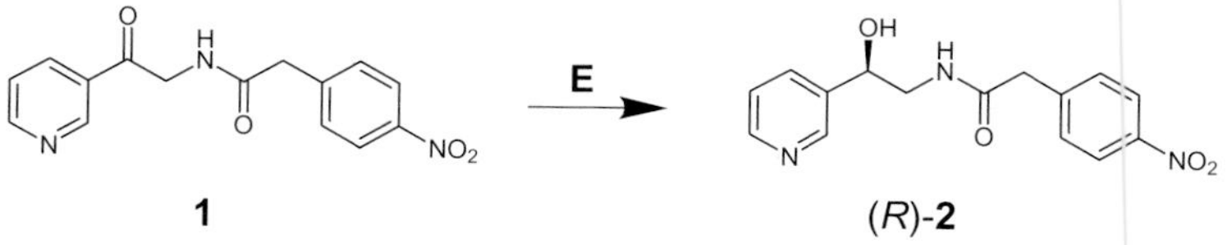

1 = 2-(4-nitro-phenyl)-*N*-(2-oxo-2-pyridin-3-ethyl)-acetamide
2 = (*R*)-*N*-(2-hydroxy-2-pyridin-3-yl-ethyl)-2-(4-nitro-phenyl)-acetamide

Merck Research Laboratories

Fig. 1.1.X.X – 1

1) Reaction conditions

[**1**]:	0.2 M, 60 g·L^{-1} [299.28 g·mol^{-1}]
pH:	5.6
T:	34 °C
medium:	two-phase: aqueous, solid
reaction type:	reduction of keto group
catalyst:	suspended whole cells
enzyme:	dehydrogenase
strain:	*Candida sorbophila*

2) Remarks

- Here the biotransformation is preferred over the chemical reduction with commercially available asymmetric catalysts (boron-based or noble-metal-based), since with the latter ones the desired elevated enantiomeric excess is not achievable.

- To prevent foaming during the biotransformation 2 mL·L^{-1} defoamer is added.

- Since the ketone has only a very low solubility in the aqueous phase, it is added as an ethanol slurry to the bioreactor. By this method no formal sterilization is possible. Heat sterilization is also not applicable because of the ketone's instability at elevated temperature. In the large scale production, 1 kg ketone is added as solution in 4 L 0.9 M H$_2$SO$_4$ to the bioreactor. This solution is sterilized prior to addition by pumping through a 0.22 μm hydrophobic Durapore microfiltration membrane (Millipore Opticap cartridge, Bedford, USA).

- After intensive investigation and optimization a glucose feed rate of 1.5 g·L^{-1} was found to support the highest initial bioreduction rate.

- The bioreduction is essentially carried out in a two-phase system, consisting of the aqueous phase and small beads made up of substrate and product.

- The downstream processing consists of multiple extraction steps with methyl ethyl ketone and precipitation induced by pH titration of the pyridine functional group (pK$_a$ = 4.66) with NaOH.

3) Flow scheme

Not published.

4) Process parameters

conversion:	> 99 %
yield:	82.5 %
ee:	> 98 %; 99.8 % after purification
chemical purity:	95 %
reactor type:	batch
reactor volume:	280 L
capacity:	multi kg
down stream processing:	extraction and precipitation
company:	Merck Research Laboratories, USA

5) Product application

- The (*R*)-amino alcohol is an important intermediate for the following synthesis of the β-3-agonist. It can be used for obesity therapy and to decrease the level of associated type II diabetes, coronary artery disease and hypertension:

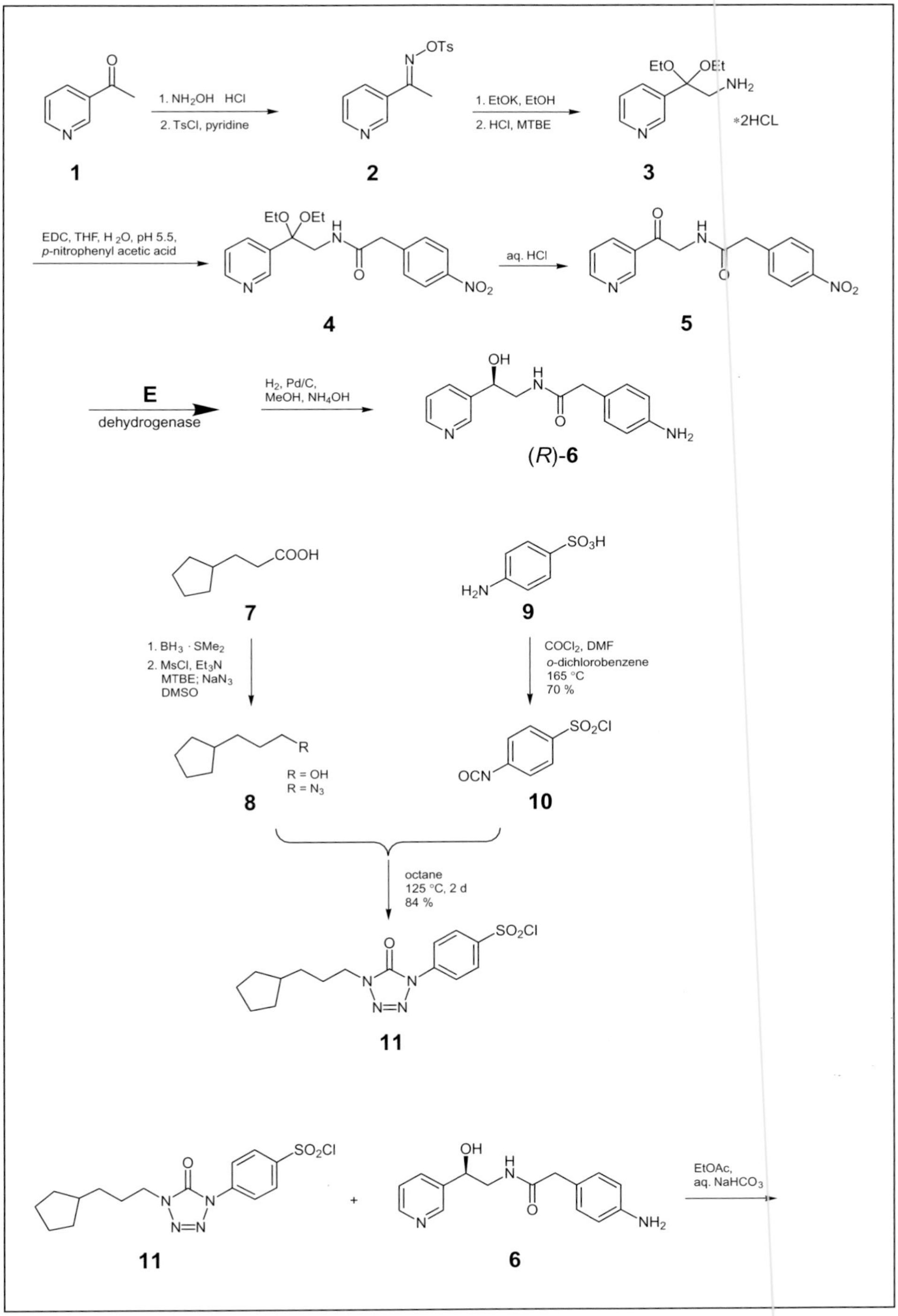
1. NH2OH HCl
2. TsCl, pyridine
1. EtOK, EtOH
2. HCl, MTBE
*2HCL
1
2
3
EDC, THF, H2O, pH 5.5,
p-nitrophenyl acetic acid
aq. HCl
4
5
E
dehydrogenase
H2, Pd/C,
MeOH, NH4OH
(R)-6
7
9
1. BH3 · SMe2
2. MsCl, Et3N
MTBE; NaN3
DMSO
COCl2, DMF
o-dichlorobenzene
165 °C
70 %
R = OH
R = N3
8
10
octane
125 °C, 2 d
84 %
11
EtOAc,
aq. NaHCO3
11
6

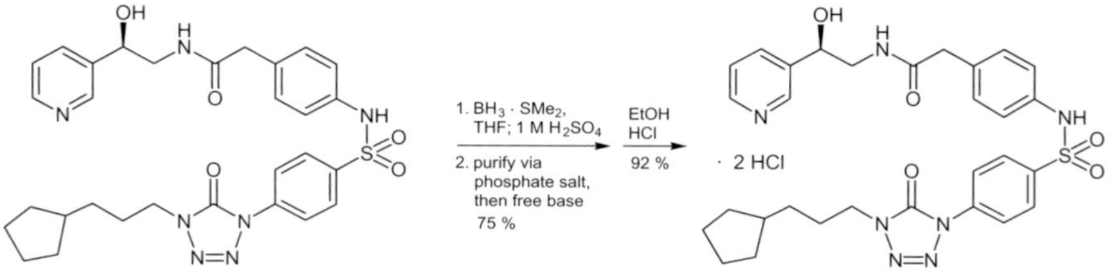

1 = 1-pyridin-3-yl-ethanone
2 = 1-pyridin-3-yl-tosyl oxime
3 = 3-pyridyl-aminomethyl ketal
4 = amide-ketal
5 = pyridyl ketone
6 = aniline-alcohol
7 = 3-cyclopentyl-propionic acid
8 = (3-azido-propyl)-cyclopentane
9 = 4-amino-benzenesulfonic acid
10 = 4-isocyanato-benzenesulfonyl chloride
11 = sulfonyl chloride
12 = sulfonylamino-acetamide
13 = di-HCl salt

Fig. 1.1.X.X – 2

6) Literature

- Chartrain, M., Chung, J., Roberge, C. (1998) *N*-(R)-(2-hydroxy-2-pyridine-3-yl-ethyl)-2-(4-nitro-phenyl)-acetamide, Merck & Co., Inc., US 5846791

- Chung, J., Ho, G., Chartrain, M., Roberge, C., Zhao, D., Leazer, J., Farr, R., Robbins, M., Emerson, K., Mathre, D., McNamara, J., Hughes, D., Grabowski, E., Reider, P. (1999) Practical chemoenzymatic synthesis of a pyridylethanolamino beta-3 adrenergic receptor antagonist – stereospecific beta-3 agonist used in obesity therapy, produced by yeast-mediated asymmetric reduction of a ketone, Tetrahedron Lett. **40** (37), 6739–6743

- Chartrain, M., Roberge, C., Chung, J., McNamara, J., Zhao, D., Olewinski, R., Hunt, R., Salmon, P., Roush, D., Yamazaki, S., Wang, T., Grabowski, E., Buckland, B., Greasham, R. (1999) Asymmetric bioreduction of (2-(4-nitro-phenyl)-*N*-(2-oxo-2-pyridin-3-yl-ethyl)-acetamide) to its corresponding (*R*)-alcohol [(*R*)-*N*-(2-hydroxy-2-pyridin-3-yl-ethyl)-2-(4-nitro-phenyl)-acetamide] by using *Candida sorbophila* MY 1833, Enzyme Microb. Technol. **25**, 489–496

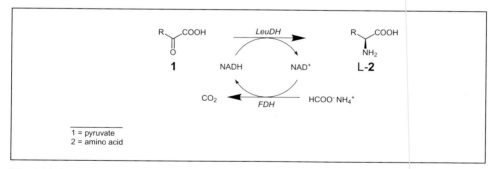

Fig. 1.4.1.9 – 1

1) Reaction conditions

[**1**]:	0.5 M, 65.07 g·L⁻¹ [130.14 g·mol⁻¹]

[**1**]: 0.5 M, 65.07 $g \cdot L^{-1}$ [130.14 $g \cdot mol^{-1}$]
pH: 8.0
T: 25 °C
medium: aqueous
reaction type: redox reaction
catalyst: solubilized enzyme
enzyme: L-leucine-NAD oxidoreductase, deaminating
strain: *Bacillus sphaericus*
CAS (enzyme): [9028–71–7]

2) Remarks

- The expensive cofactor can be easily regenerated by formate dehydrogenase (FDH) from *Candida boidinii* utilizing ammonium formate that is oxidized to CO_2 under reduction of NAD^+ to NADH:

Fig. 1.4.1.9 – 2

- The cofactor regeneration is practicably irreversible and the co-product CO_2 can be easily separated (K_{eq} = 15,000).

- Other amino acids that are synthesized by the same method:

125

Leucine dehydrogenase
Bacillus sphaericus

EC 1.4.1.9

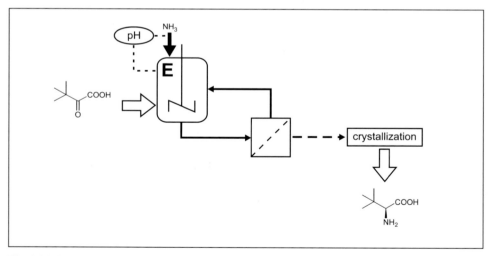

L-1   L-2   L-3   L-4

1 = neopentylglycine
2 = 3,3-dimethylpropane glycine
3 = 3-ethyl-3-methyl-propane glycine
4 = 5,5-dimethyl-butyl glycine

Fig. 1.4.1.9 – 3

- At high concentrations of the substrate trimethylpyruvate (of industrial quality), the enzymes are inhibited.

- The productivity of the process is not limited by the enzymatic reaction but by chemical reactions because at higher ammonia and keto acid concentrations the formation of by-products is increased.

- The enzymes are recycled after an ultrafiltration step.

3) Flow scheme

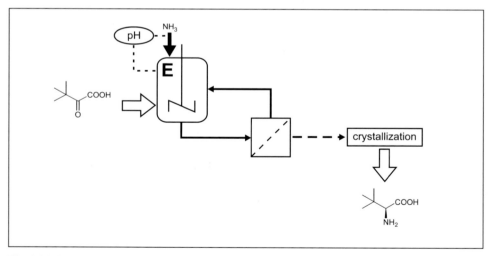

Fig. 1.4.1.9 – 4

126

4) Process parameters

yield:	74 %
reactor type:	repetitive batch with ultrafiltration (enzyme membrane reactor)
capacity:	ton scale
residence time:	2 h
space-time-yield:	638 g · L^{-1} · d^{-1}
down stream processing:	crystallization
enzyme consumption:	leucine dehydrogenase: 0.9 U · g^{-1}, formate dehydrogenase: 2.3 U · g^{-1}
company:	Degussa-Hüls AG, Hanau, Germany

5) Product application

- The amino acids are building blocks for drug synthesis, e.g. anti-tumor agents or HIV protease inhibitors.

- They can also be used for templates in asymmetric synthesis:

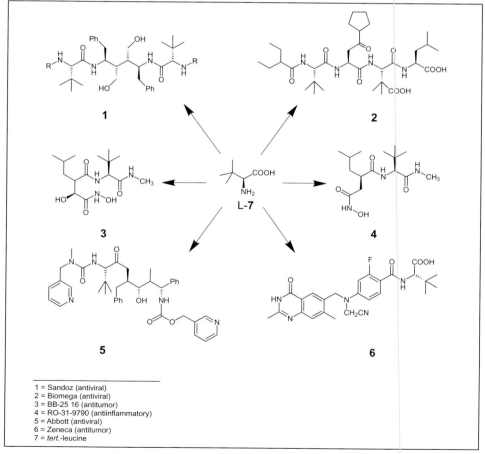

1 = Sandoz (antiviral)
2 = Biomega (antiviral)
3 = BB-25 16 (antitumor)
4 = RO-31-9790 (antiinflammatory)
5 = Abbott (antiviral)
6 = Zeneca (antitumor)
7 = *tert.*-leucine

Fig. 1.4.1.9 – 5

6) Literature

- Bommarius, A.S., Estler, M., Drauz, K. (1998) Reaction engineering of large-scale reductive amination processes; Inbio 98 Conference, Amsterdam/NL

- Kragl, U., Vasic-Racki, D., Wandrey, C. (1996) Continuous production of L-*tert*-leucine in series of two enzyme membrane reactors, Bioprocess Engineering **14**, 291–297

- Bommarius, A.S., Schwarm, M., Drauz, K. (1998) Biocatalysis to amino acid-based chiral pharmaceuticals – examples and perspectives, J. Mol. Cat. B: Enzymatic **5**, 1–11

- Wichmann, R., Wandrey, C., Bückmann, A.F., Kula, M.-R., (1981) Continuous enzymatic transformation in an enzyme membrane reactor with simultaneous NAD(H) regeneration, Biotechnol. Bioeng. **23**, 2789–2796

- Bommarius, A.S., Drauz, K., Hummel, M.-R., Wandrey, C. (1994) Some new developments in reductive amination with cofactor regeneration, Biocatalysis, **10**, 37–47

- Bradshaw, C.W., Wong, C.-H., Hummel, W., Kula, M.-R. (1991) Enzyme-catalyzed asymmetric synthesis of (S)-2-amino-4-phenylbutanoic acid and (R)-2-hydroxy-4-phenylbutanoic acid, Bioorg. Chem. **19**, 29–39

- Krix, G., Bommarius, A.S., Drauz, K., Kottenhan, M., Schwarm, M., Kula, M.-R. (1997) Enzymatic reduction of α-keto acids leading to L-amino acids or D-hydroxy acids, J. Biotechnol. **53**, 29–39

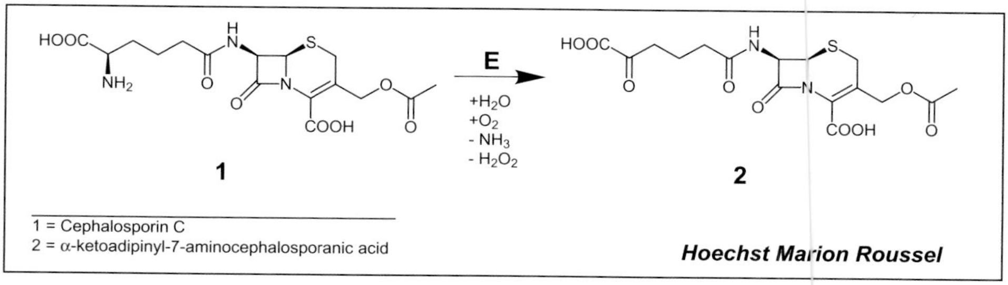

1 = Cephalosporin C
2 = α-ketoadipinyl-7-aminocephalosporanic acid

Hoechst Marion Roussel

Fig. 1.4.3.3 – 1

1) Reaction conditions

[1]:	0.02 M, 7.47 g · L^{-1} [373.38 g · mol^{-1}]
pH:	7.3
T:	25 °C
medium:	aqueous
reaction type:	oxidative deamination
catalyst:	immobilized enzyme
enzyme:	D-amino-acid oxygen oxidoreductase, deaminating (D-amino-acid oxidase)
strain:	*Trigonopsis variabilis*
CAS (enzyme):	[9000–88–8]

2) Remarks

- This step is part of the 7-aminocephalosporanic acid (7-ACA) process, see page 225.

- Ketoadipinyl-7-ACA decarboxylates *in situ* in presence of H_2O_2 that is formed by the bio-transformation step yielding glutaryl-7-ACA:

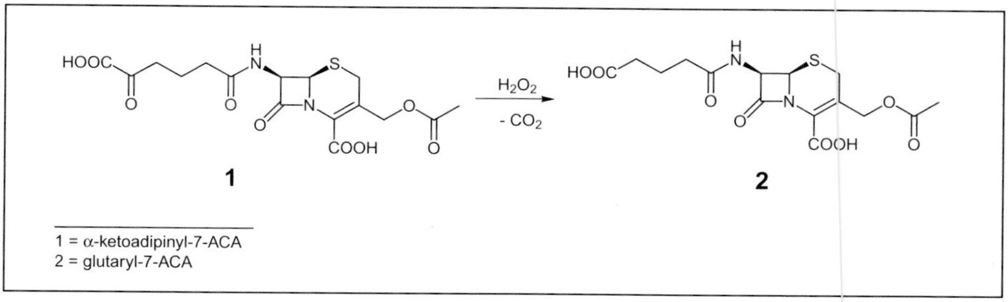

1 = α-ketoadipinyl-7-ACA
2 = glutaryl-7-ACA

Fig. 1.4.3.3 – 2

3) **Flow scheme**

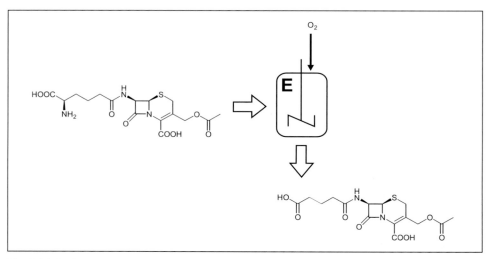

Fig. 1.4.3.3 – 3

4) **Process parameters**

reactor type:	batch
reactor volume:	10,000 L
capacity:	$200\ t \cdot a^{-1}$
residence time:	1.5 h
down stream processing:	reaction solution is directly transferred to the 7-ACA (7-aminocephalosporanic acid) production
enzyme consumption:	$1.1\ U \cdot g^{-1}$
enzyme supplier:	Hoechst-Marion Roussel, Germany
start-up date:	1996
production site:	Frankfurt, Germany
company:	Hoechst-Marion Roussel, Germany

5) **Product application**

- See 7-ACA process on page 225.

6) **Literature**

- Matsumoto, K. (1993) Production of 6-APA, 7-ACA, and 7-ADCA by immobilized penicillin and cephalosporin amidases, in: Industrial Application of Immobilized Biocatalysts (Tanaka, A, Tosa, T., Kobayashi, T. eds.), pp. 67–88, Marcel Dekker Inc., New York

- Tanaka, T. Tosa, T., Kobayashi, T. (1993) Industrial Application of Immobilized Biocatalysts, Marcel Dekker Inc., New York

- Verweij, J., Vroom, E.D. (1993) Industrial transformations of penicillins and cephalosporins, Rec. Trav. Chim. Pays-Bas **112** (2), 66–81

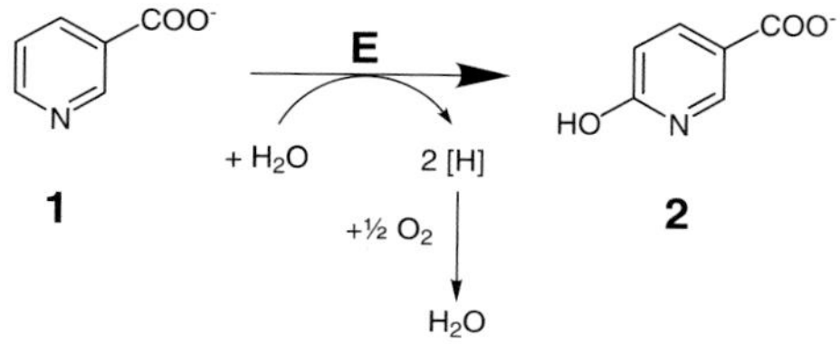

Nicotinic acid hydroxylase
Achromobacter xylosoxidans

1 = niacin = nicotinic acid = pyridine-3-carboxylate
2 = 6-hydroxynicotinate = 6-hydroxy-pyridine-3-carboxylate

Lonza AG

Fig. 1.5.1.13 – 1

1) Reaction conditions

[**1**]:	0.533 M, 65 g $\cdot$ L^{-1} [122.06 g $\cdot$ mol^{-1}]
pH:	7.0
T:	30 °C
medium:	aqueous
reaction type:	redox reaction (hydroxylation)
catalyst:	suspended whole cells
enzyme:	nicotinate: NADP$^+$6-oxidoreductase (nicotinic acid hydroxylase, nicotinate dehydrogenase)
strain:	*Achromobacter xylosoxidans*
CAS (enzyme):	[9059-03–4]

2) Remarks

- The 6-hydroxynicotinate producing strain was found by accident, when in the mother liquor of a niacin producing chemical plant precipitating white crystals of 6-hydroxynicotinate were found.

- At niacin concentrations higher than 1 % the second enzyme of the nicotinic acid pathway, the decarboxylating 6-hydroxynicotinate hydroxylase gets strongly inhibited, whereas the niacin hydroxylase operates unaffected:

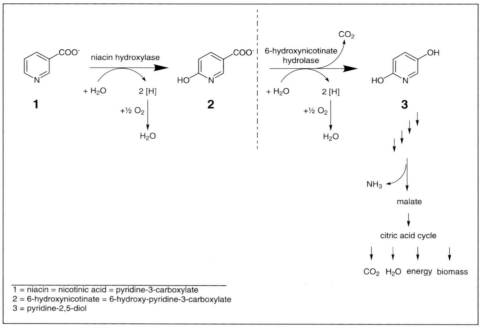

1 = niacin = nicotinic acid = pyridine-3-carboxylate
2 = 6-hydroxynicotinate = 6-hydroxy-pyridine-3-carboxylate
3 = pyridine-2,5-diol

Fig. 1.5.1.13 – 2

- The process takes place in two phases (see flow scheme):

  1) Growing of cells in a fermenter (chemostat) on niacin and subsequent storage of biomass in cooled tanks.

  2) Addition of biomass to niacin solution, incubation, separation of biomass and purification of product.

- The product is precipitated by the addition of acid.

- Alternatively, the integration of the two phases into an one reaction vessel fed-batch operation is possible (product concentration of 75 g · L^{-1} in 25 h). This procedure is not used on an industrial scale.

- Also, a continuous process was developed as 'pseudocrystal fermentation'. The substrate is added in its solid form and the product crystallizes out of the reaction solution. The process takes advantage of the fact that the Mg-salt of niacin is 100 times more soluble in H_2O at neutral pH than Mg-6-hydroxynicotinate. The pH is titrated to 7.0 with nicotinic acid. The concentration of Mg-nicotinate is regulated to 3 % using conductivity measurement techniques and direct addition of the salt. Mg-6-hydroxynicotinate is collected in a settler.

- Niacin hydroxylase works only in the presence of electron-transmitting systems such as cytochrome, flavine or NADP$^+$, and therefore air needs to be supplied to facilitate the cofactor regeneration. The oxygen-transfer rate limits the reaction.

- In contrast to the biotransformation the chemical synthesis of 6-substituted nicotinic acids is difficult and expensive due to the separation of by-products.

3) Flow scheme

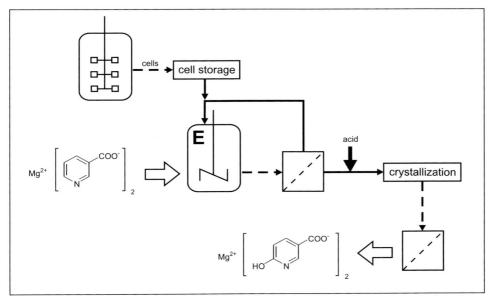

Fig. 1.5.1.13 – 3

4) Process parameters

conversion:	> 90 %
yield:	> 90 % (overall)
selectivity:	high
chemical purity:	> 99 %
reactor type:	batch
reactor volume:	12,000 L
capacity:	several tons
residence time:	12 h
down stream processing:	precipitation, centrifugation and drying
production site:	Visp, Switzerland
company:	Lonza AG, Switzerland

5) Product application

- Versatile building block chiefly in the synthesis of modern insecticides.

- By using common chemistry methods the product is converted into interesting building blocks:

133

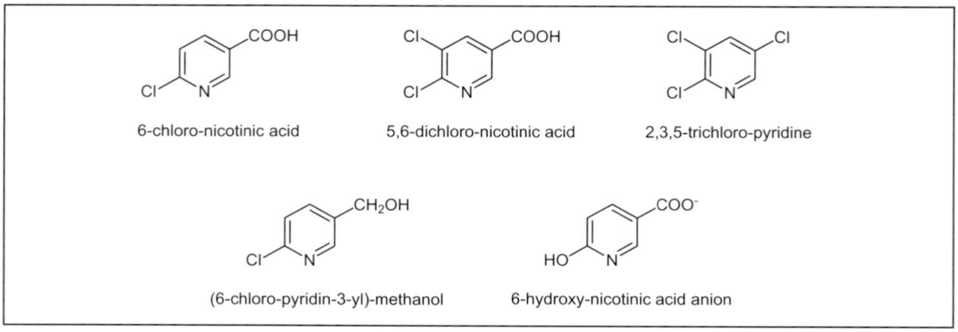

6-chloro-nicotinic acid   5,6-dichloro-nicotinic acid   2,3,5-trichloro-pyridine

(6-chloro-pyridin-3-yl)-methanol   6-hydroxy-nicotinic acid anion

Fig. 1.5.1.13 – 4

6) Literature

- Behrmann, E.J., Stanier, R.Y. (1957) The bacterial oxidation of nicotinic acid, J. Biol. Chem. **228**, 923–945

- Briauourt, D., Gilbert, J. (1973) Synthesis of pharmacological investigation concerning the series of 2-dialkylaminoalkoxy-5-pyridine carboxylic acids, Chim. Therap. **2**, 226

- Cabral, J., Best, D., Boross, L., Tramper, J. (1994) Applied Biocatalysis, Harwood Academic Publishers, Chur, Switzerland

- Glöckler, R., Roduit, J.-P. (1996) Industrial bioprocesses for the production of substituted aromatic heterocycles, Chimia, **50,** 413–415

- Gsell, L. (1989) 1-Nitro-2,2-diaminoäthylenderivate, Ciba-Geigy AG, EP 0302833

- Kieslich, K. (1991) Biotransformations of industrial use, 5th Leipziger Biotechnologiesymposium 1990, Acta Biotechnol. **11** (6) 559–570

- Kulla, H., Lehky, A. (1985) Verfahren zur Herstellung von 6-Hydroxynikotinsäure, Lonza AG, EP 0152949 A2

- Kulla, H.G. (1991) Enzymatic hydroxylations in industrial application, Chimia **45**, 81–85

- Lehky, P., Kulla, H., Mischler, S. (1995) Verfahren zur Herstellung von 6-Hydroxynikotinsäure, Lonza AG, EP 0152948 A2

- Minamida, I., Iwanaga, K., Okauchi, T. (1989) Alpha-unsaturated amines, their production and use, Takeda Chemical Industries, Ltd.,EP 0302389 A2

- Petersen, M., Kiener, A. (1999) Biocatalysis – preparation and functionalization of N-heterocycles, Green Chem. **2**, 99–106

- Sheldon, R.A. (1993) Chirotechnology, Marcel Dekker Inc., New York

- Quarroz, D., (1983) Verfahren zur Herstellung von 2-Halogenpyridinderivaten, Lonza AG, EP 0084118 A1

- Wolf, H., Becker, B., Stendel, W., Homeyer, B., (1988) Substituierte Nitroalkene, Bayer AG, EP 0292822 A2

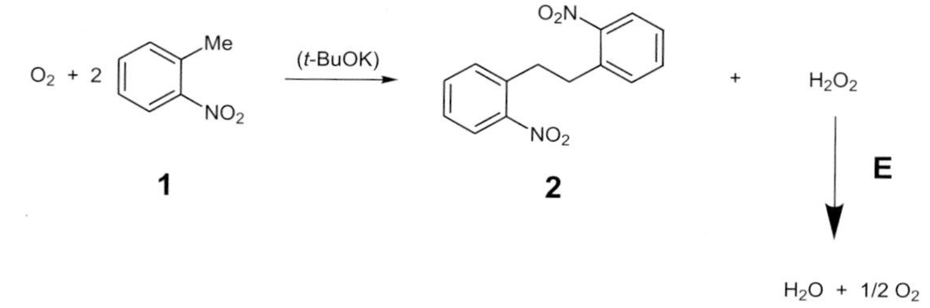

1 = nitrotoluene
2 = dinitrodibenzyl (DNDB)

Novartis

Fig. 1.11.1.6 – 1

1) Reaction conditions

pH:	6.0–9.0
medium:	3-phase-system: organic, solid, aqueous
reaction type:	redox reaction
catalyst:	solubilized
enzyme:	hydrogen-peroxide: hydrogen peroxide oxidoreductase (catalase)
strain:	*microbial source*
CAS (enzyme):	[9001-05–2]

2) Remarks

● In the synthesis of dinitrodibenzyl (DNDB) hydrogen peroxide is produced as a by-product. It is not possible to decompose H_2O_2 by the addition of heavy-metal catalysts because only incomplete conversion is reached. Additionally, subsequent process steps with DNDB are problematic due to contamination with heavy-metal catalyst.

● The reaction solution consists of three phases: 1) water, 2) organic phase and 3) solid phase consisting of DNDB.

● The catalase derived from microbial source has advantages compared to beef catalase since the activity remains constant over a broad pH range from 6.0 to 9.0; temperatures up to 50 °C are tolerated and salt concentrations up to 25 % do not affect the enzyme stability.

● The reaction is carried out in a cascade of CSTRs. After the chemical reaction in the first vessel catalase and actic acid is added to degas the product solution with nitrogen in the third vessel.

3) Flow scheme

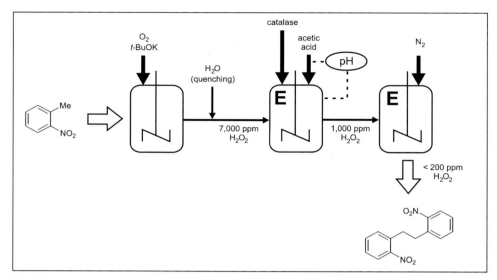

Fig. 1.11.1.6 – 2

4) Process parameters

conversion:	> 98 %
yield:	> 98 %
selectivity:	99.9 %
reactor type:	cascade of CSTR's
residence time:	> 1 h
down stream processing:	degassing with nitrogen
company:	Novartis, Switzerland

5) Product application

- The process is only relevant to remove the unwanted side product H_2O_2. The dinitrodibenzyl is used as a pharmaceutical intermediate.

6) Literature

- Onken, U., Schmidt, E., Weissenrieder, T. (**1996**) Enzymatic H_2O_2 decomposition in a three-phase suspension, Ciba-Geigy; International conference on biotechnology for industrial production of fine chemicals, 93rd event of the EFB; Zermatt, Switzerland, 29.09.1996

Oxygenase
Arthrobacter sp.

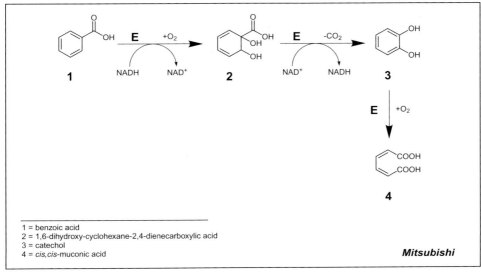

1 = benzoic acid
2 = 1,6-dihydroxy-cyclohexane-2,4-dienecarboxylic acid
3 = catechol
4 = *cis,cis*-muconic acid

Mitsubishi

Fig. 1.13.11.1 – 1

1) Reaction conditions

medium: aqueous
reaction type: redox reaction
catalyst: whole cells
enzyme: catechol-oxygen 1,2-oxidoreductase (catechase, catechol dioxygenase)
strain: *Arthrobacter* sp.
CAS (enzyme):   [9027–16–1]

2) Remarks

- Benzoic acid is continuously fed into the fermentation medium.

- The reaction solution is separated from the cells by membrane filtration (cross-flow membrane reactor).

- Since the reaction solution is colored, the first step of down stream processing is an adsorptive removal of color contaminant and the non-converted substrate benzoic acid.

- On acidification of the eluent with sulfuric acid muconic acid is precipitated and separated by an on-line filtration step.

- The residual muconic acid with a concentration of 10 % can be concentrated and precipitated by two ion-exchange steps and acidification.

3) Flow scheme

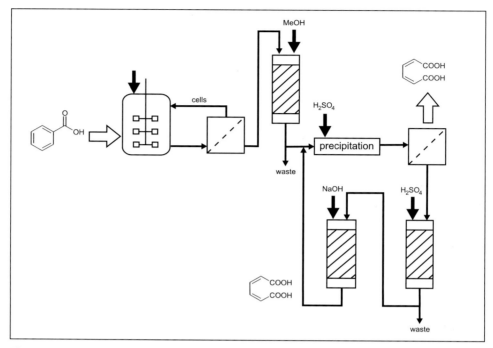

Fig. 1.13.11.1 – 2

4) Process parameters

chemical purity:	> 99 %
space-time-yield:	$70 \ g \cdot L^{-1} \cdot d^{-1}$
production site:	Yokohama, Japan
company:	Mitsubishi Kasei Corporation, Japan

5) Product application

- The product is a raw material for new resins, pharmaceuticals and agrochemicals. It can also be used as an intermediate for the synthesis of adipic acid (nylon production).

6) Literature

- Mizuno, S., Yoshikawa, N., Seki, M., Mikawa, T., Imada, Y., (1988) Microbial production of *cis,cis*-muconic acid from benzoic acid, Appl. Microbiol. Biotechnol. **28**, 20–25

- Yoshikawa, N., Ohta, K., Mizuno, S., Ohkishi, H. (1993) Production of *cis,cis*-muconic acid from benzoic acid, in: Industrial Application of Immobilized Biocatalysts (Tanaka, A., Tosa, T., Kobayashi, T., eds.), pp. 131–147, Marcel Dekker Inc., New York

- Yoshikawa, N., Ohta, K., Mizuno, S., Ohkishi, H. (1993) Production of *cis,cis*-muconic acid from benzoic acid, Bioprocess Technol. **16**, 131–147

138

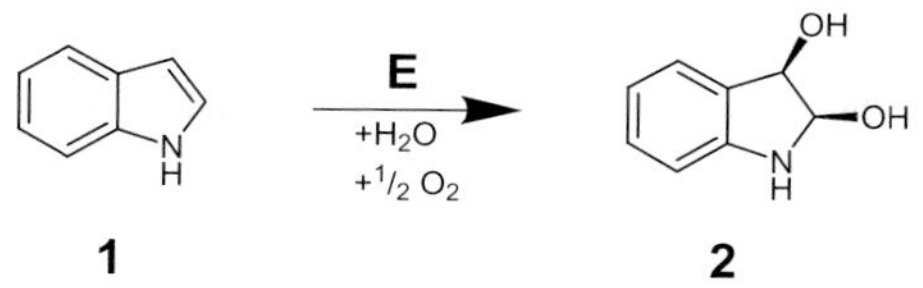

Fig. 1.13.11.11 – 1

1) Reaction conditions

medium:	aqueous
reaction type:	redox reaction
catalyst:	whole cells
enzyme:	naphthalene, NADH:oxygen oxidoreductase (naphthalene 1,2-dioxygenase, naphthalene oxygenase)
strain:	*Pseudomonas putida*
CAS (enzyme):	[9014–51–1]

2) Remarks

- This process demonstrates the potential of biological synthesis by constructing an adequate host cell.

- Several problems have to be solved for an effective synthesis of indigo:

  1) The activity of naphthalene dioxygenase is very low in *Pseudomonas putida* and even lower in the first recombinant *E. coli* strains.

  2) The half-life of the indigo oxidizing system is only about one to two hours. Indole concentrations above $400 \ mg \cdot L^{-1}$ are toxic to cells due to inactivation of the ferredoxin component of the enzyme system.

  3) Indole is too expensive for commercial application. Alternatively tryptophan can be used with a possible tryptophanase step (EC 4.1.99.1; [9024-00–4]) for synthesizing the indole in the cell. But tryprophan is also too expensive.

- To overcome all these problems several improvements were necessary:

  1) The dioxygenase system is improved by using better plasmids with stronger promoters. The actvity of the *Pseudmonas* strain can be reached.

  2) The cloning of the ferredoxin producing genes results firstly not in a more stable ferredoxin system but in a higher concentration of ferredoxin units. Secondly site-directed mutagenesis leads to a more stable ferredoxin system.

  3) The tryptophanase system is improved by site-directed mutagenesis.

- All improvements are combined in one single host to produce indigo from glucose. Further improvements of basic fermentation parameters leads to higher indole concentrations.

- In the following figures the chemical routes are compared to the biotransformation.
- The classical route of the indigo needs three steps starting from aniline:

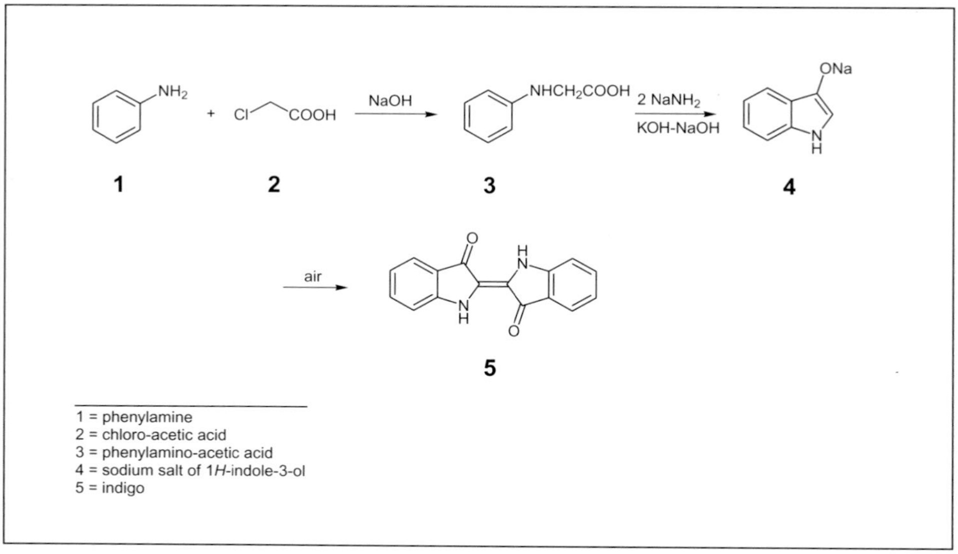
NaOH
2 NaNH2
KOH-NaOH
air
1
2
3
4
5
1 = phenylamine
2 = chloro-acetic acid
3 = phenylamino-acetic acid
4 = sodium salt of 1H-indole-3-ol
5 = indigo

Fig. 1.13.11.11 – 2

- A new catalytic route (Mitsui Toatsu Chemicals) starts also from aniline but produces less inorganic salts than the classical route:

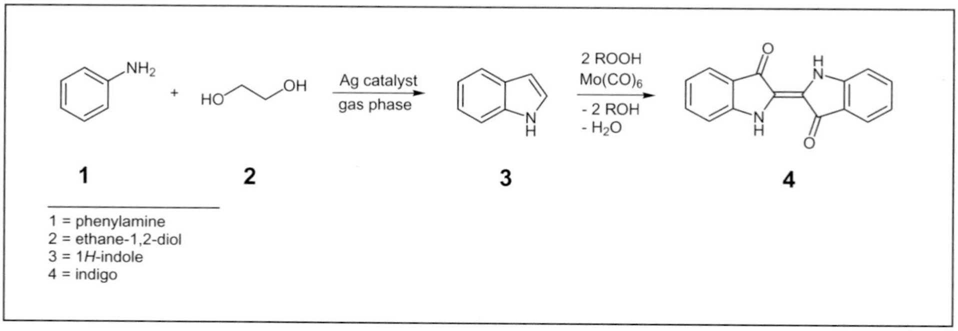
Ag catalyst
gas phase
2 ROOH
Mo(CO)6
- 2 ROH
- H2O
1
2
3
4
1 = phenylamine
2 = ethane-1,2-diol
3 = 1H-indole
4 = indigo

Fig. 1.13.11.11 – 3

- The biotransformation starts with glucose using the capability of the cells to produce tryptophan as intermediate:

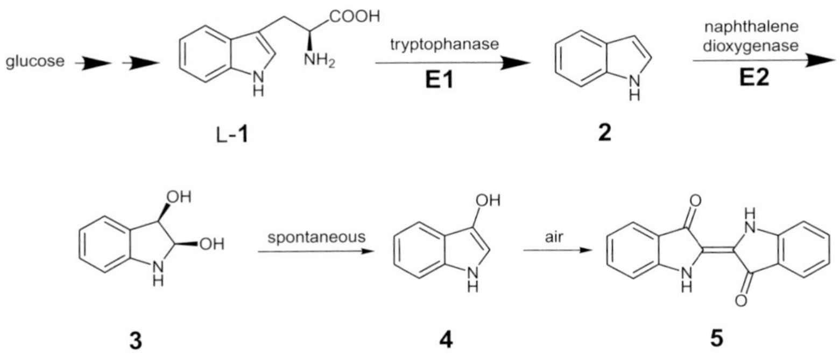

1 = L-tryptophan
2 = 1*H*-indole
3 = 2,3-dihydro-1*H*-indole-2,3-diol
4 = 1*H*-indole-3-ol
5 = indigo

Fig. 1.13.11.11 – 4

- Alternatively a toluene xylene oxidase from *Pseudomonas mendocina* can be used for the oxidation of indole to indole-3-ol:

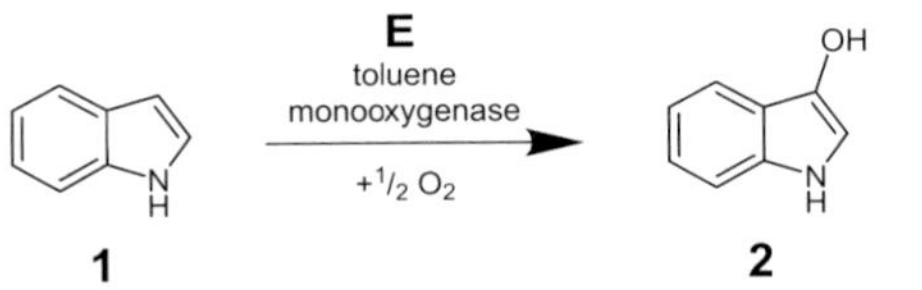

1 = 1*H*-indole
2 = 1*H*-indole-3-ol

Fig. 1.13.11.11 – 5

3) Flow scheme

Not published.

4) Process parameters

reactor type:	batch
company:	Genencor, USA

5) Product application

- Indigo is an important brilliant blue pigment extensivley used in the dyeing of cotton and woolen fabrics.

6) Literature

- Mermod, N., Harayama, S., Timmis, K.N. (1986) New route to bacterial production of indigo, Biotechnology **4**, 321–324

- Murdock, D., Ensley, B.D., Serdar, C., Thalen, M. (1993) Construction of metabolic operons catalyzing the *de novo* biosynthesis of indigo in *Escherichia coli*, Bio/Technology, **11**, 381–385

- Serdar, C.M., Murdock, D.C., Ensley, B.D. (1992) Enhancement of napthalene dioxygenase activity during microbial indigo production, Amgen Inc., US 5,173,425

- Sheldon, R.A. (1994) Consider the environmental quotient, Chemtech **3**, 38–47

- Yen, K.-M., Blatt, L.M., Karl, M.R. (1992) Bioconversion catalyzed by the toluene monooxygenase of *Pseudomonas mendocina* KR-1, Amgen Inc., WO 92/06208

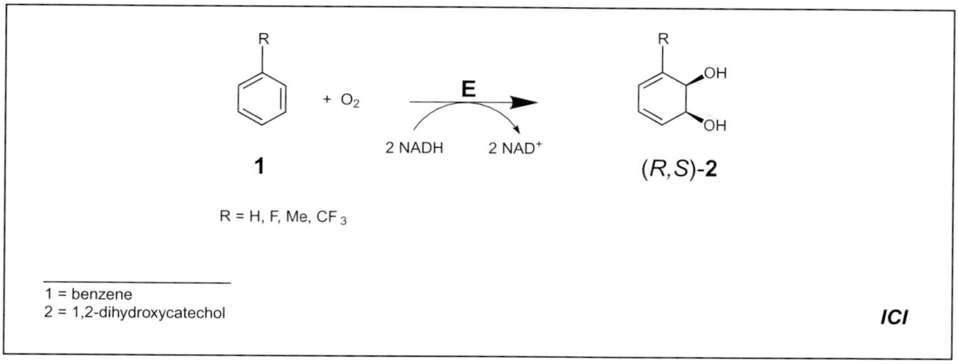

R = H, F, Me, CF$_3$

1 = benzene
2 = 1,2-dihydroxycatechol

ICI

Fig. 1.14.12.10 – 1

1) Reaction conditions

reaction type:	redox reaction
catalyst:	suspended whole cells
enzyme:	benzoate, NADH:oxygen oxidoreductase (1,2-hydroxylating)
	(benzoate 1,2-dioxygenase, benzoate hydroxylase)
strain:	*Pseudomonas putida*
CAS (enzyme):	[9059–18–1]

2) Remarks

- Benzoate dioxygenase was formerly classified as EC 1.13.99.2.

- In 1968 it was discovered that a mutant strain from *Pseudomonas putida* was lacking the activity of the dehydrogenase that normally catalyzes the dehydrogenation of *cis*-dihydro-diol to catechol. As a consequence *cis*-dihydrocatechol accumulates in the microorganism and is excreted into the medium:

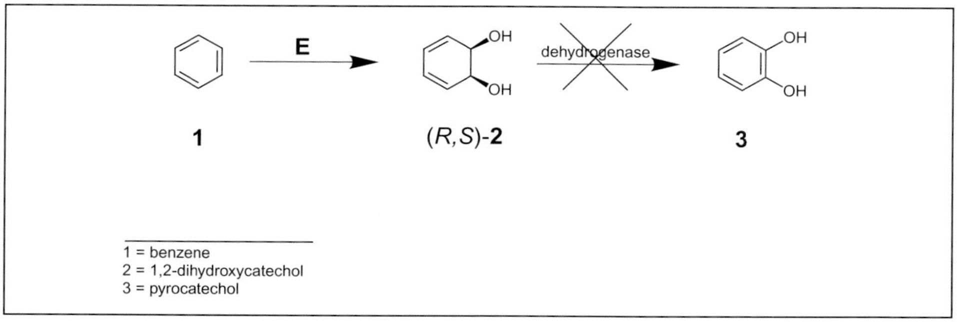

1 = benzene
2 = 1,2-dihydroxycatechol
3 = pyrocatechol

Fig. 1.14.12.10 – 2

- *Pseudomas putida* exhibits a high tolerance to aromatic substrates that are normally toxic to microorganisms.

- Since *cis*-dihydrodiols show an inhibitory effect on growth of the cells fermentation and bio-transformation are physically separated in the process.

143

- The cofactor regeneration of NADH takes place cell internally, driven by the addition of a carbon source.

3) Flow scheme

Not published.

4) Process parameters

selectivity: > 99.5 %
capacity: $t \cdot a^{-1}$
company: ICI, U.K., and others

5) Product application

- *cis*-Dihydrodiols are used as chiral intermediates in the synthesis of β-lactams that are effective as antiviral compounds.

- The product is also used as a polymerisation monomer.

6) Literature

- Evans, C., Ribbons, D., Thomas, S., Roberts, S. (1990) Cyclohexenediols and their use, Enzymatix Ltd., WO 9012798 A2

- Crosby, J. (1991) Synthesis of optically active compounds: A large scale perspective, Tetrahedron **47**, 4789–4846

- Sheldrake, G.N. (1992) Biologically derived arene *cis*-dihydrodiols as synthetic building blocks, in: Chirality in Industry (Collins, A.N., Sheldrake, G.N., Crosby, J., eds.) pp. 127–166, John Wiley & Sons, New York

- Taylor, S.C. (1983) Biochemical Process, ICI, EP 76606

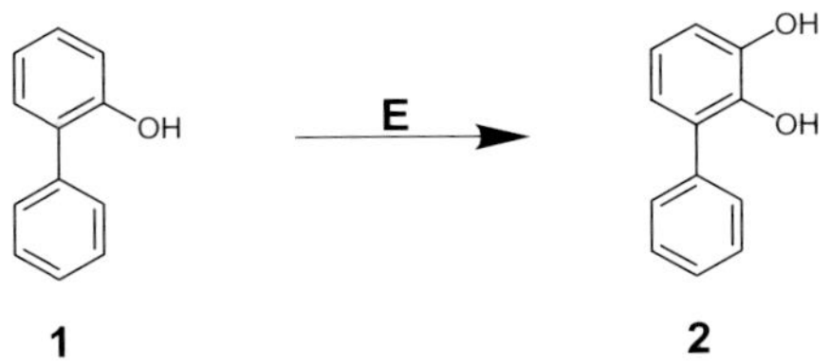

1 = 2-phenylphenol
2 = 3-phenylcatechol

Fluka

Fig. 1.14.13.44 – 1

1) Reaction conditions

[**1**]:	0.0012 M, 0.2 g · L^{-1} (total in fed batch 1,055 g) [170,21 g · mol^{-1}]
T:	30 °C
medium:	aqueous
reaction type:	redox reaction
catalyst:	suspended whole cells
enzyme:	2-hydroxybiphenyl,NADH:oxygen oxidoreductase (3-hydroxylating) (2-hydroxybiphenyl 3-monooxygenase)
strain:	*Escherichia coli* JM101
CAS (enzyme):	[118251–39–1]

2) Remarks

- 2-Phenylphenol and 3-phenylcatechol are highly toxic to whole cells. Therefore 2-phenylphenol is used in food preservation because of its fungicidal and bactericidal activity.

- The substrate is fed continuously to the reactor (0.45 g · L^{-1} · h^{-1}), establishing an actual concentration of the educt below the toxic level.

- Catechols are readily soluble in the aqueous phase (435 g · L^{-1}), which complicates extraction of the formed product.

- Dissolved 3-phenylcatechol readily polymerizes (product half-life at pH 7 and 37 °C is 10 h).

- The product is removed by continuous adsorption on the solid resin Amberlite™ XAD-4 (hydrophobic, polystyrene-based).

- The reaction mixture with the whole cells is continuously pumped through an external loop with a fluidized bed of the resin. Before circulating past the adsorbent all substrate is converted. Through the external loop 10 reactor volumes are circulated per hour.

- The growth rate of *Escherichia coli* is uneffected in the presence of XAD-4.

- The product is liberated from the resin by acidic methanol elution, followed by recrystallisation from *n*-hexane.

- By this process a series of other catechol derivatives are produced. This is easily possible by just changing the substrate in the feed (repetitive use of biocatalysts).

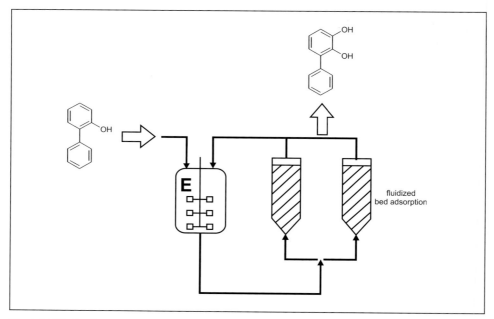

Fig. 1.14.13.44 – 2

3) Flow scheme

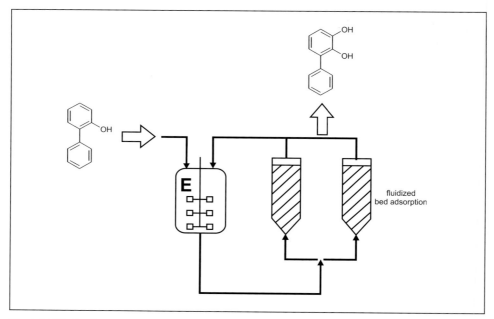

Fig. 1.14.13.44 – 3

4) Process parameters

conversion:	97 %
yield:	83 %
selectivity:	85 %
chemical purity:	77 %, 98 % after recrystallization
reactor type:	fed batch
reactor volume:	300 L

capacity:	multi kg
residence time:	10 h
space-time-yield:	$8 \text{ g} \cdot \text{L}^{-1} \cdot \text{d}^{-1}$
down stream processing:	adsorption, crystallization
company:	Fluka, Buchs, Switzerland

5) Product application

- Catechols with a substituent at the 3-position are important starting materials for the synthesis of pharmaceutical compounds (barbatusol, taxodione and L-DOPA analogues) and artificial supramolecular systems.

6) Literature

- Held, M., Panke, S., Kohler, H.-P., Feiten, H.-J., Schmid, A., Schmid, A., Wubbolts, M., Witholt, B. (1999) Solid phase extraction for biocatalytic production of toxic compounds, BioWorld **5**, 3–6

- Held, M., Schmid, A., Kohler, H.-P., Suske, W., Witholt, B., Wubbolts, M. (1999) An integrated process for the production of toxic catechols from toxic phenols based on a designer biocatalyst, Biotechnol. Bioeng. **62** (6),641–648

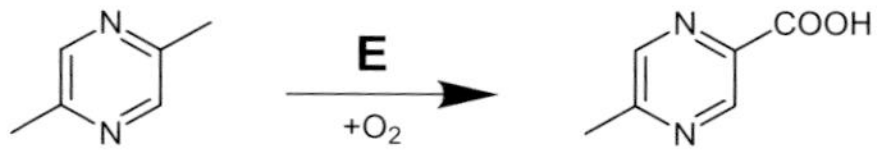

Monooxygenase
Pseudomonas putida

1 = 2,5-dimethylpyrazine
2 = 5-methylpyrazine-2-carboxylic acid

Lonza AG

Fig. 1.14.13.X – 1

1) Reaction conditions

[2]:	0.174 M, 24 g·L^{-1} [138.12 g·mol^{-1}]
pH:	7.0
T:	30 °C
medium:	aqueous
reaction type:	oxidation
catalyst:	suspended, growing whole cells
enzyme:	xylene oxygenase (monooxygenase)
strain:	*Pseudomonas putida* ATCC 33015

2) Remarks

- In contrast to growing cells the application of resting cells showed a strong accumulation of 2-hydroxymethyl-5-methylpyrazine that was only partially oxidized to 5-methylpyrazine-2-carboxylic acid.

- Selectively only one of the symmetric methyl groups is oxidized to a carboxylic acid.

- The desired enzymatic activity is expressed in the cells if 75 % *p*-xylene and 25 % 2,5-dimethylpyrazine is supplied as growth substrate in the fermentation.

- No bacterial metabolites from *p*-xylene were detected that could complicate the downstream processing.

- The biotransformation is terminated when bacterial growth enters the stationary phase.

- The concentration of the educt at the end of the reaction is below 0.1 g·L^{-1}.

- The enzyme is capable of selectively oxidizing a single methyl group of heteroarenes.

- The inhibition of the *P. putida* growth begins with a product concentration of 15 g·L^{-1} so that the process is stopped at a maximum product concentration of 24 g·L^{-1}.

- *p*-Xylene and substrate concentration in the fermenter are regulated by measuring these compounds in the exhaust air of the fermenter.

- Following alternative products could be synthesized (in parentheses yields are given):

148

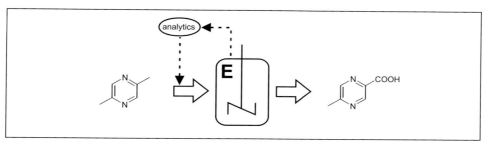

Fig. 1.14.13.X – 2

3) Flow scheme

Fig. 1.14.13.X – 3

4) Process parameters

conversion:	> 99.5 %
yield:	> 95 %
reactor type:	fed batch
reactor volume:	1,000 L
capacity:	multi $t \cdot a^{-1}$
residence time:	54 h
company:	Lonza AG, Switzerland

5) Product application

- Intermediate for the drug 5-methylpyrazine-2-carboxylic acid 4-oxide (acipimox) with antil-ipolytic activity, like nicotinic acid:

- Precursor in the synthesis of glipicide, a sulfonylurea compound with hypoglycemic activity. This generic drug is used in the control of diabetes:

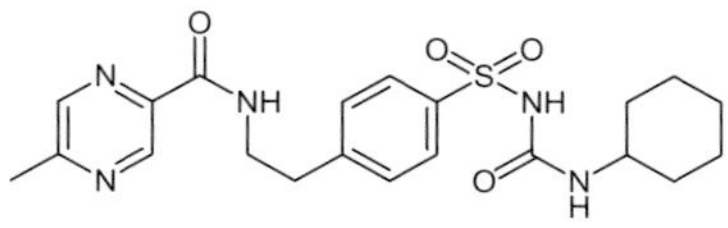

6) Literature

- Kiener, A. (1991) Mikrobiologische Oxidation von Methylgruppen in Heterocyclen, Lonza AG, EP 0442430 A2

- Kiener, A. (1992) Enzymatic oxidation of methyl groups on aromatic heterocycles: a versatile method for the preparation of heteroaromatic carboxylic acids, Angew. Chem. Int. Ed. Engl. **31**, 774–775

- Lovisolo, P. P., Briatico-Vangosa, G., Orsini, G., Ronchi, R., Angelucci, R. (1981) Pharmacological profile of a new anti-lipolytic agent: 5-methyl-pyrazine-2-carboxylic acid 4-oxide (Acipimox), Pharm. Res. Comm. **13**, 151–161

- Peterson, M., Kiener, A. (1999) Biocatalysis – Preparation and functionalization of *N*-heterocycles, Green Chem. **2**, 99–106

- Wubbolts, M.-G.,Panke, S., van Beilen, J. B., Witholt, B. (1996) Enantioselective oxidation by non-heme iron monooxygenases from *Pseudomonas*, Chimia **50**,436–437

- Zaks, A., Dodds, D. R. (1997) Application of biocatalysis and biotransformations to the synthesis of pharmaceuticals, DDT **2**, 513–531

Oxygenase
Nocardia autotropica

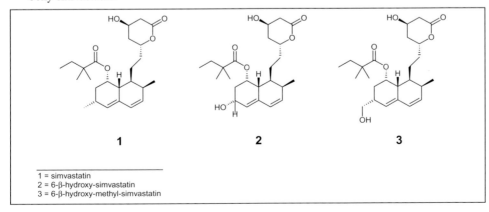

Fig. 1.14.14.1 – 1

1) Reaction conditions

[**1**]:	$< 0.05 \cdot 10^{-3}$ M, < 0.02 g·L^{-1} [416.26 g·mol^{-1}] (steady state)
[**2**]:	$1.85 \cdot 10^{-3}$ M, 0.8 g·L^{-1} [432.26 g·mol^{-1}]
pH:	6.8
T:	27 °C
medium:	aqueous
reaction type:	hydroxylation
catalyst:	suspended whole cells
enzyme:	substrate, reduced-flavoprotein: oxygen oxidoreductase (unspecific monooxygenase)
strain:	*Nocardia autotropica*
CAS (enzyme):	[62213–32–5]

2) Remarks

- Due to a strong substrate surplus inhibition, simvastatin is continuously fed to the fermenter.

- Side products are: 6-hydroxy-simvastatin, 3-hydroxy-iso-simvastatin and 3-desmethyl-3-carboxy-simvastatin:

1 = simvastatin
2 = 6-β-hydroxy-simvastatin
3 = 6-β-hydroxy-methyl-simvastatin

Fig. 1.14.14.1 – 2

151

- With a final product concentration of 0.8 g·L^{-1} and a 19,000 L reactor 15.2 kg of 6-β-hydroxy-methyl-simvastatin can be produced.

- The substrate is fed through an on-line filter sterilization system (microporous stainless steel prefilter, 1–2 µm, Pall) with a concentration of 20 g·L^{-1}. The feed tank is maintained at 45 °C to prevent precipitation.

3) Flow scheme

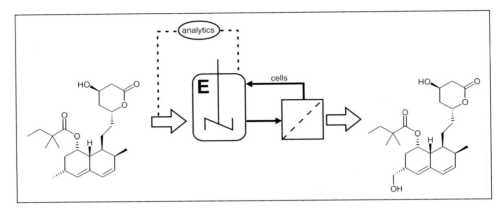

4) Process parameters

yield: 24 %
selectivity: 70 %
reactor type: fed batch
reactor volume: 19,000 L
company: Merck Sharp & Dohme, USA

5) Product application

- Simvastatin (and derivatives) belong to the family of HMG-CoA reductase inhibitors, which are, like Lovastatin, potent cholesterol-lowering therapeutic agents.

6) Literature

- Gewonyo, K., Buckland, B.C., Lilly, M.D. (1991) Development of a large-scale continuous substrate feed process for the biotransformation of simvastatin by *Nocardia sp.*, Biotech. Bioeng. **37**, 1101–1107

Monooxygenase
Nocardia corallina

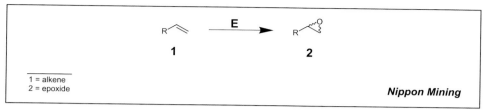

1 = alkene
2 = epoxide

Nippon Mining

Fig. 1.14.14.1 – 1

1) Reaction conditions

T:	30 °C
medium:	two phase : aqueous / organic
reaction type:	epoxidation
catalyst:	suspended whole cells
enzyme:	substrate, reduced-flavoprotein : oxygen oxidoreductase (unspecific mono-oxygenase)
strain:	*Nocardia corallina* B 276
CAS (enzyme):	[62213–32–5]

2) Remarks

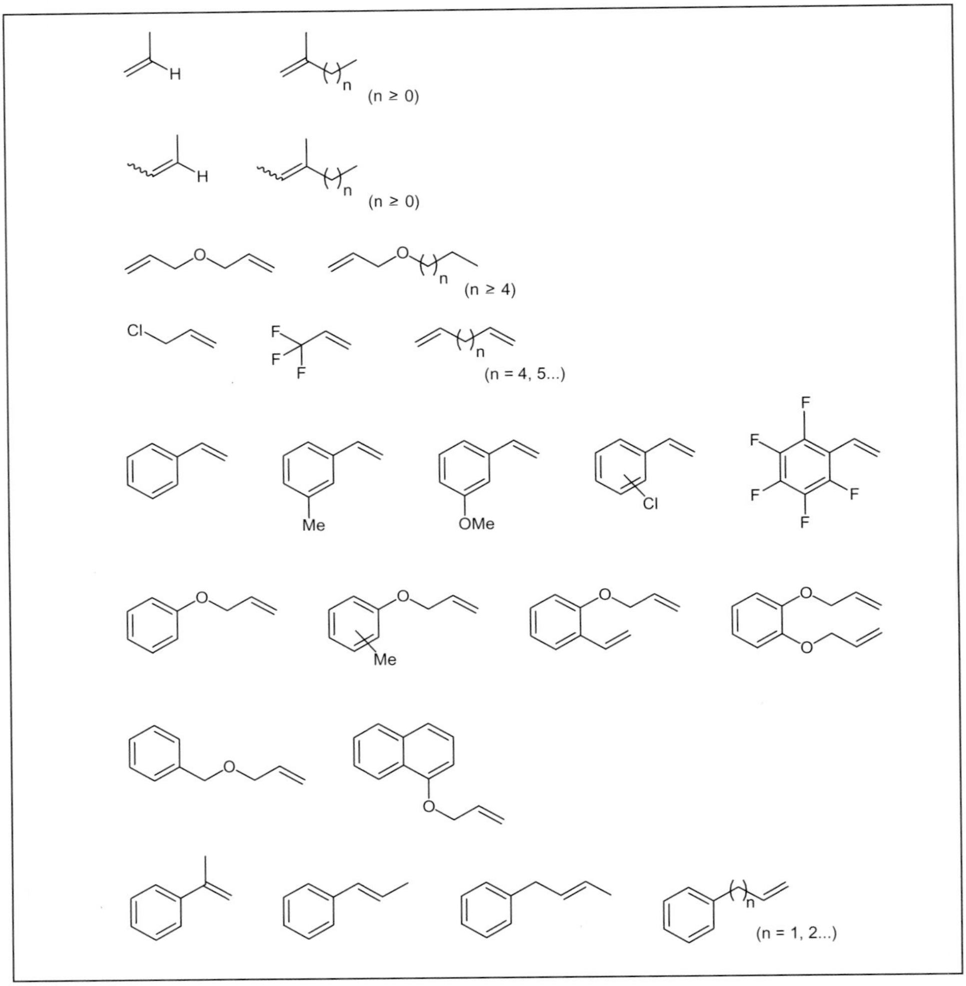

Fig. 1.14.14.1 – 2

- The alkene monooxygenase catalyzes the epoxidation of terminal and sub-terminal alkenes:

- The stereospecific epoxidation yields predominantly the (R)-enantiomer.

- The epoxidation activity is formed constitutively since the conversion takes place independent of the type of carbon source used (alkene or glucose).

- The biotransformation is carried out in a conventional fermentation system.

- The substrates have to be divided in three main classes: short-chain gaseous (C_3-C_5), short chain – liquid (C_6-C_{12}), long-chain ($C_{>12}$) olefins.

- A two-phase system is applied for C_6-C_{12} alk-1-enes with a non-toxic solvent (e.g. hexadecane or other alkanes). The addition of a solvent lowers the concentration of the inhibiting epoxide and allows continuous product extraction. Using alk-1-enes as solvent results in a competitive reaction leading preferentially to the shorter epoxides.

- For C_6-C_{12} alk-1-enes resting cells are used since the products are more toxic than in the case of longer chain epoxides where a growing cell reaction is applicable.

- For long chain alkenes it is advantageous to limit components of the medium during growth.

- Short chain, gaseous epoxides (C_3 to C_5) are very toxic to cells and product recovery is complicated.

- For these epoxides the cells have to be grown on glucose or similar carbon sources to guarantee the regeneration of the cofactor.

- The rate of aeration during fermentation has to be raised to extract the short chain toxic epoxide. The very low amounts of epoxides in the gas phase can be recovered by a special solvent extraction system. The process flow diagram of this special process is shown in the flow scheme.

3) Flow scheme

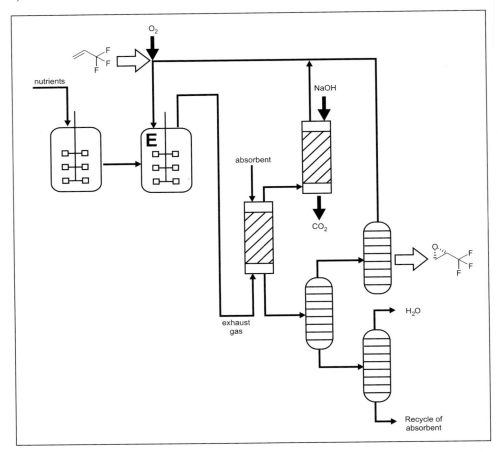

Fig. 1.14.14.1 – 3

4) Process parameters

conversion:	up to 98 %
yield:	up to 92 %
selectivity:	up to 94 %
ee:	> 80 % (up to 97 %)
reactor type:	batch with resting cells or fermenter
space-time-yield:	> 0.2 mol $\cdot$ L^{-1} $\cdot$ d^{-1}
down stream processing:	extraction and distillation
enzyme consumption:	$t_{1/2}$ > 4 d
company:	Nippon Mining, Japan

5) Product application

- From the substrates shown above the following epoxides are commercially obtained:

156

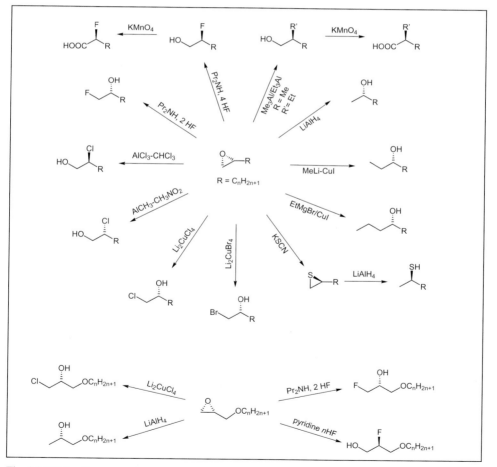

(R)-1,2-epoxyalkanes (R)-2-methyl-1,2-epoxyalkanes (R)-alkyl glycidyl ethers (R)-1,2-epoxy dec-9-ene

(R,R)-1,2,9,10-diepoxydecane pentafluorostyrene oxide 3,3,3-trifluoro-1,2-epoxypropane

Fig. 1.14.14.1 – 4

- Aliphatic epoxides can be used to synthesize various chiral intermediates for the production of ferroelectric liquid crystals:

Fig. 1.14.14.1 – 5

- 2-Methyl-1,2-epoxyalkanes are precursors for tertiary alcohols, which can be used in the synthesis of pharmaceuticals such as prostaglandins:

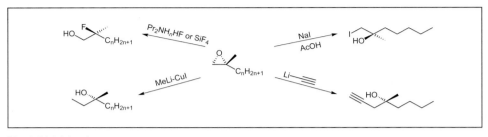

Fig. 1.14.14.1 – 6

6) Literature

- Crosby, J. (1991) Synthesis of optically active compounds: a large scale perspective, Tetrahedron **47**, 4789–4846

- Furuhashi, K. (1984) Optimization of a medium for the production of 1,2-epoxytetradecane by *Nocardia corallina* B-276, Appl. Microbiol. Biotechnol. **20**, 6–9

- Furuhashi, K., Shintani, M., Takagi, M. (1986) Effects of solvents on the production of epoxides by *Nocardia corallina* B-276, Appl. Microbiol. Biotechnol. **23**, 218–223

- Furuhashi, K. (1992) Biological routes to optically active epoxides, in: Chirality in Industry (Collins, A., Sheldrake, G.N., Crosby, J., eds.), pp. 167–186, John Wiley & Sons, London

- Gallagher, S., Cammack, R., Dalton, H. (1997) Alkene monooxygenase from *Nocardia corallina* B-276 is a member of the class of dinuclear iron proteins capable of stereospecific epoxygenation reactions, Eur. J. Biochem. **247**, 635–641.

- Hirai, T., Fukumasa, M., Nishiyama, I., Yoshizawa, A., Shiratori, N., Yokoyama, A., Yamana, M. (1991) New ferroelectric crystal having 2-methylalkanoyl group, Ferroelectrics **114**, 251–257

- Leak, D., Aikens, P., Seyed-Mahmoudian, M. (1992) The microbial production of epoxides, Tibtech **10**, 256–261

- Lilly, M.D. (1994) Advances in biotransformation processes. Eighth P.V. Danckwerts memorial lecture presented at Glaziers' Hall, London, U.K. 13 May 1993, Chem. Eng. Sci. **49**, 151–159

- Shiratori, N., Yoshizawa, A., Nishiyama, I., Fukumasa, M., Yokoyama, A., Hirai, T., Yamana, M. (1991) New ferroelectric liquid crystals having 2-fluoro-2-methyl alkanoyloxy group, Mol. Cryst. Liq. Cryst. **199**, 129–140

- Takahashi, O., Umezawa, J., Furuhashi, K., Takagi, M. (1989) Stereocontrol of a tertiary hydroxyl group via microbial epoxidation. A facile synthesis of prostaglandin ω-chains, Tetrahedron Lett. **30**, 1583–1584

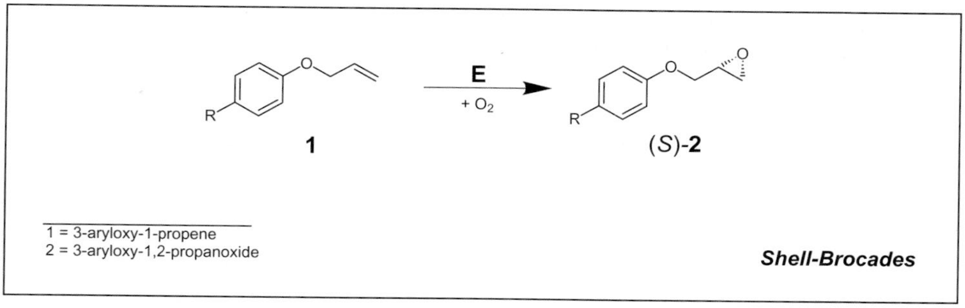

Oxidase

Pseudomonas oleovorans

Fig. 1.14.X.X – 1

1) Reaction conditions

[2]:	> 0.26 M, > 7 g · L^{-1} [267.36 g · mol^{-1}]
T:	37°C
medium:	aqueous
reaction type:	epoxidation
catalyst:	suspended whole cells
enzyme:	Oxidase
strain:	*Pseudomonas oleovarans*

2) Flow scheme

Not published.

3) Process parameters

ee:	> 99.9 %
company:	Shell AG, USA

4) Product application

- The intermediates can be converted into β-adrenergic receptor blocking agents.

- These pharmaceuticals have the common structural feature of a 3-aryloxy- or 3-heteroaryl-oxy-1-alkylamino-2-propanol.

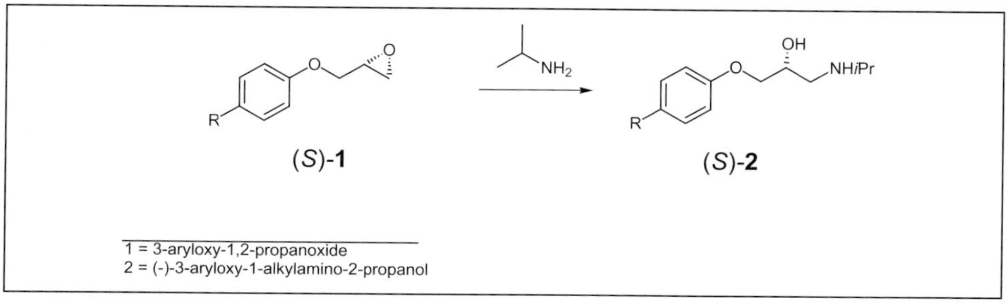

Fig. 1.14.X.X – 2

159

- Two examples are metoprolol and atenolol:

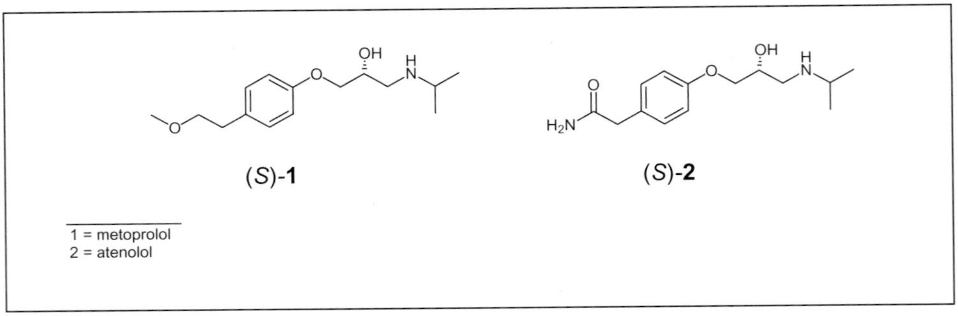

1 = metoprolol
2 = atenolol

Fig. 1.14.X.X – 3

- Since the biological activity of the enantiomers differs by more than a factor of 100, the production of the racemic mixtures of such pharmaceuticals will become outdated.

5) Literature

- Kieslich, K. (1991) Biotransformations of industrial use, Acta Biotechnol. **11**, 559–570

- Johnstone, S.L., Phillips, G.T., Robertson, B.W., Watts, P.D., Bertola, M.A., Koger, H.S., Marx, A.F. (1987) Stereoselective synthesis of *S*-(–)-β-blockers via microbially produced epoxide intermediates, Proc. Int. Symposium: Biocatalysis in organic media, 387–392, Wageningen, The Netherlands

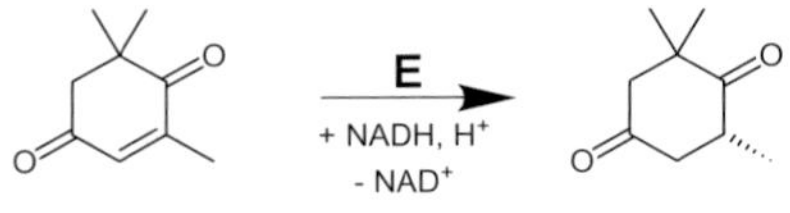

1 = oxoisophorone
2 = 2,2,6-trimethylcyclohexane-1,4-dione

Hoffmann La-Roche

Fig. 1.X.X.X – 1

1) Reaction conditions

[1]:	0.031-0.061 M, 4.74–9.33 g·L^{-1} [152.19 g·mol^{-1}]
[2]:	0.311 M, 47.9 g·L^{-1} [154.21 g·mol^{-1}]
pH:	8.0–9.0
T:	20 °C
medium:	aqueous
reaction type:	redox reaction
catalyst:	suspended whole cells
enzyme:	reductase
strain:	baker's yeast

2) Remarks

- On a 13 kg scale oxoisophorone is reacted by fermentative reduction with baker's yeast in an aqueous saccharose solution by periodical substrate addition and feeding with saccharose (fed-batch).

- Periodical addition of oxoisophorone is necessary to avoid toxic levels of oxoisophorone.

- At the suboptimal fermentation temperature of 20°C the growth rate is lowered but the product precipitates faster and the concentration in the aqueous phase is reduced, resulting in slower inactivation of the yeast. Additionally, the saccharose consumption is decreased relative to the production rate.

3) Flow scheme

Not published.

4) Process parameters

conversion:	87 %
yield:	80 %
selectivity:	92 %
ee:	< 97 %
chemical purity:	> 99 %
reactor type:	batch
residence time:	408 h
space-time-yield:	$2.8 \ \mathrm{g \cdot L^{-1} \cdot d^{-1}}$
down stream processing:	crystallization
enzyme supplier:	Klipfel (Rheinfelden, Switzerland)
company:	Hoffmann La-Roche, Switzerland

5) Product application

- The dione is an intermediate for the synthesis of natural 3-hydroxycarotenoids, e.g. zeaxanthin, cryptoxanthin and structurally related compounds:

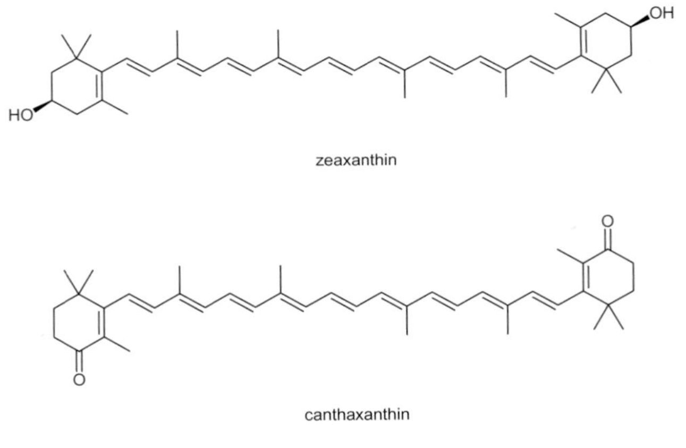

zeaxanthin

canthaxanthin

Fig. 1.X.X.X – 2

- Zeaxanthin, e.g., is the main pigment in corn and barley and can also be found in yolk.

6) Literature

- Leuenberger, H.G., Boguth, W., Widmer, E., Zell, R. (1976) 189. Synthese von optisch aktiven, natürlichen Carotinoiden und strukturell verwandten Naturprodukten. I. Synthese der chiralen Schlüsselverbindung (4R,6R)-4-Hydroxy-2,2,6-trimethylcyclohexanon, Helv. Chim. Acta **59**, 1832–1849

1 = isopropylideneglycerol
2 = isopropylideneglyceric acid

International BioSynthetics

Fig. 1.X.X.X – 1

1) Reaction conditions

[1]:	0.076 M, 10 g · L^{-1} [132.16 g · mol^{-1}]
pH:	6.8 – 7.2
T:	30 °C
medium:	aqueous
reaction type:	hydroxylation
catalyst:	whole cells
enzyme:	hydroxylase
strain:	*Rhodococcus erythropolis*

2) Remarks

- The resolution is carried out by selective microbial oxidation of the (*S*)-enantiomer.

- The chemical synthesis of (*R*)-isopropylideneglycerol starting from unnatural L-mannitol is difficult and expensive:

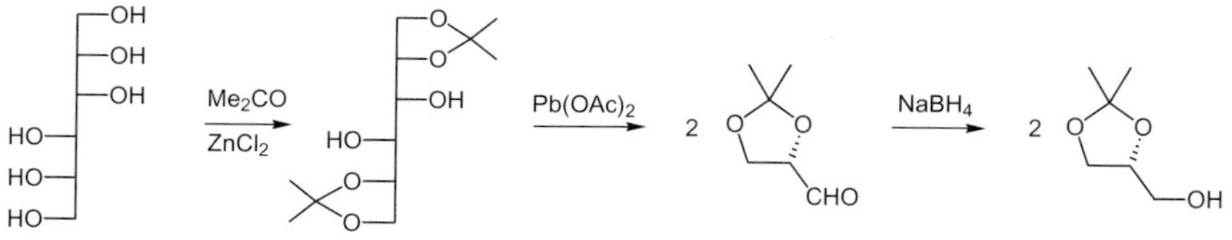

1 = mannitol
2 = isopropylideneglyceric aldehyde
3 = isopropylideneglycerol

Fig. 1.X.X.X – 2

Oxidase
Rhodococcus erythropolis

- Another chemical route starts from L-ascorbic acid:

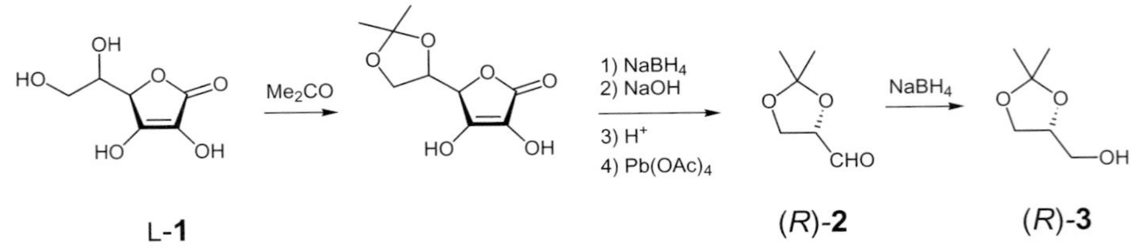

1 = ascorbic acid
2 = isopropylideneglyceric aldehyde
3 = isopropylideneglycerol

Fig. 1.X.X.X – 3

3) Flow scheme

Not published.

4) Process parameters

conversion:	50 %
ee:	> 98 % for (*R*)-**1**, > 90 % for (*R*)-**2**
reactor type:	fed batch
company:	International Bio-Synthetics

5) Product application

- (*R*)-Isopropylideneglycerol is a useful C_3-synthon in the synthesis of (*S*)-β-blockers, e.g. (*S*)-metoprolol.

- Also, the (*R*)-isopropylideneglyceric acid may be used as starting material for the synthesis of biologically active products.

6) Literature

- Crosby, J. (1991) Synthesis of optically active compounds: a large scale perspective, Tetrahedron **47**, 4789–4846

- Bertola, M. A., Koger, H. S., Phillips, G. T., Marx, A. F., Claassen, V. P. (1987) A process for the preparation of (*R*)- and (*S*)–2,2-R1,R2–1,3-dioxolane-4-methanol, Gist-Brocades NV and Shell Int. Research, EU 244912

- Baer, E., Fischer, H. (1948) J. Am. Chem. Soc. **70**, 609

- Hirth, G., Walther, W. (1985) Synthesis of the (*R*)- and (*S*)-glycerol acetonides. Determination of the optical purity, Helv. Chim. Acta **68**, 1863

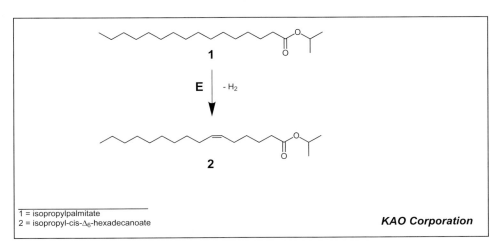

1 = isopropylpalmitate
2 = isopropyl-cis-Δ_6-hexadecanoate

KAO Corporation

Fig. 1.X.X.X – 1

1) Reaction conditions

[1]:	0.67 M, 200 g · L^{-1} [298.50 g · mol^{-1}]
[2]:	0.15 M, 45 g · L^{-1} [296.49 g · mol^{-1}]
pH:	7.0
T:	26 °C
medium:	two-phase: aqueous oil (70:30 (v:v)) emulsion
reaction type:	*cis*-desaturation (redox reaction)
catalyst:	suspended whole cells
enzyme:	desaturase
strain:	*Rhodococcus* sp. KSM-B-MT66 mutant

2) Remarks

- The dehydrogenation always takes place 9-C-atoms away from the terminal methyl group.

- With this process the production of unsaturated fatty acids is possible from low cost saturated compounds.

- The mutant KSM-B-MT66 was obtained by treatment with UV irradiation. The parent strain was originally isolated from soil samples.

- The mutant strain of *Rhodococcus* sp. *cis*-desaturates a variety of hydrocarbons and acyl fatty acids at central positions of the chains:

Fig. 1.X.X.X – 2

- Thiamine and Mg^{2+} are added for increased stability.

- The reaction is started as an O/W emulsion. Adding isopropylhexadecanoate up to 70 : 30 (v/v) leads to a phase inversion.

- The product and residual substrate in the oil phase are recovered by a hydrophobic membrane-based filtration (hydrophobic hollow-fiber module TP-113, Asahi Chemical Co., Ltd., Japan). Oil permeation flux is $5 \, L \cdot m^{-2} \cdot h^{-1}$.

- Product separation is performed by the urea-adduct method and chromatography on silica gel.

3) Flow scheme

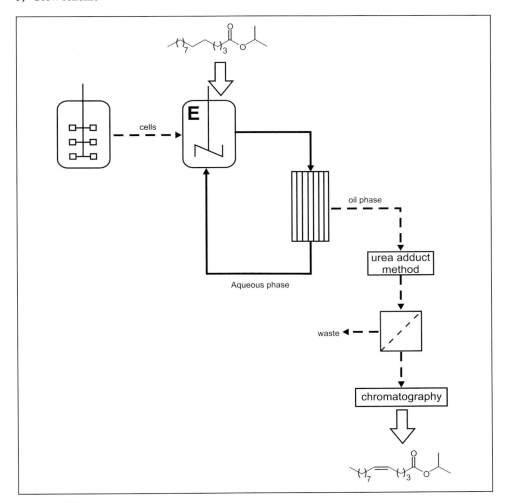

Fig. 1.X.X.X – 3

4) Process parameters

chemical purity:	97 %
reactor type:	repetitive batch
space-time-yield:	16.8 g · L^{-1} · d^{-1}
down stream processing:	chromatography, 80 % recovery
production site:	Japan
company:	Kao Corp., Tochigi, Japan

5) Product application

● The products represent intermediates for the preparation of fatty acids substituted within the aliphatic chain for use in dermatologic pharmaceuticals.

167

6) Literature

- Downing, D.T., Staruss J.S. (1974) Synthesis and composition of surface lipids of human skin, J. Invest. Dermatol. **62**, 228–244

- Kieslich, K. (1991) Biotransformations of industrial use, Acta Biotechnol. **11**, 559–570

- Morello, A.M., Downing D.T., (1976) *Trans*-unsaturated fatty acids in human skin surface lipids, J. Invest. Dermatol. **67**, 270–272

- Takeuchi, K., Koike, K., Ito, S.(1990) Production of *cis*-unsaturated hydrocarbons by strain of *Rhodococcus* in repeated batch culture with a phase-inversion, hollow-fiber system, J. Biotechnol. **14**, 179–186

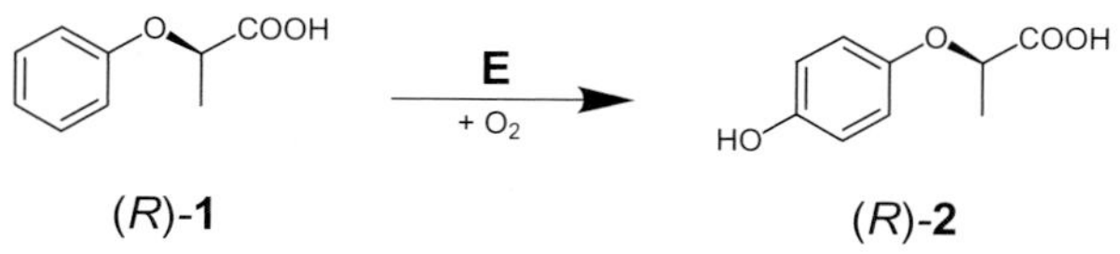

1 = 2-phenoxypropionic acid (POPS)
2 = 2-(4´-hydroxyphenoxy)propionic acid (HOPS)

BASF

Fig. 1.X.X.X – 1

1) Reaction conditions

[**1**]:	> 0.04 M, > 5 g · L^{-1} [126.17 g · mol^{-1}]
medium:	aqueous
reaction type:	hydroxylation
catalyst:	suspended whole cells
enzyme:	oxidase
strain:	*Beauveria bassiana* LU 700

2) Remarks

- The biocatalyst *Beauveria bassiana* was found by an extensive screening of microorganisms for specifically hydroxylating POPS to HPOPS regioselectively. An additional screening criteria was the tolerance to substrate concentrations > 5 g · L^{-1}.
- The production strain LU 700 was yielded by two mutations (first: UV-light, second: *N*-methyl-*N′*-nitrosoguanidine).
- The productivity was increased by optimization of the trace elements in the media by use of a genetic algorithm (cupric acid from 0.01 to 0.75 mg · L^{-1}, manganese from 0.02 to 2.4 mg · L^{-1} and ferric ions from 0.8 to 6 mg · L^{-1}).
- The hydroxylation is not growth-associated.
- The oxidase has a very broad substrate spectrum (figure 1.X.X.X – 2):

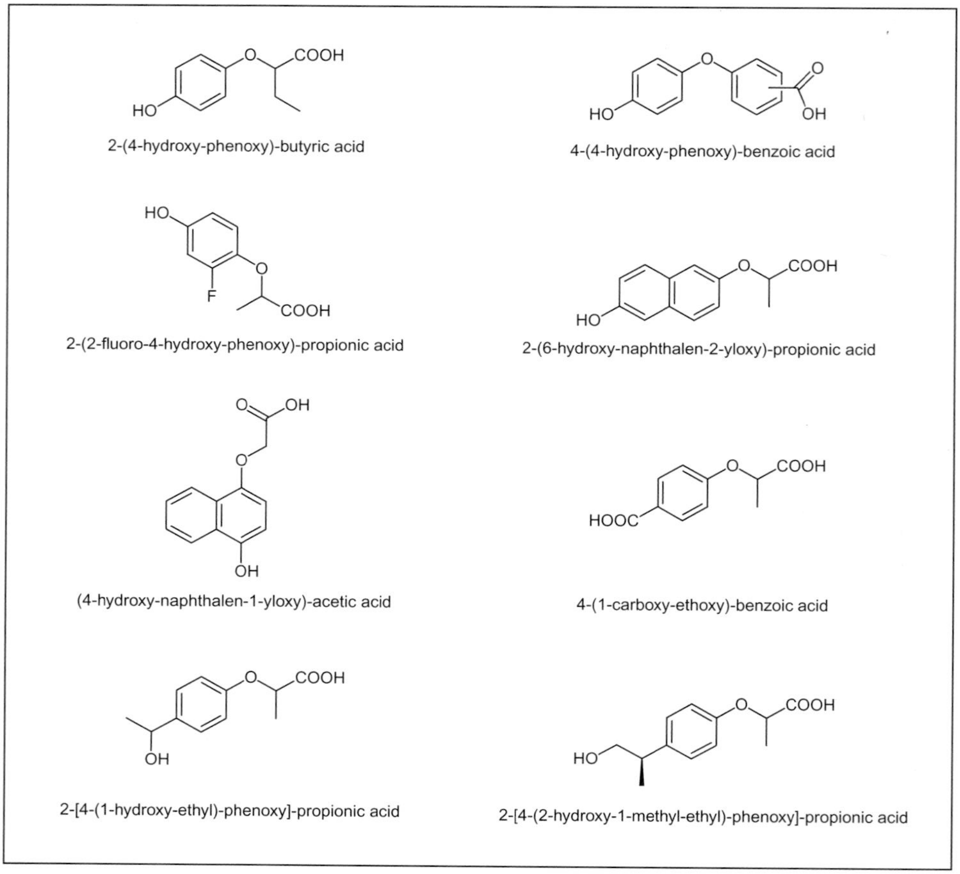

Fig. 1.X.X.X – 2

- A compound needs the structural elements of a carboxylic acid and an aromatic ring system to be a substrate for the oxidase. Hydroxylation primarily takes place in phenoxy-derivatives at the *para* position if it is free. If there is an alkyl group in the *para* position, only side-chain hydroxylation takes place. In systems with more than one ring, the most electron-rich ring is hydroxylated.

- The ee is increased during oxidation from 96 % for POPS to 98 % for HPOPS.

- The substrate (*R*)-2-phenoxypropionic acid is easily synthesized from (*S*)-2-chloropropionic acid isobutylester and phenol:

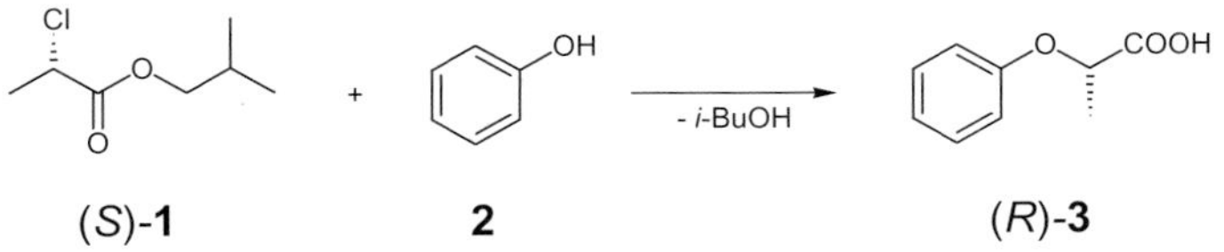

(*S*)-**1**   **2**   (*R*)-**3**

1 = 2-chloropropionic acid isobutyl ester
2 = phenol
3 = 2-phenoxypropionic acid (POPS)

Fig. 1.X.X.X – 3

3) Flow scheme

Not published.

4) Process parameters

ee:	> 98.0 %
chemical purity:	> 99.5 %
reactor type:	cstr
reactor volume:	120,000 L
space-time-yield:	$7 \text{ g} \cdot \text{L}^{-1} \cdot \text{d}^{-1}$
company:	BASF, Germany

5) Product application

- (*R*)-2-(4′-hydroxyphenoxy)propionic acid is used as an intermediate for the synthesis of enantiomerically pure aryloxyphenoxypropionic acid-type herbicides.

6) Literature

- Cooper, B., Ladner, W., Hauer, B., Siegel, H. (1992) Verfahren zur fermentativen Herstellung von 2-(4-Hydroxyphenoxy-)propionsäure, EP 0465494 B1

- Dingler, C., Ladner, W., Krei, G., Cooper, B. Hauer, B. (1996) Preparation of (*R*)-2-(4-hydroxyphenoxy)propionic acid by biotransformation, Pestic. Sci. **46**, 33–35

Cyclodextrin glycosyltransferase
Bacillus circulans

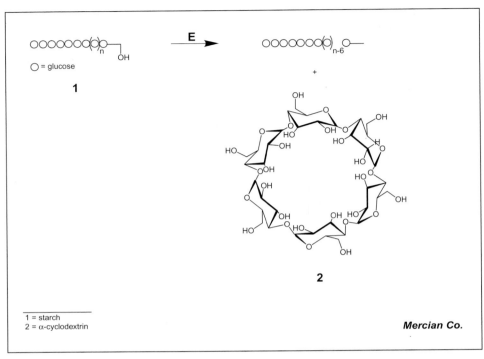

○ = glucose

1

2

1 = starch
2 = α-cyclodextrin

Mercian Co.

Fig. 2.4.1.19 – 1

1) Reaction conditions

[**1**]:	8.3 w/v % liquified starch
pH:	5.8 – 6.0
T:	55 °C
medium:	aqueous
reaction type:	hexosyl group transfer
catalyst:	solubilized enzyme
enzyme:	1,4-α-D-glucan 4-α-D-(1,4-α-D-glucano)-transferase (cyclizing) (cyclomalto-dextrin glucanotransferase)
strain:	*Bacillus circulans*
CAS (enzyme):	[9030-09–5]

2) Remarks

- Cyclodextrins are produced by cyclodextrin glycosyltransferases as a mixture of α-, β- and γ-cyclic oligosaccharides.

- The main problem that had to be overcome to establish an economic cyclodextrin production was the separation of the cyclodextrins from the reaction media. This is important due to two points:
  1) The reaction mixtures contains many by-products.
  2) Increasing cyclodextrin concentrations inhibit the enzyme.

172

Cyclodextrin glycosyltransferase
Bacillus circulans

- The separation is established by selective adsorption of α- and β-cyclodextrins on chitosan beads with appropriate ligands. α-Cyclodextrins are selectively interacted with stearic acid and β-cyclodextrins with cyclohexanepropanamide-*n*-caproic acid. The adsorption selectivity is almost 100 %. In the case of the β-cyclodextrins a capacity of 240 g·L^{-1} gel bed is reached.

- The process data given here are related to the production of α-cyclodextrins.

- Before entering the adsorption column the temperature is lowered to 30 °C for effective adsorption. At this temperature almost no cyclodextrins are formed during circulation. Before re-entering the main reactor the temperature of the solution is again adjusted to 55 °C by using the energy of the reaction solution leaving the reactor.

- To prevent adsorption of the cyclodextrin glycosyltransferase in the α-cyclodextrin adsorbent 3 w/v % NaCl are added.

3) Flow scheme

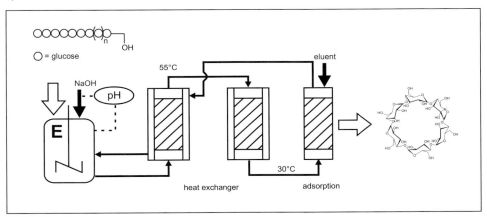

Fig. 2.4.1.19 – 2

4) Process parameters

yield:	22.3 % (α-cyclodextrin); 10.8 % (β-cyclodextrin); 5.1 % (γ-cyclodextrin)
chemical purity:	94.9 %
reactor type:	batch
down stream processing:	chromatography and crystallization
enzyme supplier:	Amano, Japan
company:	Mercian Co., Ltd., Japan

5) Product application

- Cyclodextrins serve as molecular hosts and are used in the food industry for capturing and retaining flavors. They are also used in the formulation of pharmaceuticals.

6) Literature

- Chin, P.J., Chein, S. (1989) Study of cyclodextrin production using cyclodextrin glycosyltransferase immobilized on chitosan, J. Chem. Technol. Biotechnol. **46**, 283–294

- Okabe, M., Tsuchiyama, Y., Okamoto, R. (1993) Development of a cyclodextrin production process using specific adsorbents, in: Industrial Application of Immobilized Biocatalysts (Tanaka, A., Tosa, T., Kobayashi, T., eds.) pp.109–129, Marcel Dekker Inc., New York

- Tsuchiyama, Y., Yamamoto, K., Asou, T., Okabe, M., Yagi, Y., Okamoto, R. (1991) A novel process of cyclodextrin production by use of specific adsorbents. Part I. Screening of specific adsorbents, J. Ferment. Bioeng. **71**, 407–412

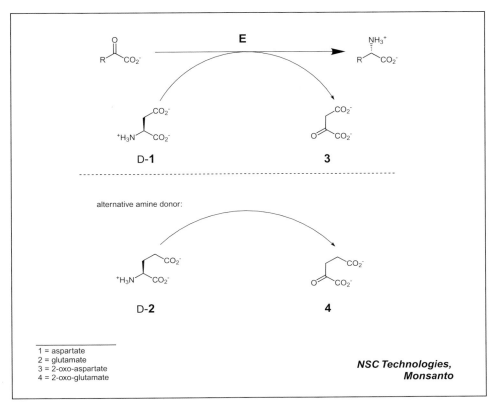

Fig. 2.6.1.21 – 1

1) Reaction conditions

pH:	8.0
T:	42 °C
medium:	aqueous
reaction type:	amino group transfer
catalyst:	whole cell
enzyme:	D-alanine; 2-oxoglutarate aminotransferase (D-aspartase transaminase)
strain:	*Bacillus* sp.
CAS (enzyme):	[37277–85–3]

2) Remarks

- The drawback of transaminases is the equilibrium conversion of about 50 %.

- Therefore in commercial application L-aspartate is often used as the amino donor for L-amino acid transaminase. The product decarboxylates to pyruvic acid and carbon dioxide. Pyruvic acid itself is also an α-keto acid and is converted by dimerisation using acetolactate synthase to acetolactate. The latter undergoes spontaneous decarboxylation to acetoin that can be easily removed and does not participate in other reactions.

175

- The production of D-amino acids proceeds in a similar way using D-aspartase or D-glutamate as amino donor.

- The α-keto acids are available by chemical or enzymatic methods. A amino acid deaminases generate the α-keto acids from inexpensive L-amino acids. Using amino acid racemase the amino-group donor is also accessible from cheap racemic mixtures of amino acids. The EC-number given above is choosen for D-alanine as precursor for the α-keto acid.

- The advantage of the process is the use of cloned strains implanted in *E. coli.* which are capable of conducting all steps of synthesis using cheap racemic educts resulting in high yields of D-amino acids.

- The following scheme shows the coupling of the enzyme systems using L-aspargine as the amino-group donor for the production of D-amino acids:

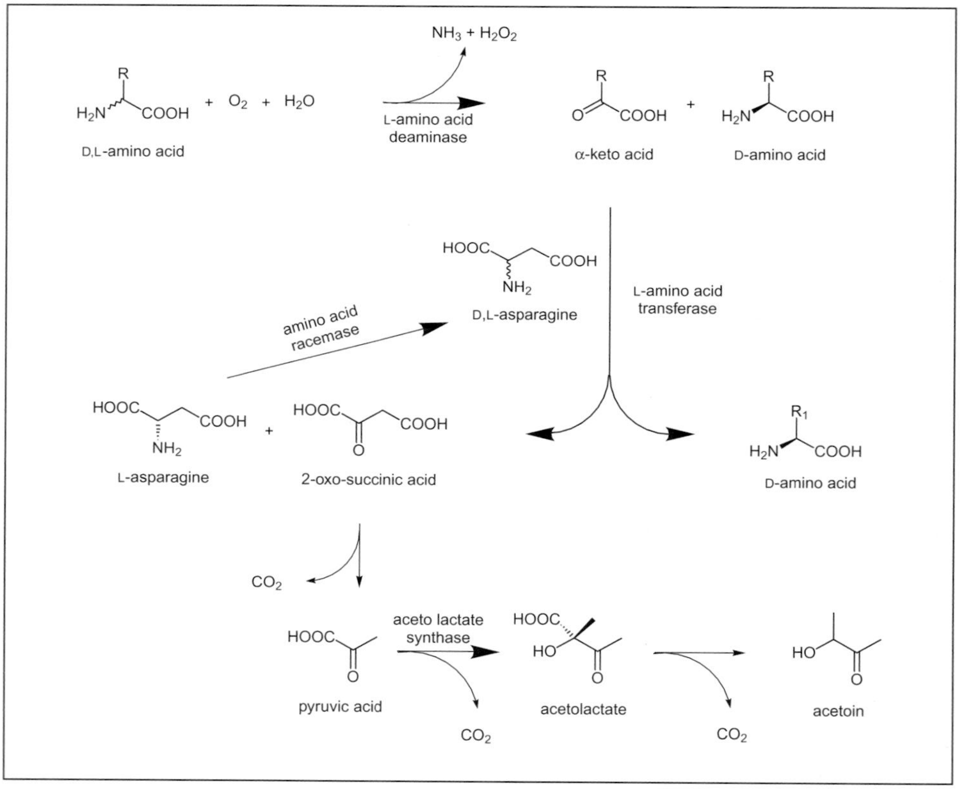

Fig. 2.6.1.21 – 2

- Possible products using this synthetic route can be:

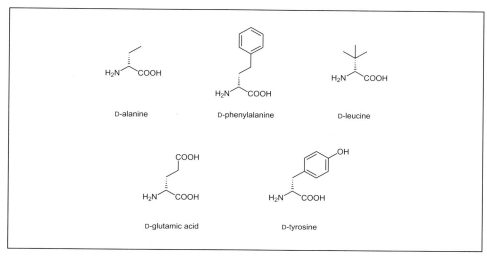

D-alanine

D-phenylalanine

D-leucine

D-glutamic acid

D-tyrosine

Fig. 2.6.1.21 – 3

3) Flow scheme

Not published.

4) Process parameters

ee:	100 %
reactor type:	batch
capacity:	multi t
down stream processing:	crystallization
start-up date:	May 1998
production site:	Georgia, San Diego, California, USA
company:	NSC Technologies, Monsanto, USA

5) Product application

- Amino acids are used as food additives and in medicine in infusion solutions.

6) Literature

- Ager, D.J., Fotheringham I.G., Laneman, S.A., Pentaleone D.P., Taylor, P.P. (1997) The large-scale synthesis of unnatural acids, Chim. Oggi, **15 (3/4)**, 11–14

- Fotheringham I.G., Pentaleone D.P., Taylor, P.P. (1997) Biocatalytic production of unnatural amino acids, mono esters, and *N*-protected derivatives, Chim. Oggi **15 (9/10)**, 33–37

- Fotheringham I.G., Taylor, P.P., Ton, J.L. (1998) Preparation of enantiomerically pure D-amino acid by direct fermentative means, Monsanto, US 5728555

177

- Fotheringham, I.G., Beldig, A., Taylor, P.P. (1998) Characterization of the genes encoding D-amino acid transaminase and glutamte racemase, two D-glutamate biosynthetic enzymes of *Bacillus Sphaericus* ATCC 10208, J. Bacteriol. **180**, 4319–4323

- Taylor, P.P., Fotheringham I.G. (1997) Nucleotide sequence of the *Bacillus licheniformis* ATCC 10716 dargene and comparison of the predicted amino acid sequence with those of other bacterial species, Biochim. Biophys. Acta **1350 (1)**, 38–40

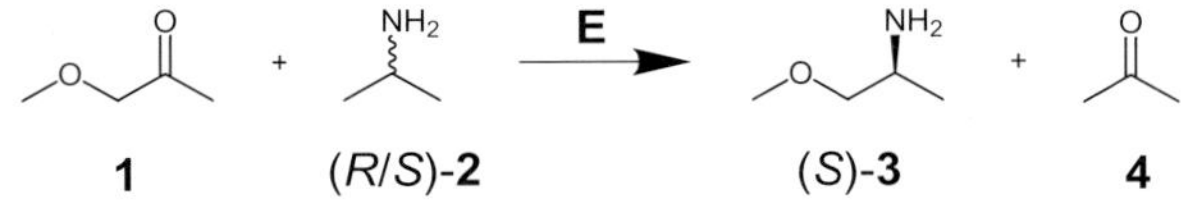

1 = 1-methoxy-propan-2-one
2 = isopropylamine
3 = methoxy isopropylamine
4 = acetone

Celgene

Fig. 2.6.1.X – 1

1) Reaction conditions

[2]:	0.7 M, 61.7 g · L^{-1} [88.11 g · mol^{-1}]
medium:	aqueous
reaction type:	amino group transfer
catalyst:	suspended whole cells
enzyme:	aminotransferase (transaminase)
strain:	*Escherichia coli*

2) Remarks

- Isopropylamine is the choice for the amino donor because it is a cheap and a kinetically attractive molecule.

- Main disadvantage of the wild-type transaminase is that the conversion is limited due to product inhibition.

- Starting with the wild-type transaminase, the gene encoding the enzyme could be optimized step by step by mutation.

- A single mutation in the gene coding the transaminase increases the possible product concentration from 0.16 M to 0.45 M.

- The screening criteria for better genes included, beside higher inhibitor concentrations, better reaction rates, higher stability and lower K_M-values.

- Since no recycling of the catalyst is integrated, the residual activity after one batch run is of no interest.

3) Flow scheme

Not published.

4) Process parameters

selectivity: 94 %
ee: > 99 %
reactor type: batch
enzyme consumption: 0.15 $kg_{catalyst\ powder} \cdot kg^{-1}_{product}$ (t > 6 h)
enzyme supplier: Celgene, USA
company: Celgene, USA

5) Product application

- Possible products derived from (*S*)-methoxyisopropylamine are several agrochemicals, e. g.:

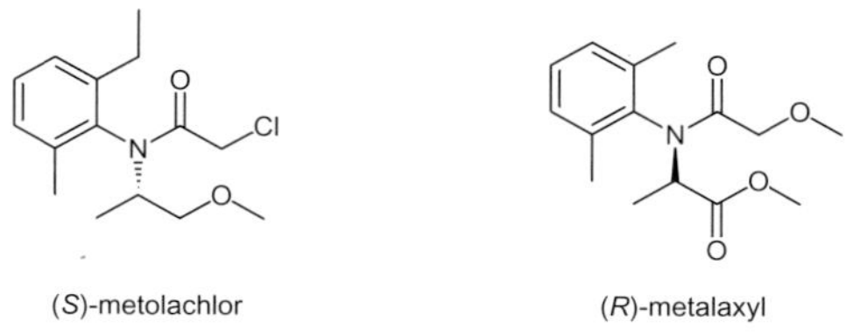

(*S*)-metolachlor   (*R*)-metalaxyl

Fig. 2.6.1.X – 2

- Ciba spent about 10 years to find a chemical catalyst for the production of an enantiomer-enriched (*S*)-metolachlor. But only an ee of 79 % is possible resulting in a reduced application rate of only 38 %. Other companies developed different approaches leading to higher product costs than the biotransformation route.

6) Literature

- Matcham, G.W. (1997) Chirality and biocatalysis in agrochemical production, INBIO Europe 97 Conference, Spring Innovations Ltd, Stockport, UK

- Matcham, G.W., Lee, S. (1994) Process for the preparation of chiral 1-aryl-2-aminopropanes, Celgene Corporation, US 5,360,724

- Stirling, D., Matcham, G.W., Zeitlin, A.L. (1994) Enantiomeric enrichment and stereoselective synthesis of chiral amines, Celgene Corporation, US 5,300,437

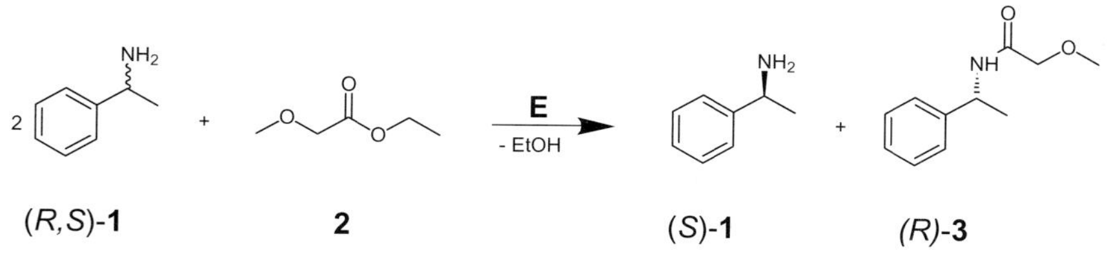

Fig. 3.1.1.3 – 1

1) Reaction conditions

[(R/S)-**1**]:	1.65 M, 200 g·L^{-1} [121.18 g·mol^{-1}] in MTBE (= methl-*tert*-butylether)
[(S)-**1**]:	1.4 M, 170 g·L^{-1} [121.18 g·mol^{-1}] in MTBE
pH:	8.0–9.0
T:	25 °C
medium:	MTBE-ethylmethoxyacetate
reaction type:	carboxylic ester hydrolysis
catalyst:	immobilized enzyme
enzyme:	triacylglycerol acylhydrolase (triacylglycerol lipase)
strain:	*Burkholderia plantarii*
CAS (enzyme):	[9001–62–1]

2) Remarks

- The lipase is immobilized on polyacrylate.

- The lowering in activity caused by the use of organic solvent can be offset (about 1,000 times and more) by freeze-drying a solution of the lipase together with fatty acids (e.g. oleic acid).

- The E-value of the reaction is above 500.

- The (*R*)-phenylethylmethoxy amide can be easily hydrolyzed to get the (*R*)-phenylethylamine:

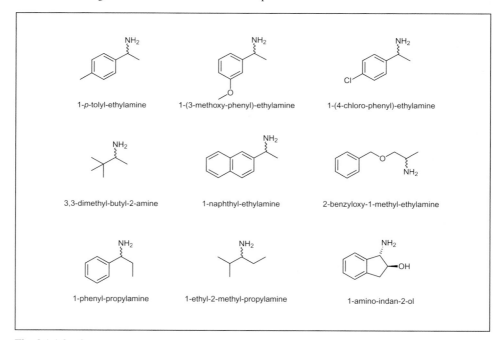

```
1 = phenylethylmethoxyamide
2 = 1-phenylethylamine
```

Fig. 3.1.1.3 – 2

- The (*S*)-enantiomer can be racemized using a palladium catalyst.
- The following amines can also be used in this process:

Fig. 3.1.1.3 – 3

3) Flow scheme

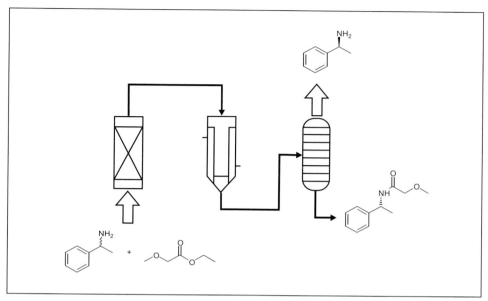

Fig. 3.1.1.3 – 4

4) Process parameters

conversion:	50 %
yield:	> 90 %
ee:	> 99 % (*S*); 93 % (*R*)
reactor type:	plug-flow reactor or batch
capacity:	> 100 t · a^{-1}
residence time:	5–7 h
down stream processing:	distillation or extraction
production site:	Ludwigshafen, Germany
company:	BASF, Germany

5) Product application

- Products are intermediates for pharmaceuticals and pesticides.

- They can also be used as chiral synthons in asymmetric synthesis.

6) Literature

- Balkenhohl, F., Hauer, B., Lander, W., Schnell, U., Pressler, U., Staudemaier H.R. (1995) Lipase katalysierte Acylierung von Alkoholen mit Diketenen, BASF AG, DE 4329293

- Balkenhohl, F., Ditrich, K., Hauer, B., Lander, W. (1997) Optisch aktive Amine durch Lipase-katalysierte Methoxyacetylierung, J. prakt. Chem. **339**, 381–384

- Reetz, M.T., Schimossek, K. (1996) Lipase-catalyzed dynamic kinetic resolution of chiral amines: use of palladium as the racemization catalyst, Chima **50**, 668

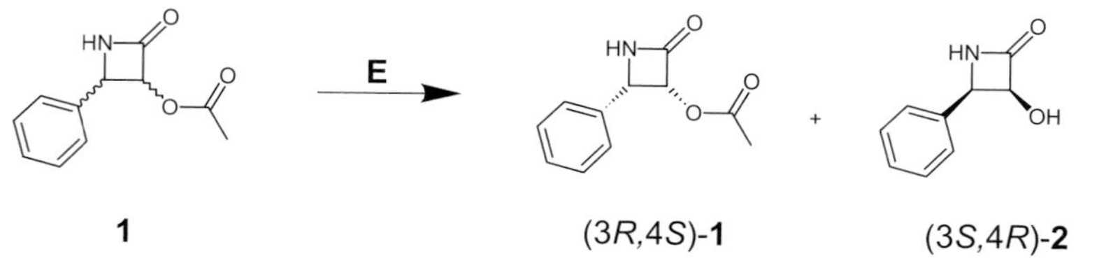

Lipase
Pseudomonas cepacia

1 = cis-azetidinone acetate
2 = cis-azetidinone

Bristol-Myers Squibb

Fig. 3.1.1.3 – 1

1) Reaction conditions

[**1**]:	0.049 M, 10 g · L^{-1} [205.21 g · mol^{-1}]
pH:	7.0
T:	29 °C
medium:	aqueous
reaction type:	carboxylic ester hydrolysis
catalyst:	immobilized enzyme
enzyme:	triacylglycerol acylhydrolase (lipase, triacylglycerol lipase)
strain:	*Pseudomonas cepacia*
CAS (enzyme):	[9001–62–1]

2) Remarks

● The enzyme is immobilized by adsorption to polypropylene beads. It can be recycled many times.

● The immobilized enzyme is reused for ten cycles without any loss in activity, productivity or optical purity of the product.

● The rate of hydrolysis is determined to be 0.12 g · L^{-1} · h^{-1}. It remains constant over ten cycles.

● At the end of the reaction the temperature is lowered to 5 °C and the agitation from 200 rpm to 50 rpm. The product (*3R,4S*)-azetidinone acetate precipitates from the reaction mixture.

● The immobilized enzyme floats due to its hydrophobicity on top of the fermenter and is separated by draining.

3) Flow scheme

Not published.

4) Process parameters

yield:	> 96 %
ee:	> 99.5 %
reactor type:	repetitive batch
reactor volume:	150 L
capacity:	1.2 kg · batch^{-1}
down stream processing:	precipitation
enzyme supplier:	Amano Int., Japan, as well as Bristol-Myers Squibb, USA
company:	Bristol-Myers Squibb, USA

5) Product application

- (*3R,4S*)-Azetidinone acetate is an intermediate for the synthesis of paclitaxel (tradename of the product is taxol):

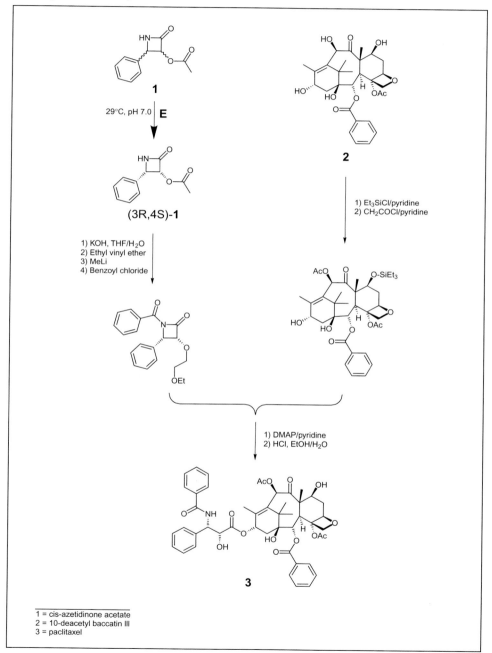

1 = cis-azetidinone acetate
2 = 10-deacetyl baccatin III
3 = paclitaxel

Fig. 3.1.1.3 – 2

- It is indicated for the treatment of various cancers.

- The main problem until now was that the source of taxol is the bark of the Pacific yew *Taxus brevifolia* that contains taxol at a very low level (< 0.02 wt.-%).

- The alternate semisynthetic pathway in which the tetracyclic diterpene 10-deacetyl baccatin can be isolated from the leaves of the European yew *Taxus baccata* (0,1 wt.-%) leads directly to paclitaxel.

6) Literature

- Holton, R.A. (1992) Method for preparation of taxol using ß-lactam, Florida State University, US 5,175,315

- Patel, R.N., Szarka, L.J., Partyka, R.A. (1993) Enzymatic process for resolution of enantiomeric mixtures of compounds useful as intermediates in the preparation of taxanes, E.R. Squibb Sons Inc., EP 552041

- Patel, R.N., Banerjee, A., Ko, R.Y., Howell, J.M., Li, W.-S., Comezoglu, F.T. (1994) Enzymic preparation of (3*R-cis*)-3-(acetyloxy)-4-phenyl-2-azetidinone: a taxol side-chain synthon, Biotechnol. Appl. Biochem. **20**, 23–33

- Zaks, A., Dodds, D.R. (1997) Application of biocatalysis and biotransformations to the synthesis of pharmaceuticals; Drug Discovery Today **2**, 513–531

Lipase
Pseudomonas cepacia

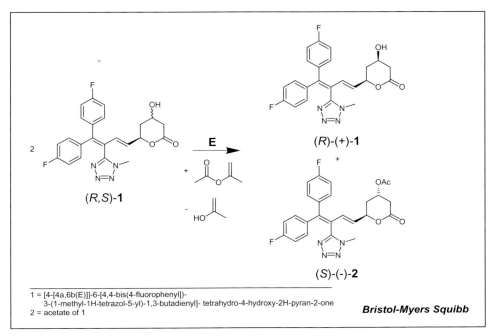

1 = [4-[4a,6b(E)]]-6-[4,4-bis(4-fluorophenyl])-
3-(1-methyl-1H-tetrazol-5-yl)-1,3-butadienyl]- tetrahydro-4-hydroxy-2H-pyran-2-one
2 = acetate of 1

Bristol-Myers Squibb

Fig. 3.1.1.3 – 1

1) Reaction conditions

[1]:	0.009 M, 4 g·L^{-1} [438.43 g·mol^{-1}]
T:	37 °C
medium:	toluene
reaction type:	acetylation
catalyst:	immobilized enzyme
enzyme:	triacylglycerol acylhydrolase (triacylglycerol lipase)
strain:	*Pseudomonas cepacia*
CAS (enzyme):	[9001–62–1]

2) Remarks

- The lipase is immobilized on a polypropylene support (Accurel PP).

- The reaction suspension additionally contains 2 % (v/v) isopropenyl acetate, 0.05 % distilled water and 1 % (w/v) lipase (Amano PS-30).

- Isopropenyl acetate is used as acyl donor, since here acetone is produced. If vinyl acetate is applied as acyl donor then acetaldehyde is formed, which is much more incompatible with lipases than acetone.

- Immobilized lipase is reused for 5 times in a repetitive batch.

- The (*S*)-acetate can be hydrolyzed in a subsequent reaction by the same lipase in an aqueous/organic two-phase system. The cleaved lactone is resynthesized using pivaloyl chloride.

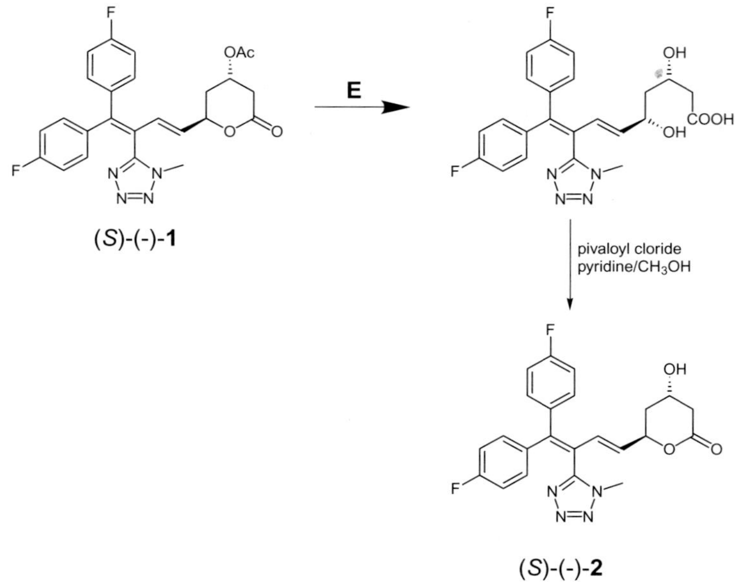

1 = isopropenyl acetate
2 = [4-[4a,6b(E)]]-6-[4,4-bis(4-fluorophenyl)]-
    3-(1-methyl-1H-tetrazol-5-yl)-1,3-butadienyl]- tetrahydro-4-hydroxy-2H-pyran-2-one

Fig. 3.1.1.3 – 2

3) Flow scheme

Not published.

4) Process parameters

yield:	48 %
ee:	98.5 %
chemical purity:	99.5 %
reactor type:	repetitive batch
reactor volume:	640 L
capacity:	multi kg
residence time:	20 h
down stream processing:	filtration, distillation, crystallization
enzyme supplier:	Amano, Japan
company:	Bristol-Myers Squibb, USA

5) Product application

- The product is a hydroxymethyl glutaryl coenzyme A (HMG-CoA) reductase inhibitor and a potential anticholesterol drug candidate.

6) Literature

- Patel, R.N. (1992) Enantioselective enzymatic acetylation of racemic [4-[4α,6β(*E*)]]-6-[4,4-bis(4-fluorophenyl])-3-(1-methyl-1H-tetrazol-5-yl)-1,3-butadienyl]-tetrahydro-4-hydroxy-2-H-pyran-2-one, Appl. Microbiol. Biotechnol. **38**, 56–60

- Patel, R.N. (1997) Stereoselective biotransformations in synthesis of some pharmaceutical intermediates, Adv. Appl. Microbiol. **43**, 91–140

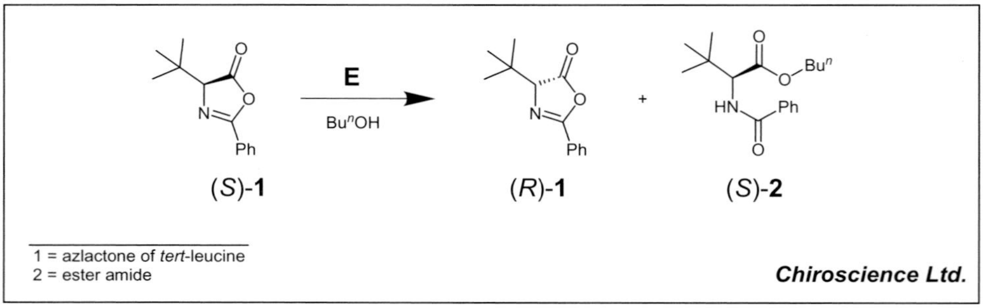

Fig. 3.1.1.3 – 1

1 = azlactone of *tert*-leucine
2 = ester amide

Chiroscience Ltd.

1) Reaction conditions

medium:	organic solvent
reaction type:	carboxylic ester hydrolysis
catalyst:	immobilized enzyme
enzyme:	triacylglycerol acylhydrolase (triacylglycerol lipase)
strain:	*Mucor miehei*
CAS (enzyme):	[9001–62–1]

2) Remarks

- At Degussa, the synthesis of (S)-*tert*-leucine is carried out as an asymmetric reductive amination of the prochiral keto acid (dehydrogenase and cofactor recycling); see page 300. Chiroscience uses the hydrolase-catalyzed resolution of azlactones instead.

- Azlactones easily racemize and can be prepared by cyclodehydration of an N-acylated amino acid with acetic anhydride.

- Although the opening of the ring with water can also be achieved non-enzymatically, the reaction in presence of water is very unselective. An alcohol in a water-free system has to be used instead.

- Even if the process is formally described as a kinetic resolution it is ultimately a deracemization. The azlactone is racemized *in situ* resulting in an overall yield of 90 % for the chiral amino acid. The following figure shows the reaction steps:

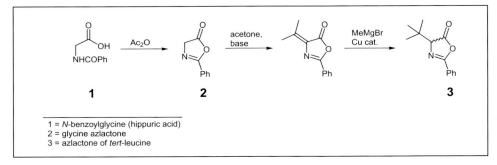

Fig. 3.1.1.3 – 2

- A high substrate concentration promotes racemization of the (*R*)-lactone and improves yield and ee.

- Since the substrate can also be synthesized via the glycine azlactone this methodology can be applied to other *tert*-alkyl glycine derivatives. The following figure shows the route to the azlactone derivative of *tert*-leucine:

Fig. 3.1.1.3 – 3

- The concentration of the substrate could be increased to 20 %, and conditions were established whereby the reaction was completed within 24 h.

- The deprotection of the ester amide to the amino acid could be carried out with potassium hydroxide without racemization (the cleavage by acid hydrolysis leads to partial racemization caused by transient recyclization to the stereochemically labile azlactone).

3) Flow scheme

Not published.

4) Process parameters

yield:	>90 %
ee:	>97 %
residence time:	24 h
company:	Chiroscience Ltd., U.K.

5) Product application

- The product is useful as a lipophilic, hindered component of peptides. Cleavage by peptidases is disfavoured because of its bulky nature, resulting in peptides of improved metabolic stability.

- The amino acids are also useful building blocks for a number of chiral auxiliaries and ligands where the presence of the bulky *tert*-butyl group makes these compounds particularly effective for asymmetric synthesis:

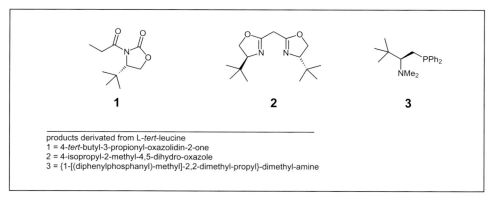

1   **2**   **3**

products derivated from L-*tert*-leucine
1 = 4-*tert*-butyl-3-propionyl-oxazolidin-2-one
2 = 4-isopropyl-2-methyl-4,5-dihydro-oxazole
3 = {1-[(diphenylphosphanyl)-methyl]-2,2-dimethyl-propyl}-dimethyl-amine

Fig. 3.1.1.3 – 4

6) Literature

- Taylor, S.J.C., McCague, R. (1997) Dynamic resolution of an oxazolinone by lipase biocatalysis: Synthesis of *(S)-tert*-leucine, in: Chirality In Industry II (Collins, A.N., Sheldrake, G.N.,Crosby, J., eds.),pp. 201–203, John Wiley & Sons,New York

- Turner, N.J., Winterman, J.R., McCague, R., Parratt, J.S., Taylor, S.J.C. (1995) Synthesis of homochiral L-(*S*)-*tert*-leucine via a lipase catalysed dynamic resolution process, Tetrahedron Lett. **36** (7)**,** 1113–1116

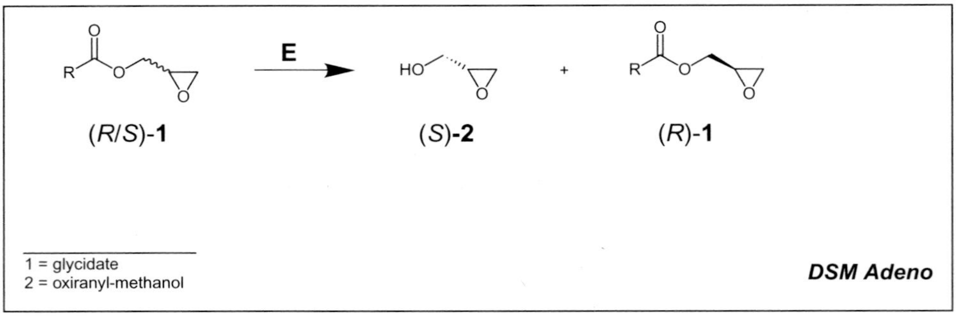

Lipase
Porcine pancreas

EC 3.1.1.3

(R/S)-**1**   (S)-**2**   +   (R)-**1**

1 = glycidate
2 = oxiranyl-methanol

DSM Adeno

Fig. 3.1.1.3 – 1

1) Reaction conditions

reaction type:	carboxylic ester hydrolysis
catalyst:	immobilized enzyme
enzyme:	triacylglycerol acylhydrolase (triacylglycerol lipase, porcine pancreas lipase = PPL)
strain:	porcine pancreas (organism)
CAS (enzyme):	[9001–62–1]

2) Remarks

- This is the oldest biocatalytic process at DSM, still operated campaign-wise.

3) Flow scheme

Not published.

4) Process parameters

yield:	> 85 %
ee:	> 99.9 %
reactor type:	batch
company:	DSM, The Netherlands

196

5) Product application

- The propane oxirane can be converted into (*S*)-beta blockers:

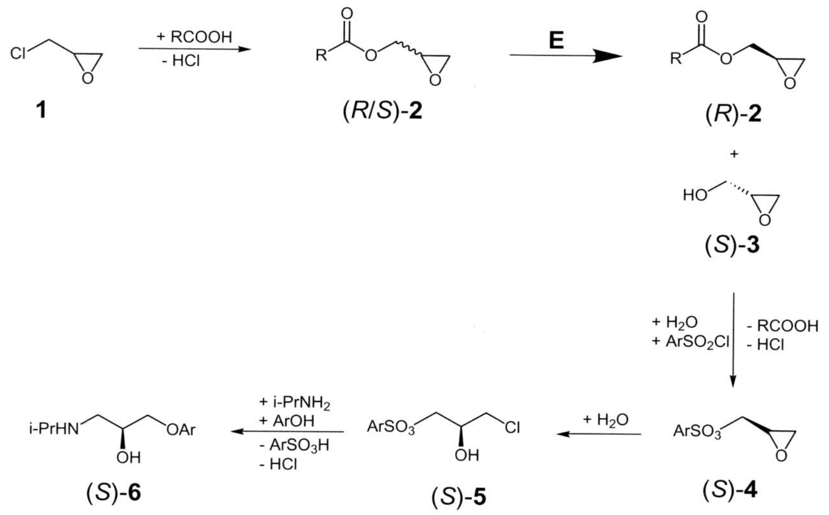

1 = 3-chlorpropaneoxirane
2 = glycidate
3 = oxiranyl-methanol
4 = 2-ôxiranyl-ethanesulfonic acid phenyl ester
5 = 3-chloro-2-hydroxy-propane-1-sulfonic acid phenyl ester
6 = beta blocker

Fig. 3.1.1.3 – 2

- Typical examples of beta blockers are shown in the following figure:

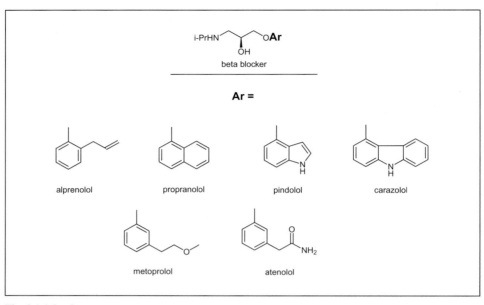

Fig. 3.1.1.3 – 3

6) Literature

- Sheldon, R.A. (1993) Chirotechnology, Marcel Dekker Inc., New York
- Elferink, V.H.M. (1995) Progress in the application of biocatalysis in the industrial scale manufacture of chiral molecules, Chiral USA 96, 11th International Spring Innovations Chirality Symposium, Boston, pp. 79–80

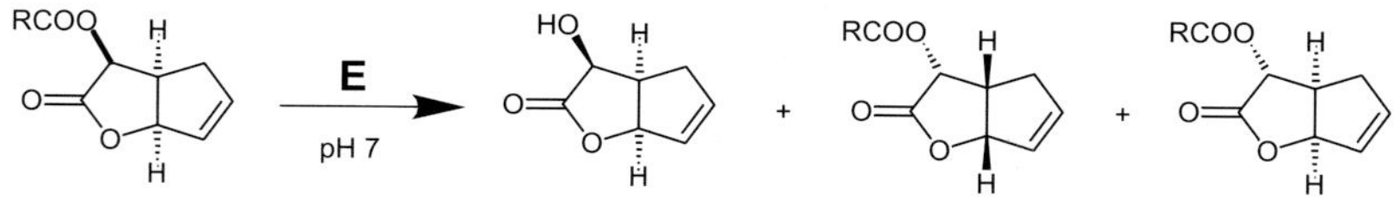

1 = 4-hydroxy-2-oxabicyclo[3.3.0]oct-7-en-3-one butyrate ester (R = propyl)
  (crude mixture of *endo*- and *exo*-butyrate esters)
2 = 4-*endo*-hydroxy-2-oxabicyclo[3.3.0]oct-7-en-3-one
3 = 4-*exo*-hydroxy-2-oxabicyclo[3.3.0]oct-7-en-3-one butyrate ester

Chiroscience Ltd.

Fig. 3.1.1.3 – 1

1) Reaction conditions

pH:	7.0
T:	25 °C
medium:	aqueous
reaction type:	carboxylic ester hydrolysis
catalyst:	solubilized enzyme
enzyme:	triacylglycerol acylhydrolase (lipase, triacylglycerol lipase)
strain:	*Pseudomonas fluorescens*
CAS (enzyme):	[9001–62–1]

2) Remarks

- The hydrolysis is preferred over the transesterification in this case since the reaction rate and enantioselectivity of the acylation are drastically reduced.

- Continuous addition of NaOH to the reaction mixture during hydrolysis is necessary to maintain neutral pH.

- If a hydrophobic ester (e.g. butyrate) is used, the ester can be extracted into the organic phase (heptane), while the alcohol remains in the aqueous phase.

- The butyrate ester is insoluble in water, so that after centrifugation and separation of the aqueous phase the alcohol can be easily extracted and purified by crystallization.

- About ten percent of the ester are lost because of pH-dependent ring-opening to an unstable carboxylic salt.

3) Flow scheme

Not published.

4) Process parameters

yield:	22 %
ee:	> 92 % (after recrystallization >99 %)
reactor type:	batch
capacity:	multi-kg
residence time:	75 h
down stream processing:	extraction, crystallization
production site:	Cambridge, UK
company:	Chiroscience Ltd, UK

5) Product application

- The product can be used as intermediate for the anti-HIV agent carbovir:

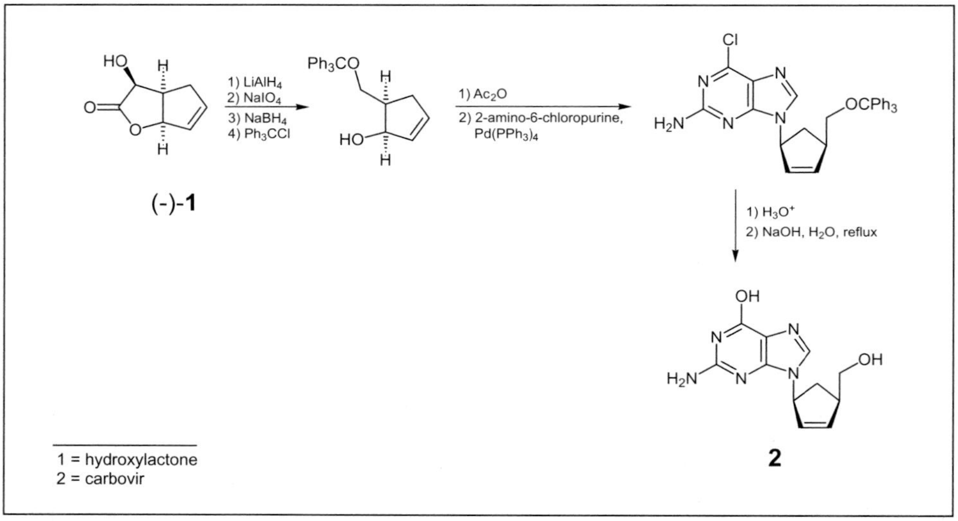

1 = hydroxylactone
2 = carbovir

Fig. 3.1.1.3 – 2

- As (–)-hydroxylactone it can be used as intermediate for the synthesis of hypocholesteremic reagents and the antifungal agent brefeldin A:

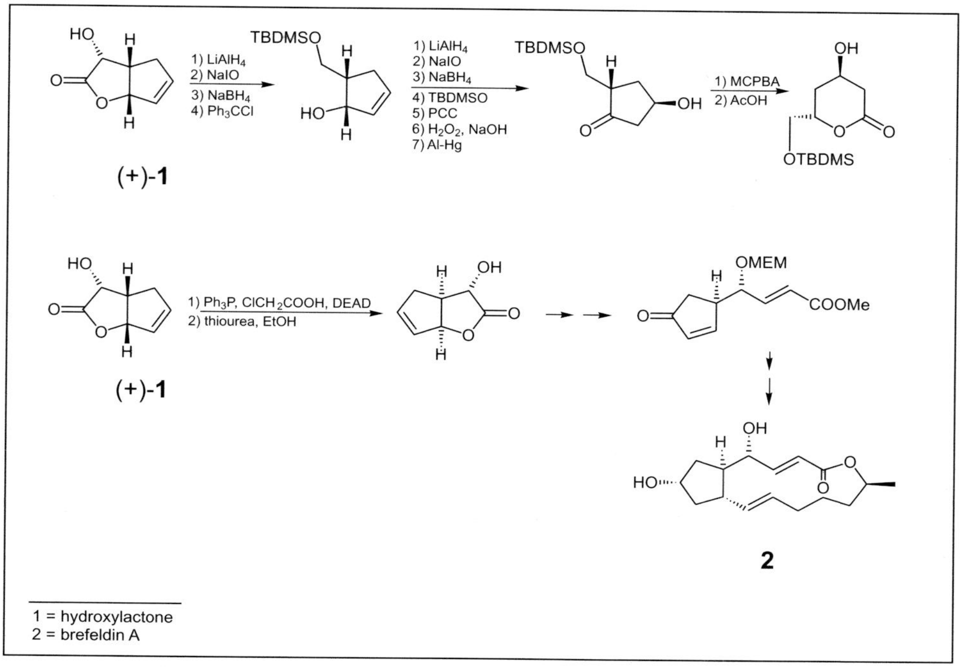

1 = hydroxylactone
2 = brefeldin A

Fig. 3.1.1.3 – 3

6) Literature

- Evans, C.T., Roberts, S.M., Shoberu, K.A., Sutherland, A.G. (1992) Potential use of carbocylic nucleosides for the treatment of AIDS: Chemo-enzymatic syntheses of the enantiomers of Carbovir; Chem. Soc. Perkin Trans. **1**, 589–592

- MacKeith, R.A., McCague, R., Olivo, H.F., Palmer, C.F., Roberts, S.M. (1993) Conversion of (–)-4-hydroxy-2-oxabicyclo[3.3.0]oct-7-en-3-one into the anti-HIV agent Carbovir, J. Chem. Soc. Perkin Trans. 1 **3**, 313–314

- MacKeith R.A., McCague R., Olivo H.F., Roberts S.M., Taylor S.J., Xiong H. (1994) Enzyme-catalysed kinetic resolution of 4-endo-hydroxy-2-oxabicyclo[3.3.0]oct-7-en-3-one and employment of the pure enantiomers for the synthesis of anti-viral and hypocholesteremic agents, Bioorg. Med. Chem. **2**, 387–394

- Taylor, S.J.C., Mc Cague, R. (1997) Resolution of a versatile hydroxylactone synthon 4-endo-hydroxy-2-oxabicyclo[3.3.0]oct-7-en-3-one by lipase deesterification, in: Chirality In Industry II (Collins, A. N., Sheldrake, G. N. and Crosby, J., eds.), pp. 190–193, John Wiley & Sons, New York

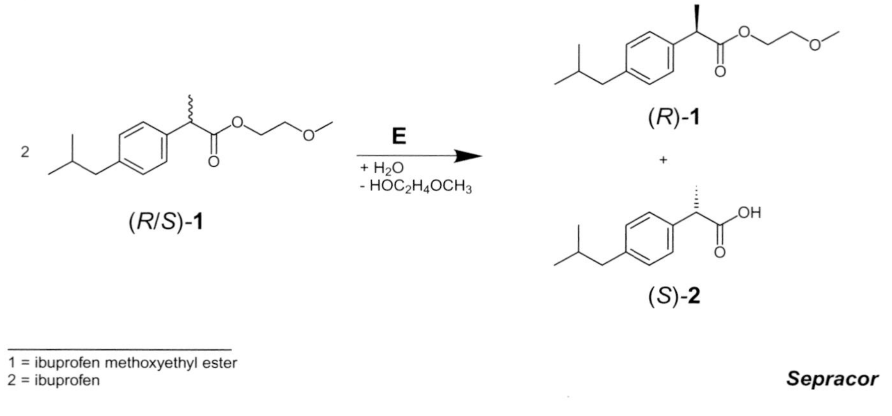

Lipase
Candida cylindracea

1 = ibuprofen methoxyethyl ester
2 = ibuprofen

Sepracor

Fig. 3.1.1.3 – 1

1) Reaction conditions

pH:	5.0
T:	20 °C
medium:	multiphase: aqueous / organic / solid
reaction type:	carboxylic ester hydrolyis
catalyst:	immobilized enzyme
enzyme:	triacylglycerol acylhydrolase (triacylglycerol lipase, lipase)
strain:	*Candida cylindraceae*
CAS (enzyme):	[9001–62–1]

2) Remarks

- Although the lipase shows good activity over a broad pH range, a low pH value has to be employed because the enzyme is deactivated by ibuprofen. At low pH-values the low solubility of ibuprofen ester can be used to prevent deactivation.

- The main problem of the enzymatic synthesis is the low solubility of substrates in water. In this case the ester solubility is below 1 mM. To circumvent problems of handling big volumes of water, a membrane based concept is realized.

- A hollow fibre membrane is used, where the lipase is immobilized (non-covalently, entrapped) in the pores of the membrane. The hydrophobic ibuprofen methoxyethylester is delivered solubilized in the organic phase to the outside of the asymmetric membrane. After conversion, the ibuprofen acid is extracted by the aqueous phase into the lumen of the hollow fibres.

- The advantage of this reactor setup is that the membrane stabilizes the aqueous/organic interface providing a high surface area for contact between the organic and aqueous phases without dispersing one phase into the other.

- In combination with another membrane module adjusted to a higher pH, the product can be easily separated from the unconverted ester, which can be easily recycled to the first membrane system. These techniques allows a low ibuprofen concentration at low pH leading to high catalyst stability.

- An alternative process starting from racemic 2-arylpropionitrile by the action of nitrilase was investigated by Asahi Chemical Ind. Co.

3) Flow scheme

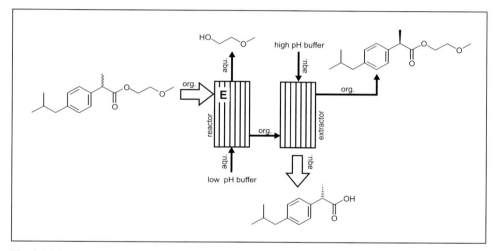

Fig. 3.1.1.3 – 2

4) Process parameters

ee:	96 %
reactor type:	membrane reactor in batch mode
capacity:	multi kg
space-time-yield:	$3.6 \ g \cdot d^{-1} \cdot g_{Lipase}^{-1}$ (related to membrane area: $18 \ g \cdot d^{-1} \cdot m^{-2}$)
down stream processing:	extraction and distillation
enzyme activity:	$5 \ g \cdot m^{-2}$
enzyme consumption:	$t_{1/2} = 30 \ d$
enzyme supplier:	Genzyme Corp., USA
company:	Sepracor, USA

5) Product application

- Ibuprofen is an important nonsteroidal antiinflammatory drug.

- For a long time ibuprofen was sold as a racemic mixture. The activity *in vitro* of the (*S*)-enantiomer is 100 times that of the (*R*)-enantiomer.

6) Literature

- Adams, S.S., Bresloff, P., Mason, C.G. (1976) Pharmacological differences between the optical isomers of ibuprofen: evidence for metabolic inversion of the (–)-isomer, J. Pharm. Pharmacol. **28**, 256–257.

- Cesti, P., Piccardi, P. (1988) Biotechnological preparation of optically active α-arylalkanoic acids, Montedison S.p.A., Italy, EP 195 717

- Lopez, J.L., Wald, S.A., Matson, S.L., Quinn, J.A. (1990) Multiphase membrane reactors for separating stereoisomers, Ann. N. Y. Acad. Sci., **613**, 155–166

- Mc Conville F.X., Lopez, J.L., Wald, S.A. (1990) Enzymatic resolution of ibuprofen in a multiphase membrane reactor, in: Biocatalysis (Abramowicz, D.A., ed.) pp.167–177, van Nostrand Reinhold, New York

- Sheldon, R.A. (1993) Chirotechnology, Marcel Dekker Inc., New York

- Sih, C. J. (1987) Process for preparing (S)-α-methylarylacetic acids from a mixture of their esters by enantiospecific hydrolysis with microbial lipase, Wisconsin Alumni Research Foundation, USA, EP 227 078

- Yamamoto, K., Ueno, Y., Otsubo, K., Kawakami, K., Komatsu, K.I. (1990) Production of (S)-(+)-ibuprofen from a nitrile compound by *Acinetobacter sp.* Strain AK226, Appl. Environ. Microbiol. **56**, 3125–3129

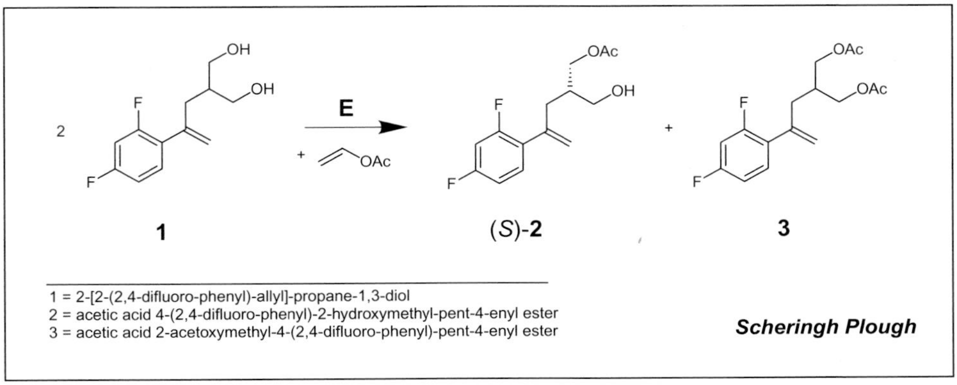

Lipase
Candida antarctica

1 = 2-[2-(2,4-difluoro-phenyl)-allyl]-propane-1,3-diol
2 = acetic acid 4-(2,4-difluoro-phenyl)-2-hydroxymethyl-pent-4-enyl ester
3 = acetic acid 2-acetoxymethyl-4-(2,4-difluoro-phenyl)-pent-4-enyl ester

Scheringh Plough

Fig. 3.1.1.3 – 1

1) Reaction conditions

[1]:	0.876 M, 200 g·L^{-1} [228.24 g·mol^{-1}]
T:	0 °C
medium:	vinyl acetate in acetonitrile
reaction type:	carboxylic ester hydrolysis
catalyst:	immobilized enzyme (Novozyme 435)
enzyme:	triacylglycerol acylhydrolase
strain:	*Candida antarctica*
CAS (enzyme):	[9001–62–1]

2) Remarks

- The lipase from *Candida antarctica* catalyzes the pro-*S* acetylation of the diol.

- Beside 74 % of the (*S*)-monoacetate, 26 % of the diacetate is formed.

- As solvent acetonitrile is selected, since the subsequent iodocyclization is also carried out in acetonitrile (see product application). Therefore, the reaction solution can be directly transferred to the chemical step after separating the lipase by filtration. To do so, it is important to reach a very high conversion, since the racemic diol also reacts in the iodocyclization. With the diacetate no reaction takes place.

- As acylating agents 1.25 equivalents of vinyl acetate are used.

3) Flow scheme

Not published.

4) Process parameters

conversion: 99 %
yield: 74 %
selectivity: 73 %
ee: > 99 %
enzyme supplier: Novo Industry, Denmark
company: Schering-Plough, USA

5) Product application

• The product is used as an improved azole antifungal. The complete synthesis is shown in the next figure:

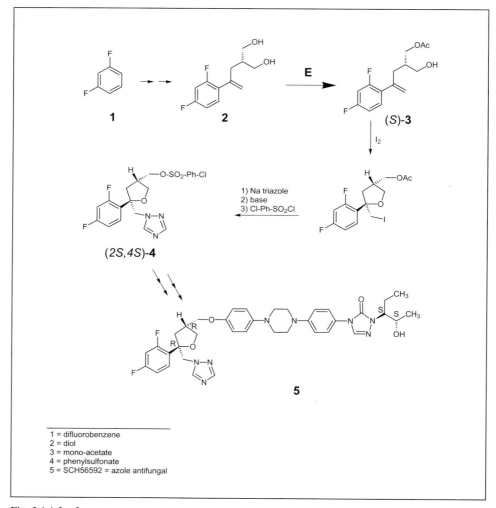

1 = difluorobenzene
2 = diol
3 = mono-acetate
4 = phenylsulfonate
5 = SCH56592 = azole antifungal

Fig. 3.1.1.3 – 2

- It shows activity against systemic *Candida* and pulmonary *Aspergillus* infections (phase II, clinical).

- The increased activity in comparison to other azole antifungals results from the tetrahydrofuran ring that replaces the 1,3-dioxolane ring present in other azole drugs.

6) Literature

- Morgan, B., Dodds, D.R., Zaks, A., Andrews, D.R., Klesse, R. (1997) Enzymatic desymmetrization of prochiral 2-substituted-1,3-propanediols: A practical chemoenzymatic synthesis of a key precursor of SCH51048, a broad-spectrum orally active antifungal agent, J. Org. Chem. **62**, 7736–7743

- Morgan, B., Stockwell, B.R., Dodds, D., Andrews, D.R., Sudhakar, A.R. (1997) Chemoenzymatic approaches to SCH 56592, a new azole antifungal, J. Am. Oil Chem. Soc. **74**, 1361–1370

- Pantaleone, D. (1999) Biotransformations: "Green" processes for the synthesis of chiral fine chemicals, in: Handbook of Chiral Chemicals (Ager, D., ed.) pp. 245–286, Marcel Dekker Inc., New York

- Saksena, A.K., Girijavallabhan, V.M., Pike, R.E., Wang, H., Lovey, R.G., Liu, Y.-T., Ganguly, A.K., Morgan, W.B., Zaks, A. (1995) Process for preparing intermediates for the synthesis of antifungal agents, Schering Corporation, US 5,403,937

- Zaks, A., Dodds, D.R. (1997) Application of biocatalysis and biotransformations to the synthesis of pharmaceuticals, Drug Discovery Today **2** (6), 513–531

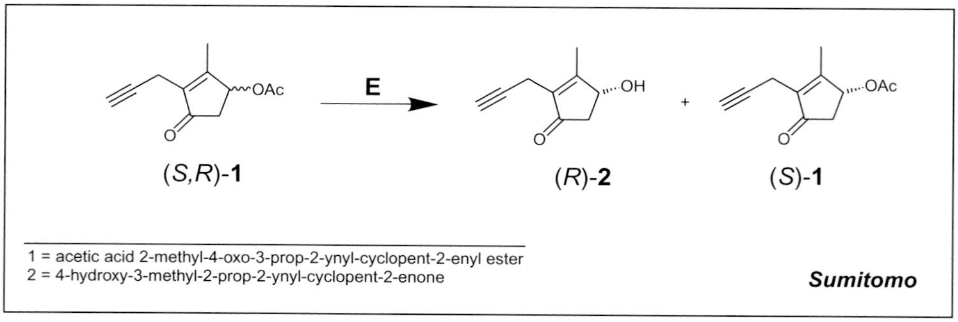

Fig. 3.1.1.3 – 1

1) Reaction conditions

[1]:	4.16 M, 800 g · L^{-1} [192.21 g · mol^{-1}]
pH:	7.0
T:	40 °C
medium:	two-phase: aqueous / organic
reaction type:	carboxylic ester hydrolysis
catalyst:	suspended enzyme
enzyme:	triacylglycerol acylhydrolase (triacylglycerol lipase)
strain:	*Arthrobacter* sp.
CAS (enzyme):	[9001–62–1]

2) Remarks

- Subsequent to the lipase-catalyzed hydrolysis, the cleaved alcohol is sulfonated in the presence of the acylated compound with methanesulfonyl chloride. The hydrolysis of the sulfonated enantiomer in the presence of small amounts of calcium carbonate takes place under inversion of the chiral center as opposed to the hydrolysis of the acylated enantiomer, which is carried out under retention of the chiral center. By this means, an enantiomeric excess of 99.2 % and a very high yield is achieved for the (*R*)-alcohol:

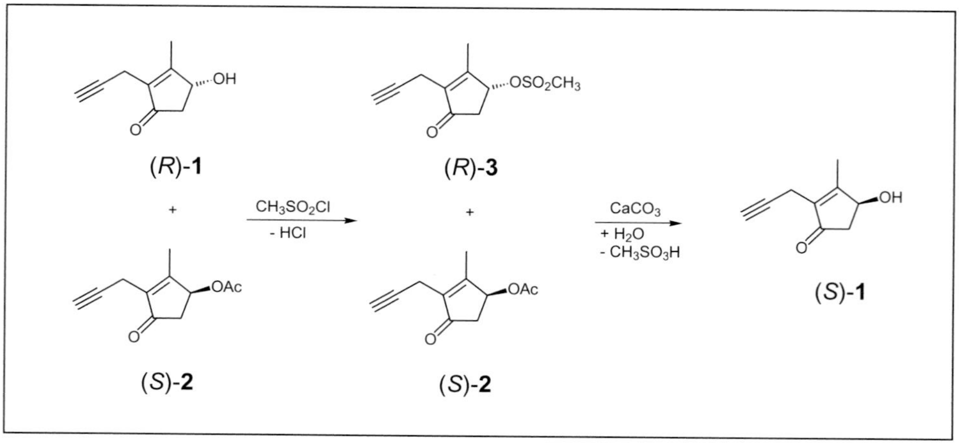

Fig. 3.1.1.3 – 2

- For this resolution an E-value of 1,300 was determined.

3) Flow scheme

Not published.

4) Process parameters

conversion:	49.9 %
ee:	99.2 % (alcohol)
reactor type:	batch
company:	Sumitomo Chemical Co., Japan

5) Product application

- The (S)-alcohol is used as an intermediate in the synthesis of pyrethroids, which are used as insecticides. They show excellent insecticidal activities and a low toxicity in mammals.

6) Literature

- Hirohara, H., Nishizawa M. (1998) Biochemical synthesis of several chemical insecticide intermediates and mechanism of action of relevant enzymes, Biosci. Biotechnol. Biochem. **62**, 1–9

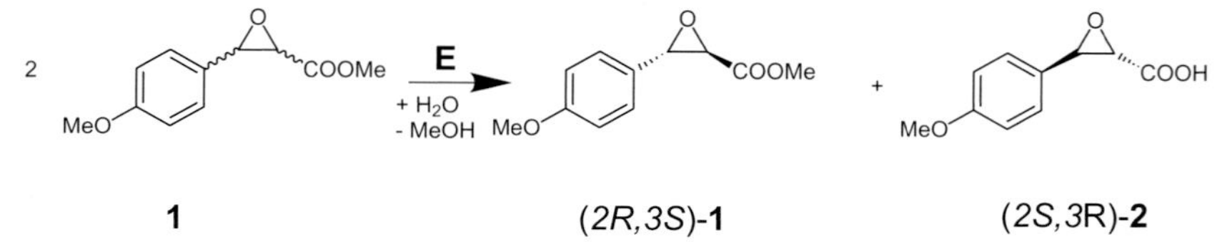

1 = 3-(4-methoxyphenyl) glycidic acid methyl ester = MPGM
2 = 3-(4-methoxyphenyl) glycidic acid

Tanabe Seiyaku Co., Ltd.

Fig. 3.1.1.3 – 1

1) Reaction conditions

[**1**]:	< 0.6 M, < 125 g $\cdot$ L^{-1} [208.21 g $\cdot$ mol^{-1}]
pH:	8.5 (aqueous solution in lumen loop)
T:	22 °C
medium:	two-phase-system, aqueous / toluene
reaction type:	carboxylic ester hydrolysis
catalyst:	immobilized enzyme
enzyme:	triacylglycerol acylhydrolase (triacylglycerol lipase)
strain:	*Serratia marescens* Sr41 8000
CAS (enzyme):	[9001–62–1]

2) Remarks

- A hydrophilic hollow fibre membrane is used (polyacrylonitrile) as reactor unit (see flow scheme).

- The lipase is immobilized onto a spongy layer by pressurized adsorption.

- The lipase does not attack *(2R,3S)*-(4-methoxyphenyl)glycidic acid methyl ester which acts as a competitive inhibitor.

- The formed acid (hydrolyzed (+)-methoxyphenylglycidate) is unstable and decarboxylates to give 4-methoxyphenylacetaldehyde; this aldehyde strongly inhibits and deactivates the enzyme. It can be removed continuously by filtration as bisulfite adduct. The bisulfite acts also as buffer to maintain constant pH during synthesis.

- The lipase is also inhibited by Co^{2+}, Ni^{2+}, Fe^{2+}, Fe^{3+} and EDTA, but can be activated by Ca^{2+}, Li^+.

- Apparent V_{max} for hydrolysis of (+)MPGM is 1.7 U $\cdot$ mg$_{Protein}^{-1}$.

- The substate specificity of the hydrolase from *S. marcescens* is shown in the following table:

Substrate	R	Specific activity (U*mg⁻¹)[a]	Substrate	R	Specific activity (U*mg⁻¹)[b]
MeO—⟨⟩—epoxide-COOR	—CH₃	570		—CH₃	61
	—C₂H₅	320		—C₃H₇	450
	—C₃H₇	510		—C₅H₁₁	91
	—C₄H₉	640	R—COOMe	—C₇H₁₅	620
	—C₅H₁₁	650		—C₉H₁₉	570
				—C₁₁H₂₃	380
R—⟨⟩—epoxide-COOMe	—CH₃	640		—C₁₃H₂₇	210
	—H	560		—C₁₅H₃₁	91
MeO—⟨⟩—CH(OH)CH(OH)COOMe		350		—CH₃	270
				—C₃H₇	3300
			CH₂—O—COR	—C₅H₁₁	2500
			CH—O—COR	—C₇H₁₅	3500
R—⟨⟩—CH=CH—COOMe	—OCH₃	0	CH₂—O—COR	—C₉H₁₉	1500
	—CH₃	0		—C₁₁H₂₃	140
	—H	0		—C₁₃H₂₇	120
				—C₁₅H₃₁	56

a)/ b) for conditions of assays see lit. H. Matsumae and T. Shibatani

Fig. 3.1.1.3 – 2

3) Flow scheme

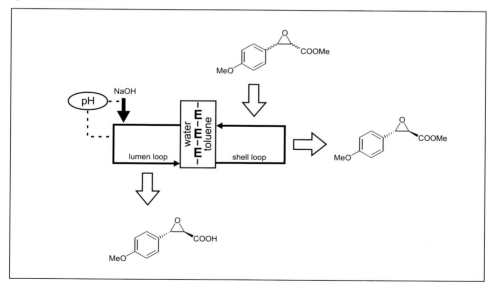

Fig. 3.1.1.3 – 3

4) Process parameters

yield:	40–45 %
ee:	99.9 %
reactor type:	batch, hollow fibre reactor (Sepracor Inc. Massachusetts, USA)
capacity:	40 kg (–)-MPGM $\cdot$ m^{-2} $\cdot$ a^{-1}
down stream processing:	crystallization
enzyme activity:	$1.6 \cdot 10^5$ U $\cdot$ m^{-2}
start-up date:	1993
production site:	Japan and other sites
company:	Tanabe Seiyaku Co. Ltd., Japan and DSM, the Netherlands

5) Product application

- The product is an intermediate in the synthesis of diltiazem.

- Diltiazem hydrochloride is a coronary vasodilator and a calcium channel blocker (for anti-anginal and anti-hypertensive actions) and is produced worldwide in excess of 100 t $\cdot$ a^{-1}.

- In comparison to the chemical route only 5 steps (instead of 9 steps) are necessary. The kinetic resolution is carried out in an earlier step during the synthesis resulting in reduction of waste:

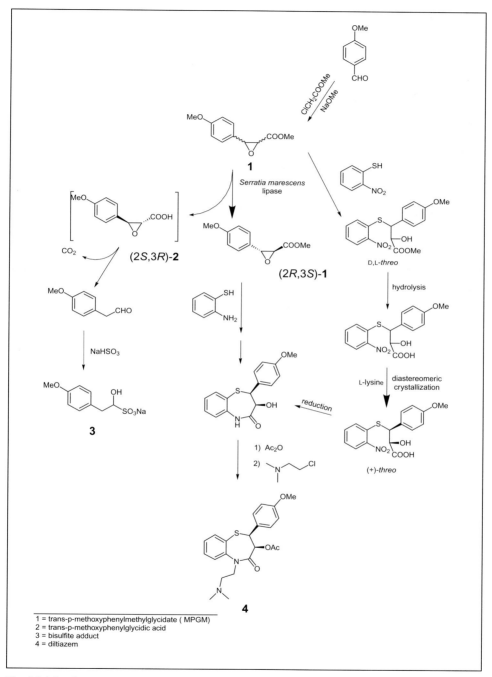

Fig. 3.1.1.3 – 4

1 = trans-p-methoxyphenylmethylglycidate (MPGM)
2 = trans-p-methoxyphenylglycidic acid
3 = bisulfite adduct
4 = diltiazem

6) Literature

- Elferink, V.H.M. (1996) Progress in the application of biocatalysts in the industrial scale manufacture of chiral molecules, Proc. 11th Int. Spring Innovations Chirality Symposium, USA

- Kierkels, J.G., Peeters, W.P.H. (1993) Process for the enzymatic preparation of optically active transglycidic acid esters, DSM NV, EP 602740 A1

- López, J.L., Matson, S.L. (1991) Liquid-liquid extractive membrane reactors, Bioproc. Technol. 11, 27–66

- Matson, S.L. (1987) Method and apparatus for catalyst containment in multiphase membrane reactor systems, PCT WO 87/02381, PCT US 86/02089

- Matsumane, H., Furui, M., Shibatani, T., Tosa, T. (1994) Production of optically active 3-phenylglycidic acid ester by the lipase from *Serratia marcescens* in a hollow-fiber membrane reactor, J. Ferment. Bioeng. **78**, 59–63

- Matsumane, H., Shibatani, T. (1994) Purification and characterization of the lipase from *Serratia marcescens* Sr41 8000 responsible for asymmetric hydrolysis of 3-phenylglycidic acid esters, J. Ferment. Bioeng. **77**, 152–158

- Tosa, T., Shibatani, T. (1995) Industrial application of immobilized biocatalysts in Japan, Ann. N. Y. Acad. Sci. **750**, 364–375

- Zaks, A., Dodds, D.R. (1997) Application of biocatalysis and biotransformations to the synthesis of pharmaceuticals, Drug Discovery Today, **2**, 513–531

β-Galactosidase
Saccharomyces lactis

EC 3.2.1.23

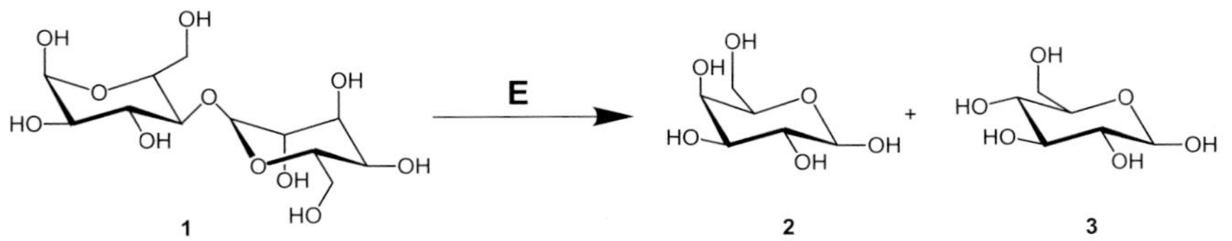

1 2 3

1 = lactose
2 = galactose
3 = glucose

Sumitomo Chemical Industries
Snow Brand Milk Products
Central del Latte
and others

Fig. 3.2.1.23 – 1

1) Reaction conditions

[**1**]:	~ 5 % in milk [342.30 g · mol^{-1}]
T:	35 °C
medium:	milk
reaction type:	O-glycosyl bond hydrolysis
catalyst:	immobilized enzyme
enzyme:	β-D-galactoside galactohydrolase (lactase, hydrolact, β-galactosidase)
strain:	*Aspergillus oryzae*, *Saccharomyces lactis*, and others
CAS (enzyme):	[9031–11–2]

2) Remarks

- Milk with hydrolyzed lactose is much sweeter since the sweetening powers of lactose, glucose and galactose are 20, 70, and 58 % respectively.
- Different processes for hydrolyzing lactose in milk are established by different companies.
- Snamprogretti's enzyme is immobilized in the microcavities of fibers made from cellulose triacetate. The advantage of fibers as support is the enormous surface area. The fibers are stretched between two metallic bars in a column.
- Snow Brand Milk Products, Japan, uses a horizontal rotary column reactor. Here the fibrous immobilized β-galactosidase is placed on a wire mesh cylinder.

3) Flow scheme

Not published.

4) Process parameters

conversion: 70 – 81 %
reactor type: plug-flow reactor
reactor volume: typical size: 1,500 to 250,000 L
capacity: e.g. 8,000 L · d⁻¹ (Central del Latte, Italy)
enzyme supplier: Snamprogretti and others
company: Sumitomo Chemical Industries, Japan; Snow Brand Milk Products Co., Ltd., Japan; Central del Latte, Italy; and others

5) Product application

• Lactose needs to be removed before consumption by babies and people (e.g. Asians and Italians) who are not able to produce or do not have enough β-galactosidase activity. These people are called lactose intolerant.

6) Literature

• Cheetham, P.S.J. (1994) Case studies in applied biocatalysis – From ideas to products, in: Applied Biocatalysis (Cabral, J.M.S., Best, D., Boross, L., Tramper, H., eds.) pp. 47–108, Harwood Academic Publishers, Chur

• Honda, Y., Kako, M., Abiko, K., Sogo, Y. , (1993) Hydrolysis of lactose in milk, in: Industrial Application of Immobilized Biocatalysts (Tanaka, A., Tosa, T., Kobayashi, T., eds.) pp.109–129, Marcel Dekker Inc., New York

• Marconi, W., Morisi, F., (1979) Industrial application of fiber-entrapped enzymes, in: Appl. Biochem. Bioeng. 2, Enzyme Technology, (Wingard, L.B., Katchalski-Katzir, E., Goldstein, L. eds.) pp. 219–258, Academic Press, New York

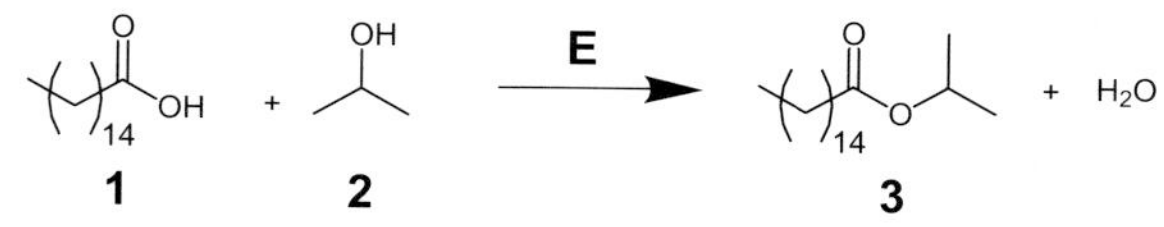

Fig. 3.1.1.3 – 1

1 = palmitic acid
2 = isopropanol
3 = isopropyl palmitate

UNICHEMA Chemie BV

1) Reaction conditions

[**1**]:	3.1 M, 800 g · L^{-1} [256.43 g · mol^{-1}]
pH:	7.0
T:	60 °C
medium:	2-propanol
reaction type:	carboxylic ester hydrolysis
catalyst:	immobilized enzyme
enzyme:	triacylglycerol acylhydrolase (triacylglycerol lipase, lipase)
strain:	*Candida antarctica*
CAS (enzyme):	[9001–62–1]

2) Remarks

- The problem during ester synthesis is the produced water, which leads to equilibrium conditions meaning forward and backward reaction have the same rates.

- Two possible process layouts are published:

  1) The reaction water is removed by azeotropic distillation (alcohol/water) at 0.26 bar. 2-propanol is continuously fed to the reactor (58 g · h^{-1}) to replace the distilled one. The catalyst can be easily removed by filtration.

  2) Alternatively the reaction water is removed during esterification by pervaporation at only 80 °C. The reaction solution with the lower water level is cooled to 65 °C and passed to the second reactor unit. After a second pervaporation step the water content is lowered to 0.2 wt.-%.

- The production can be adapted to other alcohols and acids. By the same process isopropyl myristate is produced from myristic acid (H$_3$C-(CH$_2$)$_{12}$-COOH).

217

3) Flow scheme

1) Process with azeotropic distillation:

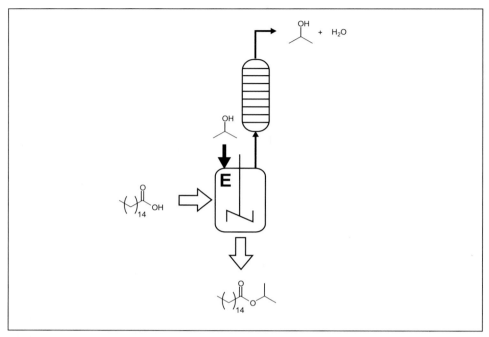

Fig. 3.1.1.3 – 2

2) Process with pervaporation:

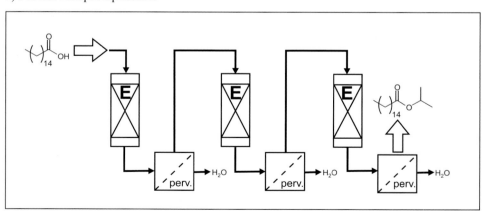

Fig. 3.1.1.3 – 3

4) Process parameters

yield:	99 %
chemical purity:	> 99 %
reactor type:	batch
residence time:	14 h
enzyme supplier:	Novo Industry, Denmark
company:	UNICHEMA Chemie BV, the Netherlands

5) Product application

- Isopropyl palmitate and isopropyl myristate are used in the preparation of soaps, skin creams, lubricants and grease.

6) Literature

- Hills, G.A., McCrae, A.R., Poulina, R.R. (1990) Ester preparation, Unichema Chemie BV, EP0383405

- Kemp, R.A., Macrae, A.R. (1992) Esterification process, Unichema Chemie BV, EP 0506159

- McCrae, A.R., Roehl, E.-L., Brand, H.M. (1990) Bio-Ester – Bio-Esters, Seifen-Öle-Fette-Wachse **116** (6), 201–205

Lactonase
Fusarium oxysporum

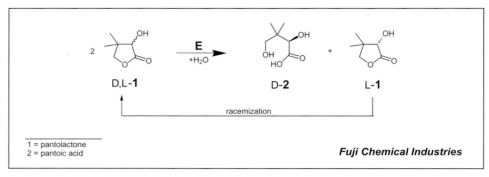

D,L-**1**   D-**2**   L-**1**

racemization

1 = pantolactone
2 = pantoic acid

Fuji Chemical Industries

Fig. 3.1.1.25 – 1

1) Reaction conditions

[**1**]:	2.69 M, 350 g · L^{-1} [130.14 g · mol^{-1}]
pH:	6.8 – 7.2
T:	30 °C
medium:	aqueous
reaction type:	carboxylic ester hydrolysis
catalyst:	immobilized whole cells
enzyme:	1,4-lactone hydroxyacylhydrolase (γ-lactonase)
strain:	*Fusarium oxysporum*
CAS (enzyme):	[37278–38–9]

2) Remarks

- The reverse reaction, the lactonization of aldonic acid, is catalyzed under acidic conditions. The reverse reaction does not take place with aromatic substrates.

- The lactonase from *Fusarium oxysporum* has a very broad substrate spectrum:

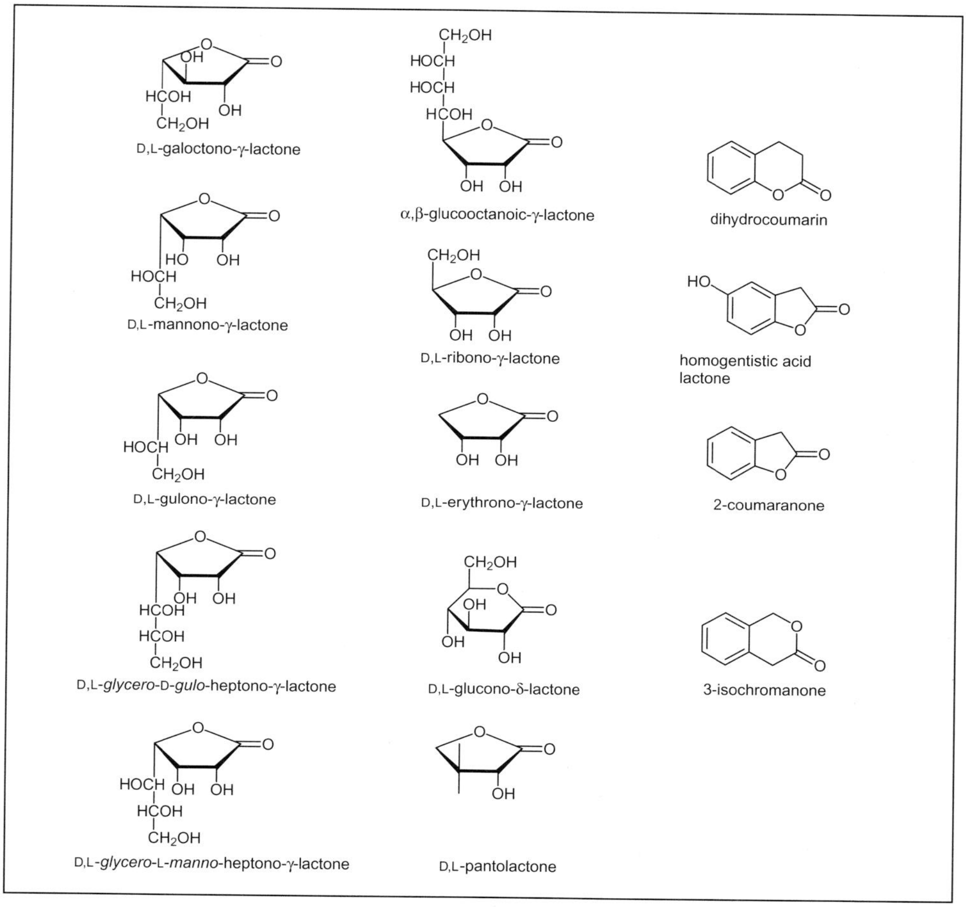

D,L-galoctono-γ-lactone

D,L-mannono-γ-lactone

D,L-gulono-γ-lactone

D,L-*glycero*-D-*gulo*-heptono-γ-lactone

D,L-*glycero*-L-*manno*-heptono-γ-lactone

α,β-glucooctanoic-γ-lactone

D,L-ribono-γ-lactone

D,L-erythrono-γ-lactone

D,L-glucono-δ-lactone

D,L-pantolactone

dihydrocoumarin

homogentistic acid lactone

2-coumaranone

3-isochromanone

Fig. 3.1.1.25 – 2

- For the synthesis whole cells are immobilized in calcium alginate beads and used in a fixed bed reactor.

- The immobilized cells retain more than 90 % of their initial activity even after 180 days of continuous use.

- At the end of the reaction L-pantolactone is extracted and reracemized to D,L-pantolactone that is recycled into the reactor. The D-pantoic acid is chemically lactonized to D-pantolactone and extracted:

Lactonase
Fusarium oxysporum

EC 3.1.1.25

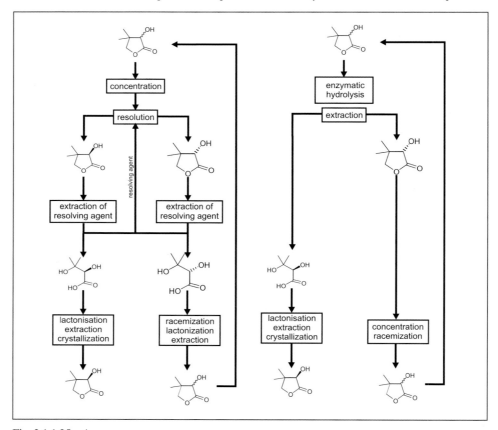

Fig. 3.1.1.25 – 3

- The biotransformation skips several steps that are necessary in the chemical resolution process:

Fig. 3.1.1.25 – 4

- By using the lactonase from *Brevibacterium protophormia* L-lactones are available:

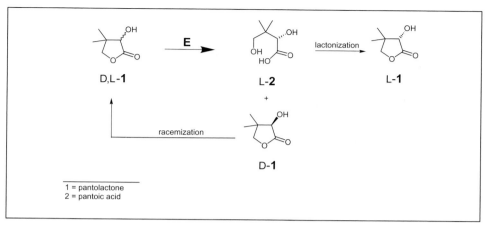

1 = pantolactone
2 = pantoic acid

Fig. 3.1.1.25 – 5

3) Flow scheme

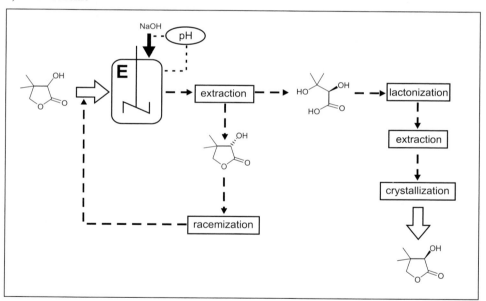

Fig. 3.1.1.25 – 6

4) Process parameters

conversion:	90–95 %
ee:	90–97 %
reactor type:	plug-flow reactor
residence time:	21 h
down stream processing:	lactonization, extraction and crystallization
production site:	Takaoka, Toyama Prefecture, Japan
company:	Fuji Chemical Industries, Japan

223

5) Product application

- The pantoic acid is used as a vitamin B_2-complex.

- D- and L-pantolactones are used as chiral intermediates in chemical synthesis.

6) Literature

- Simizu, S., Ogawa, J., Kataoka, M., Kobayashi. M. (1997) Screening of novel microbial enzymes for the production of biologically and chemically useful compounds, in: New Enzymes for Organic Synthesis (Scheper, T., ed.) pp. 45–88, Springer, New York

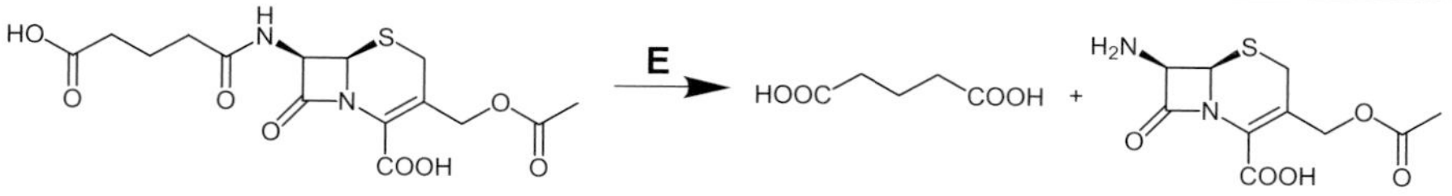

Glutaryl amidase
Escherichia coli

Fig. 3.1.1.41 – 1

1) Reaction conditions

pH:	8.3
T:	30 °C
medium:	aqueous
reaction type:	carboxylic ester hydrolysis
catalyst:	immobilized enzyme
enzyme:	ω-amidodicarboxylate amidohydrolase (ω-amidase, α-keto acid-ω-amidase)
strain:	*Escherichia coli*
CAS (enzyme):	[9025–19–8]

2) Remarks

- Second step of the 7-aminocephalosporanic acid process, for first step see page 129.

- The glutaryl amidase is immobilized on a spherical carrier.

- Here an existing chemical process has been replaced by an enzymatic process due to environmental considerations:

Glutaryl amidase
Escherichia coli

EC 3.1.1.41

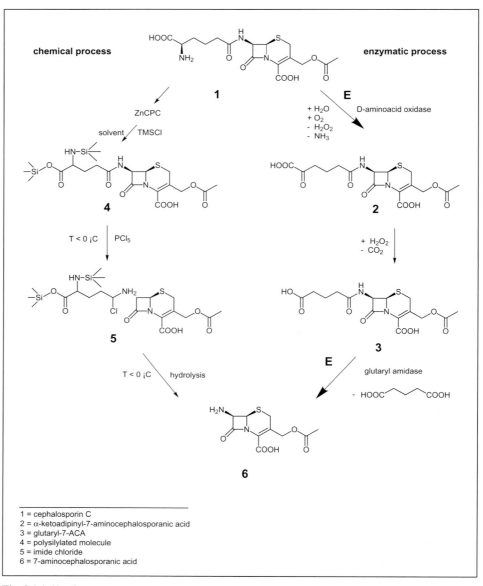

chemical process

enzymatic process

ZnCPC

+ H$_2$O
+ O$_2$
− H$_2$O$_2$
− NH$_3$

D-aminoacid oxidase

solvent / TMSCl

1

E

4

2

T < 0 ¡C | PCl$_5$

+ H$_2$O$_2$
− CO$_2$

5

3

E

glutaryl amidase

T < 0 ¡C | hydrolysis

− HOOC ⟋⟍ COOH

6

1 = cephalosporin C
2 = α-ketoadipinyl-7-aminocephalosporanic acid
3 = glutaryl-7-ACA
4 = polysilylated molecule
5 = imide chloride
6 = 7-aminocephalosporanic acid

Fig. 3.1.1.41 – 2

- In the first step the zinc salt of cephalosporin C (ZnCPC) is produced, followed by the protection of the functional groups (NH_2 and COOH) with trimethylchlorosilane. The imide chloride is synthesized in the subsequent step at $0\,°C$ with phosphorous pentachloride. Hydrolysis of the imide chloride yields 7-ACA. By replacement of this synthesis with the biotransformation the usage of heavy-metal salts ($ZnCl_2$), chlorinated hydrocarbons and precautions for highly flammable compounds can be circumvented. The waste-gas emission is reduced from 7.5 to 1.0 kg. Mother liquors requiring incineration are reduced from 29 to 0.3 t. Residual zinc that is recovered as $Zn(NH_4)PO_4$ is completely reduced from 1.8 to 0 t.

- The absolute costs of environmental protection are reduced by 90 % per ton of 7-ACA.

3) Flow scheme

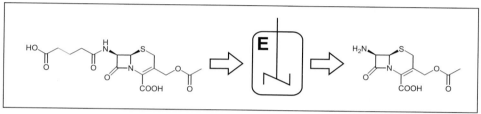

Fig. 3.1.1.41 – 3

4) Process parameters

reactor type:	batch
reactor volume:	10,000 L
capacity:	$200\ \text{t} \cdot \text{a}^{-1}$
residence time:	1.5 h
down stream processing:	crystallization
enzyme supplier:	Hoechst Marion Roussel, Germany
start-up date:	1996
production site:	Frankfurt, Germany
company:	Hoechst Marion Roussel, Germany

5) Product application

- 7-ACA is an intermediate for semi-synthetic penicillins and cephalosporins.

6) Literature

- Aretz, W. (1998) Hoechst Marion Roussel, personal communication

- Christ, C. (1995) Biochemical production of 7-aminocephalosporanic acid, in: Ullmann's Encyclopedia of Industrial Chemistry, Vol. **B8** (Arpe, H.-J. ed.) pp. 240–241, VCH Verlagsgesellschaft, Weinheim

- Verweij, J., Vroom, E.D. (1993) Industrial transformations of penicillins and cephalosporins, Rec. Trav. Chim. Pays-Bas **112** (2), 66–81

- Matsumoto, K. (1993) Production of 6-APA, 7-ACA, and 7-ADCA by immobilized penicillin and cephalosporin amidases, in: Industrial Application of Immobilized Biocatalysts (Tanaka, A., Tosa, T., Kobayashi, T. eds.) pp. 67–88, Marcel Dekker Inc., New York

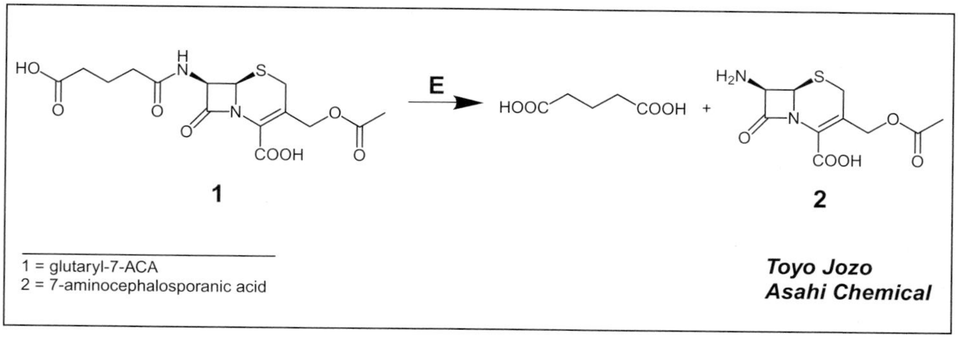

Glutaryl amidase
Pseudomonas sp.

1 = glutaryl-7-ACA
2 = 7-aminocephalosporanic acid

Toyo Jozo
Asahi Chemical

Fig. 3.1.1.41 – 1

1) Reaction conditions

[1]:	0.026 M, 10 g · L⁻¹ [386.38 g · mol⁻¹]
pH:	7.5 – 8.5
T:	30 °C
medium:	aqueous
reaction type:	carboxylic ester hydrolysis
catalyst:	immobilized enzyme
enzyme:	ω-amidodicarboxylate amidohydrolase (ω-amidase, α-keto acid-ω-amidase)
strain:	*Pseudomonas* GK-16
CAS (enzyme):	[9025–19–8]

[1]: $0.026\ M,\ 10\ g \cdot L^{-1}\ [386.38\ g \cdot mol^{-1}]$

2) Remarks

- Second step of the 7-aminocephalosporanic acid process, for the first step see page 129.

- The glutaryl amidase is immobilized by adsorption onto a porous styrene anion-exchange resin and subsequent crosslinking with 1 % glutaraldehyde.

- The liberated glutaric acid is an inhibitor of the glutaryl amidase and additionally lowers the pH. Therefore the pH is controlled by an autotitrator.

- The process is started at 15 °C. To compensate enzyme deactivation during the production, it is gradually increased to 25 °C. After 70 cycles the enzyme is replaced.

- The reaction solution circulates at the rate of 10,000 L · h⁻¹.

- Here an existing chemical process has been replaced by an enzymatic process due to environmental reasons.

- For general remarks regarding the biocatalytic synthesis of 7-ACA and the comparison to the chemical route see page 225.

3) Flow scheme

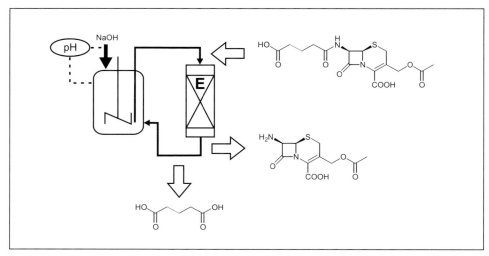

Fig. 3.1.1.41 – 2

4) Process parameters

yield:	95 %
reactor type:	repetitive batch, fixed bed reactor
reactor volume:	1,000 L
capacity:	90 t · a^{-1}
residence time:	4 h
start-up date:	1973
company:	Asahi Chemical Industry Co., Ltd., Japan, and Toyo Jozo, Japan

5) Product application

- 7-ACA is an intermediate for semisynthetic penicillins and cephalosporins.

6) Literature

- Christ, C. (1995) Biochemical production of 7-aminocephalosporanic acid, in: Ullmann's Encyclopedia of Industrial Chemistry, Vol. **B8** (Arpe, H.-J. ed.) pp. 240–241, VCH Verlagsgesellschaft, Weinheim

- Matsumoto, K. (1993) Production of 6-APA, 7-ACA, and 7-ADCA by immobilized penicillin and cephalosporin amidases, in: Industrial Application of Immobilized Biocatalysts (Tanaka, A, Tosa, T., Kobayashi, T. eds.) pp. 67–88, Marcel Dekker Inc., New York

- Tsuzuki, K., Komatsu, K., Ichikawa, S., Shibuya, Y. (1989) Enzymatic synthesis of 7-aminocephalosporanic acid (7-ACA), Nippon Nogei Kagaku Kaishi **63** (12), 1847

- Verweij, J., Vroom, E.D. (1993) Industrial transformations of penicillins and cephalosporins, Rec. Trav. Chim. Pays-Bas **112** (2), 66–81

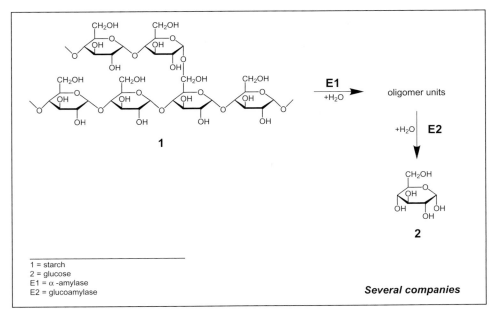

1 = starch
2 = glucose
E1 = α-amylase
E2 = glucoamylase

Several companies

Fig. 3.2.1.1/3.2.1.3 – 1

1) Reaction conditions

pH:	liquefying: 6.0 – 6.5 / glucosidation: 4.2
T:	115°C – 95 °C / 60 °C
medium:	aqueous
reaction type:	O-glucosyl-bond hydrolysis (endo / exo)
catalyst:	solubilized enzyme
enzyme:	1,4-α-D-glucan glucanohydrolyse (α-amylase, glycogenase) / glucan 1,4-a-glucosidase (glucoamylase, amyloglucosidase)
strain:	*Bacillus licheniformis / Aspergillus niger*
CAS (enzyme):	[9000–90–2] / [9032-08-0]

2) Remarks

- The process is a part of the production of high fructose corn syrup (see page 387).

- After several improvements this process provides an effective way for an important, low-cost sugar substitute derived from grain.

- At various stages enzymes are employed in this process.

- The corn kernels are softened by treatment with sulfur dioxide and lactic acid bacteria to separate oil, fiber and proteins. The enzymatic steps are cascaded to yield the source product for the invertase process after liquefaction in continuous cookers, debranching and filtration.

- Since starches from different natural sources have different compositions the procedure is not unique. The process ends if all starch is completely broken down to limit the amount of oligomers of glucose and dextrins. Additionally, recombination of molecules has to be prevented.

- The thermostable enzymes can be used up to 115 °C. The enzymes need Ca^{2+} ions for stabilization and activation. Since several substances in corn can complex cations, the cation concentration is increased requiring a further product purification causing the neccessity to refine the product.

3) Flow scheme

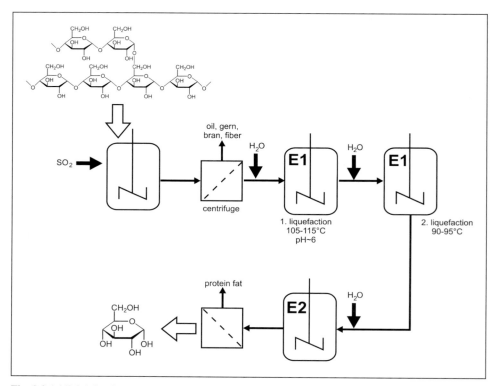

Fig. 3.2.1.1/3.2.1.3 – 2

4) Process parameters

conversion:	> 95 %
yield:	depending on natural source
selectivity:	> 99 %
reactor type:	continuously operated stirred tank reactor
capacity:	> 10,000,000 t · a^{-1} (worldwide)
residence time:	2–3 h / 48–72 h
down stream processing:	filtration
production site:	world wide
company:	several companies

5) Product application

- The product is a feed stock for high fructose syrup (see page 387).

6) Literature

- Gerhartz, W.(1990) Enzymes in Industry: Production and Application, VCH, Weinheim

- Holm, J., Bjoerck, I., Ostrowska, S., Eliasson, A.C., Asp, N.G., Larsson, K., Lundquist, I. (1983) Digestibility of amylose-lipid complexes *in vitro* and *in vivo*, Stärke, **35**, 294–297.

- Holm, J., Bjoerck, I., Eliasson, A.C. (1985) Digestibility of amylose-lipid complexes *in vitro* and *in vivo*, Prog. Biotechnol., **1**, 89–92

- Kainuma, K.(1998) Applied glycoscience-past, present and future, Foods Food Ingredients J. Jpn., **178**, 4–10

- Labout, J.J.M. (1985) Conversion of liquefied starch into glucose using a novel glucoamylase system, Stärke, **37**, 157–161

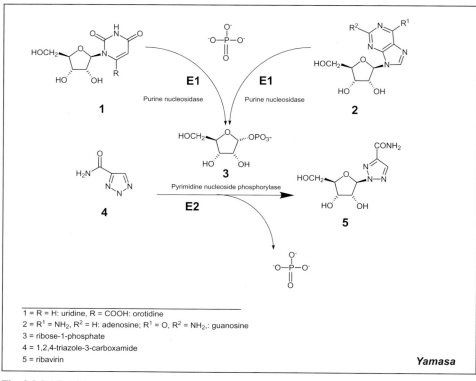

Fig. 3.2.2.1/2.4.2.2 – 1

1) Reaction conditions

T:	60 °C
medium:	aqueous
reaction type:	N-glycosyl bond hydrolysis / pentosyl group transfer
catalyst:	solubilized enzyme
enzyme:	purine nucleosidase / purine – nucleoside phosphorylase
strain:	*Erwinia carotovora*
CAS (enzyme):	[9025–44–9] / [9055–35–0]

2) Remarks

- The enzymes were isolated and purified from *Erwinia carotovora*.

- Alternatively biocatalysts from *Brevibacterium acetylicum* and *Bacillus megaterium* can be used.

3) Flow scheme

Not published.

4) Process parameters

company: Yamasa, Japan

5) Product application

- Ribavirin is an antiviral drug.

6) Literature

- Shirae, H., Yokozeki, K., Uchiyama, M., Kubota, K. (1988) Production of Ribavirin from purine nucleosides by *Brevibacterium acetylicum*, Agric. Biol. Chem. **52** (7), 1777–1784

- Shirae, H., Yokozeki, K. (1991) Purification and properties of purine nucleoside phosphorylase from *Brevibacterium acetylicum* ATCC 954, Agric. Biol. Chem. **55** (2), 493–499

- Shirae, H., Yokozeki, K. (1991) Purine nucleoside phosphorylase from *Erwinia carotovora* AJ 2992, Agric. Biol. Chem. **55** (7), 1849–1857

- Zaks, A., Dodds, D.R. (1997) Application of biocatalysis and biotransformations to the synthesis of pharmaceuticals, Drug Discovery Today **2** 6, 513–531

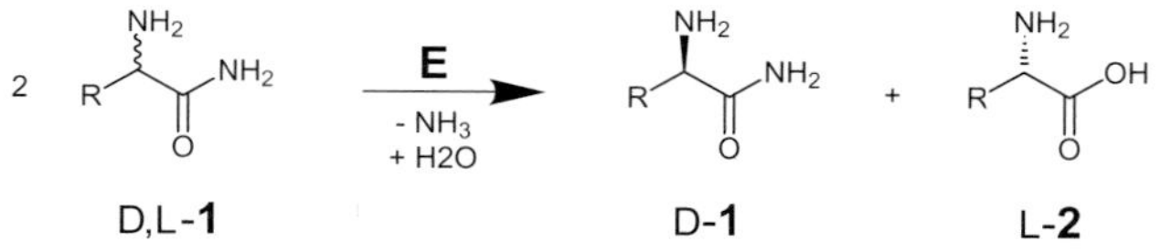

Aminopeptidase
Pseudomonas putida

Fig. 3.4.11.1 – 1

1) Reaction conditions

[1]:	up to 20 g · L^{-1}
pH:	8.0 – 10.0
medium:	aqueous
reaction type:	carboxylic acid amide hydrolysis
catalyst:	suspended whole cells
enzyme:	α-aminoacyl-peptide hydrolase (cytosol aminopeptidase, aminopeptidase, leucyl peptidase)
strain:	*Pseudomonas putida* ATCC 12633
CAS (enzyme):	[9001–61–0]

2) Remarks

- The substrates for this biotransformation can be readily obtained from the appropriate aldehyde via the Strecker synthesis.

- Conversion to the racemic amide in one step is possible under alkaline conditions in the presence of a catalytic amount of ketone with yields above 90 %.

- After the enzymatic step benzaldehyde is added so that the Schiff base of the D-amide is precipitated and can be easily isolated by filtration. An acidification step leads to the D-amino acid.

- The L-amino acid can be recycled by racemization so that a theoretical yield of 100 % is possible.

- The same process can be used for the synthesis of L-amino acids by racemizing the Schiff base of the D-amide in a short time using small amounts of base in organic solvents.

- The following figure shows all the synthetic routes:

236

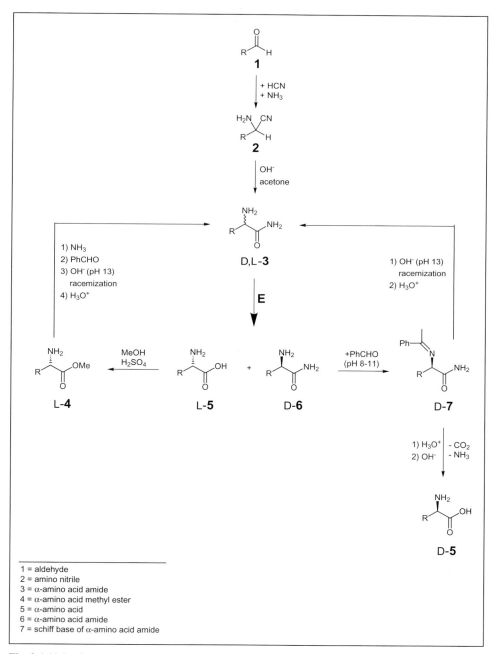

1 = aldehyde
2 = amino nitrile
3 = α-amino acid amide
4 = α-amino acid methyl ester
5 = α-amino acid
6 = α-amino acid amide
7 = schiff base of α-amino acid amide

Fig. 3.4.11.1 – 2

- The whole cell catalyst of *Pseudomonas putida* accepts a wide range of substrates:

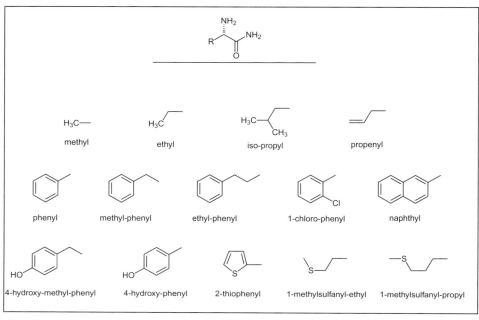

Fig. 3.4.11.1 – 3

- Addition of Mn^{2+} to the purified enzyme (up to 20 mM) resulted in a 12-fold increase in activity, whereas Cu^{2+} and Ca^{2+} inhibit the enzyme at a concentration of 1 mM.

- The next table shows the K_M and V_{max} values for different substrates:

substrate	K_M / mM	V_{max} / $U \cdot mg^{-1}_{protein}$
phenyl	65	1,565
methyl-phenyl	15	80
iso-propyl	130	110

Fig. 3.4.11.1 – 4

- Using *in vivo* protein engineering not only mutant strains of *Pseudomonas putida* exhibiting L-amidase and also D-amidase but also amino acid amide racemase activities were obtained. Using these mutants a convenient synthesis of α-H-amino acids with 100 % yield would be possible with a single cell system.

- It is noteworthy that only α-H-substrates can be used. By screening techniques a new bio-catalyst of the strain *Mycobacterium neoaurum* was found, which is capable of converting α-substituted amino acid amides. The next figure shows possible substrates for *Mycobacterium neoaurum*:

Fig. 3.4.11.1 – 5

3) Flow scheme

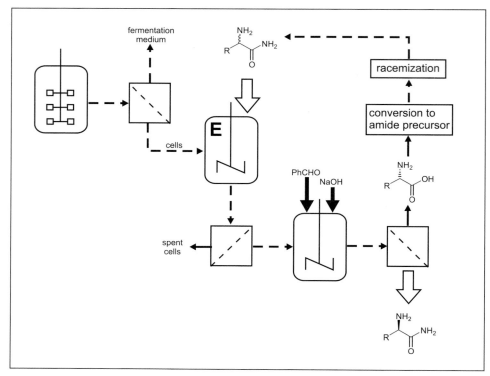

Fig. 3.4.11.1 – 6

4) Process parameters

ee:	> 99 %
reactor type:	batch
capacity:	ton scale
enzyme activity:	$338{,}000 \ \mathrm{U \cdot g_{protein}^{-1}}$
enzyme supplier:	Novo Nordisk, Denmark
start-up date:	1988
company:	DSM, The Netherlands

5) Product application

- The α-H-amino acids are intermediates in the synthesis of antibiotics, injectables, food and feed additives.

- Examples are the antibiotics ampicillin, amoxicillin (see page 290), the sweetener aspartame (see page 270), several ACE-inhibitors and the pyrethroid insecticide fluvalinate:

Aminopeptidase
Pseudomonas putida

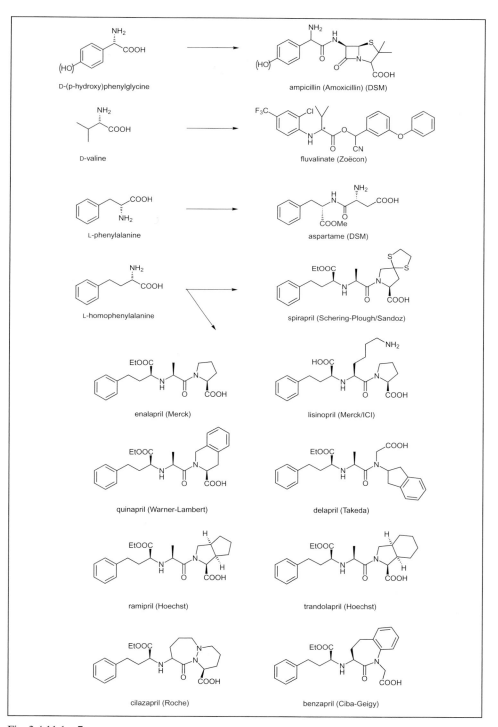

Fig. 3.4.11.1 – 7

- The antihypertensive drug L-methyl-DOPA (see page 342) and the herbicide arsenal can be synthesized using α-alkyl amino acids:

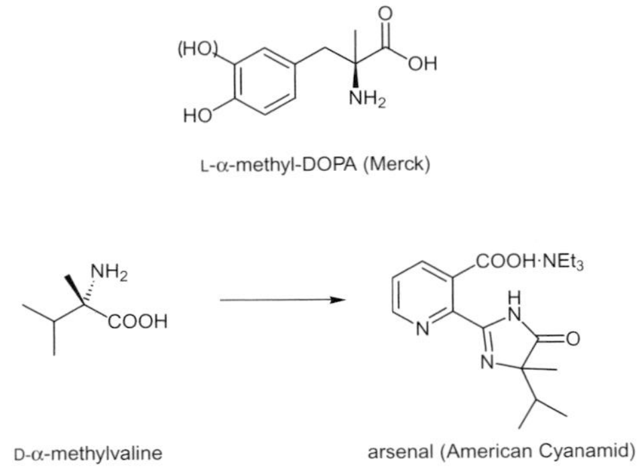

L-α-methyl-DOPA (Merck)

D-α-methylvaline

arsenal (American Cyanamid)

Fig. 3.4.11.1 – 8

6) Literature

- Kamphuis, J., Meijer, E.M., Boesten, W.H.J., Sonke, T., van den Tweel, W.J.J., Schoemaker, H.E. (1992) New developments in the synthesis of natural and unnatural amino acids, in: Enzyme Engineering XI, Vol. 672 (Clark, D.S., Estell, D.A., eds.), pp. 510–527, Ann. N. Y. Acad. Sci.

- Crosby, J. (1991) Synthesis of optically active compounds: A large scale perspective, Tetrahedron **47**, 4789–4846

- Sheldon, R.A. (1993) Chirotechnology, Marcel Dekker, New York

- Schoemaker, H.E., Boesten, W.H.J., Kaptein, B., Hermes, H.F.M., Sonke, T., Broxterman, Q.B., van den Tweel, W.J.J., Kamphuis, J.(1992) Chemo-enzymatic synthesis of amino acids and derivatives, Pure & Appl. Chem. **64**, 1171–1175

- Van den Tweel, W.J.J., van Dooren, T.J.G.M., de Jonge, P.H., Kaptein, B., Duchateau, A.L.L., Kamphuis, J. (1993) *Ochrobacterium anthropi* NCIMB 40321: a new biocatalyst with broad-spectrum L-specific amidase activity, Appl. Microbiol. Biotechnol. **39**, 296–300

- Kamphuis, J., Hermes, H.F.M. van Balken, J.A.M., Schoemaker, H.E., Boesten W.H.J., Meijer, E.M., (1990) Chemo-enzymatic synthesis of enantiomerically pure alpha-H and alpha-alkyl alpha-amino acids and derivates, in: Amino acids: Chemistry, Biology, Medicine; (Lubec, G., Rosenthal, G.A., eds.) pp. 119–125, ESCOM Science Pupl., Leiden

- Meijer, E.M., Boesten W.H.J., Schoemaker, H.E., van Balken J.A.M. (1985) Use of biocatalysts in the industrial production of speciality chemicals, in: Biocatalysis in Organic Synthesis, (Tramper, J., Van der Plas, H.C., Linko, P.), , eds.) pp. 135–156 Elsevier, Amsterdam, The Netherlands

Carboxypeptidase B
Pig Pancreas

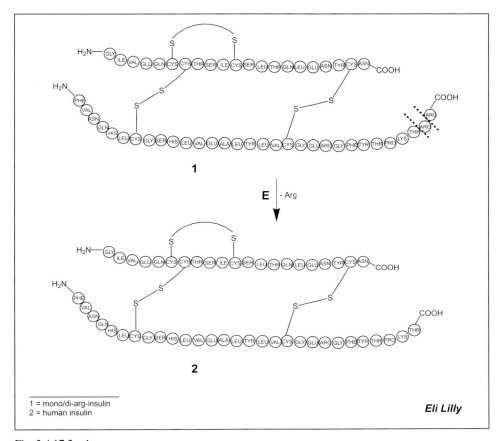

Fig. 3.4.17.2 – 1

1) Reaction conditions

T:	30–35 °C
medium:	aqueous
reaction type:	carboxylic acid amide hydrolysis
catalyst:	solubilized enzyme
enzyme:	peptidyl-L-arginine hydrolyase (carboxypeptidase)
strain:	pig pancreas
CAS (enzyme):	[9025–24–5]

2) Remarks

- See preparation of educt on page 249.

- For an overview of alternative synthetic path see page 245.

Carboxypeptidase B
Pig Pancreas

3) Flow scheme

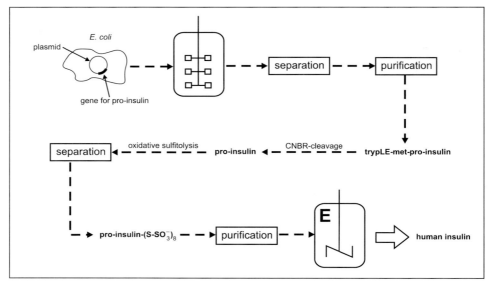

Fig. 3.4.17.2 – 2

4) Process parameters

conversion: > 99 %
yield: > 95 %
selectivity: > 99 %
down stream processing: chromatography
company: Eli Lilly, USA

5) Product application

- For details about insulin see process of Hoechst Marion Roussel on page 245.

6) Literature

- Frank, B.H., Chance, R.E. (1983) Two routes for producing human insulin utilizing recombinant DNA technology, Münch. Med. Wschr. **125**, 14–20

- Jørgensen, L.N., Rasmussen, E., Thomsen, B. (1989) HM(ge), Novo's biosynthetic insulin; Med. View. **III**, No. 4, 1–7

- Ladisch, M.R., Kohlmann, K.L. (1992) Recombinant human insulin, Biotechnol. Prog. **8**, 469–478

Carboxypeptidase B
Pig Pancreas

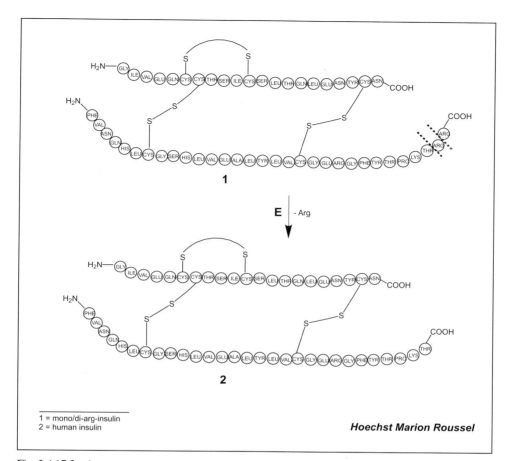

Fig. 3.4.17.2 – 1

1) Reaction conditions

[1]:	$0.4 \text{ g} \cdot \text{L}^{-1}$
[2]:	$0.36 \text{ g} \cdot \text{L}^{-1}$
pH:	8.0
T:	30–35 °C
medium:	aqueous
reaction type:	carboxylic acid amide hydrolysis
catalyst:	solubilized enzyme
enzyme:	peptidyl-L-arginine hydrolyase (carboxypeptidase)
strain:	Pig pancreas
CAS (enzyme):	[9025–24–5]

2) Remarks

- See preparation of educt on page 251.

- Historically, insulin has been purified from animal tissues. Frozen bovine or porcine pancreas are diced and in a multi-step procedure of acidification, neutralization and concentration and after several chromatography steps insulin can be isolated.

- Nowadays there are four main routes to produce insulin:
  1) Extraction from human pancreas,
  2) chemical synthesis from individual amino acids,
  3) conversion of porcine insulin to human insulin and
  4) fermentation of genetically engineered microorganisms.

- The last fermentation procedure can be divided into four main syntheses:

  1) Synthesis of chain A and chain B in separated genetically modified *E. coli* strains. The chains are combined by several chemical steps and the crude insulin is purified (Genentech):

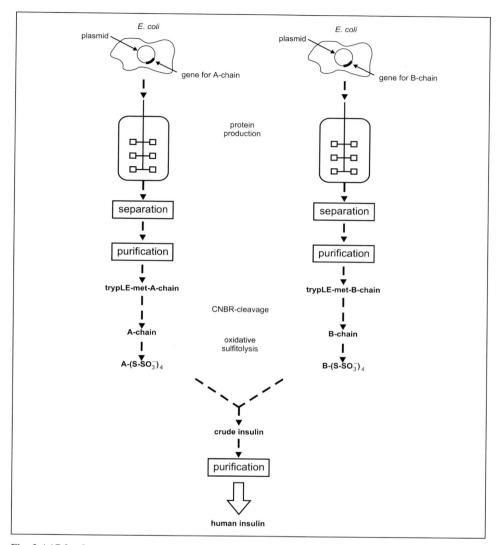

Fig. 3.4.17.2 – 2

2) Production of TrpLE-met-pro-insulin by genetically modified *E. coli* and further synthetic and purification steps (Eli Lilly, see process on page 243.)

3) Production of pre-pro-insulin by genetically modified *E. coli* (HMR, process shown above).

4) Production of a pro-insulin which is converted to insulin by transpeptidation (Novo Nordisk, see page 253).

- Insulin belongs to the first mammal proteins that were synthesized with identical amino acid sequence using recombinant DNA-technology. Today the world capacity is about $2 \text{ t} \cdot \text{a}^{-1}$.

3) Flow scheme

Not published.

4) Process parameters

conversion:	>99.9 %
yield:	> 90 %
selectivity:	> 90 %
reactor type:	batch
reactor volume:	7,500 L
capacity:	$> 0.5 \, t \cdot a^{-1}$
residence time:	4–6 h
space-time-yield:	$1.7 \, g \cdot L^{-1} \cdot d^{-1}$
down stream processing:	chromatography
enzyme activity:	$20 \, U \cdot g^{-1}_{enzyme}$
enzyme supplier:	Calbiochem, USA
start-up date:	1998
production site:	Frankfurt, Germany
company:	Hoechst Marion Roussel, Germany

5) Product application

- In 1921 insulin could be isolated from dog pancreas.

- The sequence was identified by Sanger in 1955.

- Insulin regulates the blood sugar level. It is used for the treatment of diabetis mellitus. Up to 5 % of the population of the western world suffers from Diabetis.

- A glucose level below 0.25 mM leads to problems in energy supply to the brain. Results can be coma, irreversible brain damage and death.

- Since insulin is decomposed by proteases it cannot be applied orally. A new application method is inhalation of modified insulin with faster effects due to exchange of single hydrophobic amino acid with several hydrophilic amino acids. The continuous application after measurement of the blood sugar level is the most comfortable and efficient way.

6) Literature

- Frank, B.H., Chance, R.E. (1983) Two routes for producing human insulin utilizing recombinant DNA technology, Münch. Med. Wschr. **125**, 14–20

- Jørgensen, L.N., Rasmussen, E., Thomsen, B. (1989) HM(ge), Novo's biosynthetic insulin; Med. View. **III**, No. 4, 1–7

- Ladisch, M.R., Kohlmann, K.L. (1992) Recombinant human insulin, Biotechnol. Prog. **8**, 469–478

Trypsin
Pig Pancreas

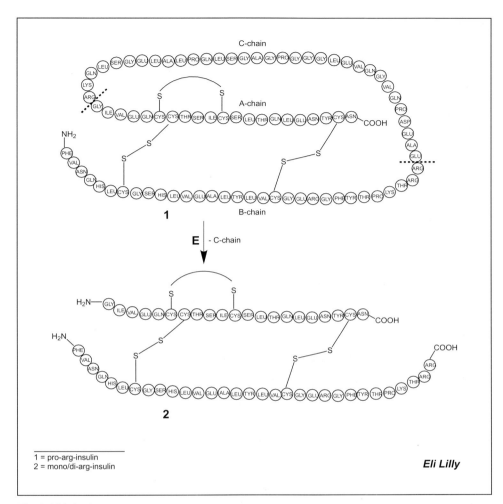

1 = pro-arg-insulin
2 = mono/di-arg-insulin

Eli Lilly

Fig. 3.4.21.4 – 1

1) Reaction conditions

pH: < 7.0
T: 6 °C
medium: aqueous
reaction type: carboxylic acid amide hydrolysis
catalyst: solubilized enzyme
enzyme: tryptase (trypsin)
strain: pig pancreas
CAS (enzyme):   [9002-07–7]

Trypsin
Pig Pancreas

EC 3.4.21.4

2) Remarks

- The precursor pro-insulin is directly produced by fermentation of recombinant *E. coli* (see second step of process on page 243).

- See page 245 for an overview of possible syntheses of insulin.

3) Flow scheme

- See page 243.

4) Process parameters

yield: > 70 %
reactor type: batch
company: Eli Lilly, USA

5) Product application

- See process of Hoechst Martion Roussel on page 245.

6) Literature

- Frank, B.H., Chance, R.E. (1983) Two routes for producing human insulin utilizing recombinant DNA technology, Münch. Med. Wschr. **125**, 14–20

- Jørgensen, L.N., Rasmussen, E., Thomsen, B. (1989) HM(ge), Novo's biosynthetic insulin; Med. View. **III**, No. 4, 1–7

- Ladisch, M.R., Kohlmann, K.L. (1992) Recombinant human insulin, Biotechnol. Prog. **8**, 469–478

Trypsin
Pig Pancreas

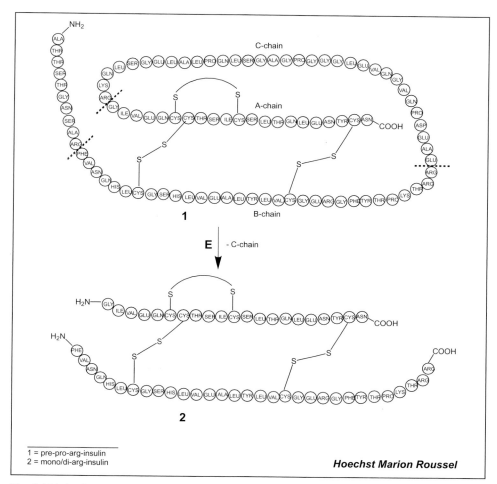

Fig. 3.4.21.4 – 1

1) Reaction conditions

[1]:	$0.8 \text{ g} \cdot \text{L}^{-1}$
pH:	8.3
T:	6 °C
medium:	aqueous
reaction type:	carboxylic acid amide hydrolysis
catalyst:	solubilized enzyme
enzyme:	tryptase (trypsin)
strain:	Pig pancreas
CAS (enzyme):	[9002-07–7]

2) Remarks

- The precursor pre-pro insulin (PPI) is directly produced by fermentation of *E. coli* using the recombinant DNA-technology.

3) Flow scheme

Not published.

4) Process parameters

conversion:	> 99.9 %
yield:	> 65 %
selectivity:	> 65 %
reactor type:	batch
reactor volume:	10,000 L
capacity:	$> 0.5 \ t \cdot a^{-1}$
residence time:	6 h
space-time-yield:	$2.1 \ g \cdot L^{-1} \cdot d^{-1}$
enzyme activity:	$80 \ U \cdot L_{reaction \ solution}^{-1}$
enzyme supplier:	Calbiochem, USA
start-up date:	1998
production site:	Frankfurt, Germany
company:	Hoechst Marion Roussel, Germany

5) Product application

- See next step of process on page 245.

6) Literature

- Frank, B.H., Chance, R.E. (1983) Two routes for producing human insulin utilizing recombinant DNA technology, Münch. Med. Wschr. **125**, 14–20

- Jørgensen, L.N., Rasmussen, E., Thomsen, B. (1989) HM(ge), Novo's biosynthetic insulin; Med. View. **III**, No. 4, 1–7

- Ladisch, M.R., Kohlmann, K.L. (1992) Recombinant human insulin, Biotechnol. Prog. **8**, 469–478

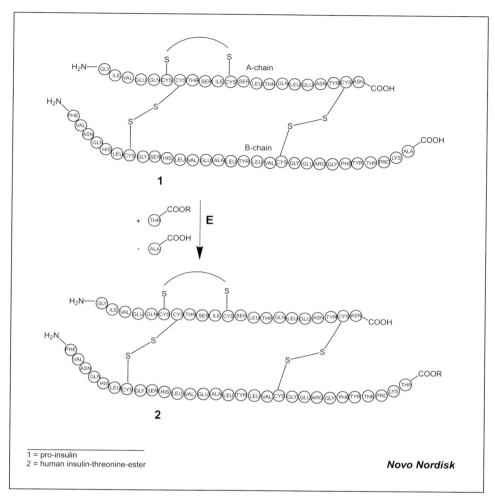

Fig. 3.4.21.4 – 1

1) Reaction conditions

pH:	< 7.0
T:	6 °C
medium:	acetic acid with low watercontent
reaction type:	carboxylic acid amide hydrolysis
catalyst:	solubilized enzyme
enzyme:	tryptase (trypsin)
strain:	pig pancreas
CAS (enzyme):	[9002-07–7]

2) Remarks

- The precursor pro-insulin is directly produced by fermentation of *Saccharomyces cerevisiae* using recombinant DNA-technology. The fermenters with a volume of 80 m³ run continuously for 3–4 weeks by adding substrate at the same speed the broth is drawn.

- The difference between the synthetic route of Hoechst Marion Russel and Eli Lilly is that a threonine ester is synthesized by a transpeptidation that can be easily hydrolyzed to insulin.

- To prevent the possible cleavage beside trypsin at position 22 of the B-chain, the following conditions are applied: low water concentrations by adding organic solvents, surplus of threonin ester, low temperature and low pH.

- See page 245 for an overview of possible synthesis of insulin.

3) Flow scheme

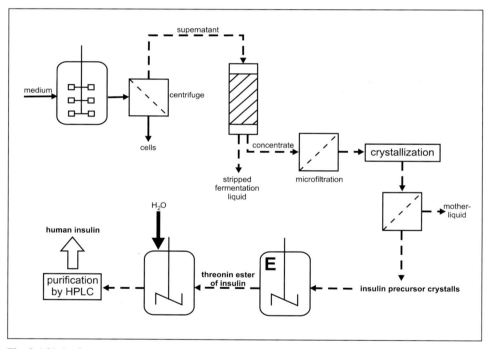

Fig. 3.4.21.4 – 2

4) Process parameters

conversion:	> 99.9 %
yield:	> 97 %
selectivity:	> 97 %
reactor type:	batch
down stream processing:	chromatography
company:	Novo Nordisk, Denmark

5) Product application

- The threoninester of human insulin can be converted into human insulin by simple hydrolysis with subsequent purification steps (see also flow scheme):

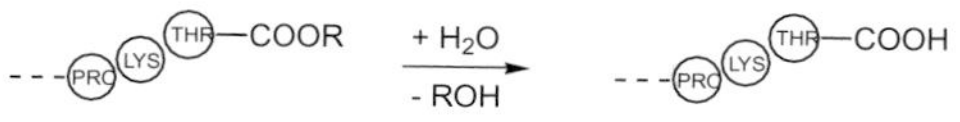

1 = human insulin-threonine-ester
2 = human insulin

Fig. 3.4.21.4 – 3

- For details about insulin see the process by Hoechst Marion Russel on page 245.

6) Literature

- Frank, B.H., Chance, R.E. (1983) Two routes for producing human insulin utilizing recombinant DNA technology, Münch. Med. Wschr. **125**, 14–20

- Jørgensen, L.N., Rasmussen, E., Thomsen, B. (1989) HM(ge), Novo's biosynthetic insulin; Med. View. **III**, No. 4, 1–7

- Ladisch, M.R., Kohlmann, K.L. (1992) Recombinant human insulin, Biotechnol. Prog. **8**, 469–478

- Novo Nordisk (1981) Process for preparing insulin esters, GB 2069502

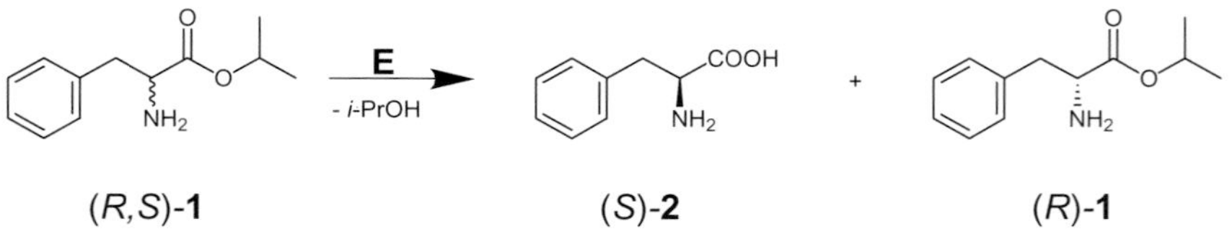

Subtilisin
Bacillus licheniformis

EC 3.4.21.62

1 = phenylalanine-isopropylester
2 = phenylalanine

Coca-Cola

Fig. 3.4.21.62 – 1

1) Reaction conditions

[**1**]:	1.2 M, 248.76 g · L^{-1} [207.3 g · mol^{-1}]
[pH]:	7.5
T:	25 °C
medium:	aqueous two-phase
reaction type:	peptide bond hydrolysis
catalyst:	solubilized enzyme
enzyme:	subtilisin Carlsberg (hydrolase)
strain:	*Bacillus licheniformis*
CAS (enzyme):	[9014-01–1]

2) Remarks

- The EC number was formerly 3.4.4.16 and 3.4.21.14.

- The availability of (*S*)-phenylalanine at low cost is critical to the manufacture of the sweet dipeptide (*S*)-aspartyl-(*S*)-phenylalanine methyl ester (aspartame).

- The non-converted enantiomer is continuously extracted via a supported liquid membrane (=SLM) that is immobilized in a microporous membrane. The liquid membrane consists of 33 % *N,N*-diethyldodecanamide in dodecane.

256

Subtilisin
Bacillus licheniformis

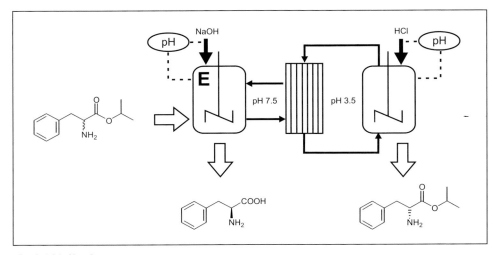

Fig. 3.4.21.62 – 2

- The uniqueness of this process is the continuous extraction with a second aqueous phase of pH 3.5. The hydrolyzation product, the amino acid, is charged at pH 7.5 and is not extracted into the acidic aqueous phase.

- The non-converted (*R*)-amino ester can be racemized by refluxing in anhydrous toluene in the presence of an immobilized salicylaldehyde catalyst.

3) Flow scheme

Fig. 3.4.21.62 – 3

4) Process parameters

yield:	73 % (= 36,6 % with reference to racemic mixture)
ee:	95 %
reactor type:	cstr with continuous extraction of non-converted enantiomer
residence time:	0.2 h
space-time-yield:	$14 \text{ g} \cdot \text{L}^{-1} \cdot \text{d}^{-1}$
enzyme supplier:	Sigma Chem. Co., USA
company:	Coca-Cola, USA

5) Product application

- (S)-Phenylalanine is used as an intermediate for the synthesis of aspartame.

6) Literature

- Mirviss, S. (1987) Racemization of amino acids, US 4713470

- Ricks, E.E., Estrada-Valdes, M.C., McLean, T.L., Iacobucci, G.A. (1992) Highly enantioselective hydrolysis of (R,S)-phenylalanine isopropyl ester by subtilisin Carlsberg. Continuous synthesis of *(S)*-phenylalanine in a hollow fibre/liquid membrane reactor, Biotechnol. Prog. **8**,197–203

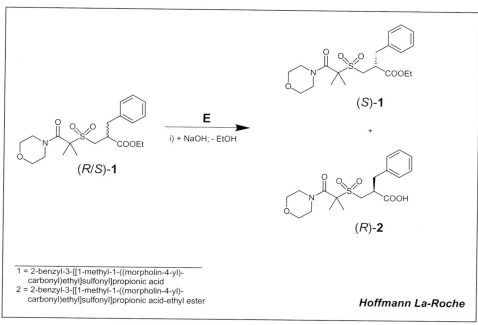

Fig. 3.4.21.62 – 1

1) Reaction conditions

[1]:	0.198 M, 80.68 g · L⁻¹ [407.45 g · mol⁻¹]
pH:	7.5
T:	40 °C
medium:	two-phase: aqueous/organic
reaction type:	ester bond hydrolysis
catalyst:	solubilized enzyme
enzyme:	subtilisin Carlsberg (hydrolase)
strain:	*Bacillus licheniformis*
CAS (enzyme):	[9014-01–1]

2) Remarks

- The EC number was formerly 3.4.4.16 and 3.4.21.14.

- On a 17 kg scale (*R,S*)-**1** was reacted as an 8 % aqueous suspension using solid Optimase® M 440 (6.5 % with respect to (*R,S*)-**1**, pH 7.5, 40 °C, 69 h) to yield 43 % of (*S*)-**2**.

- Alternatively, (*R,S*)-**1** is reacted as a 10 % solution in an aqueous buffer / *tert*-butyl methyl ether mixture using Protease® L 660 (10 % with respect to (*R,S*)-**1**, pH 7.5, 40°C) to give 46 % of (*S*)-**2**.

- Optimase® M 440 is a granular subtilisin preparation used in detergent formulations.

- Protease® L 660 is a stabilized, water-miscible, liquid food-grade subtilisin preparation used in detergent formulations.

- The major component of both the preparations is subtilisin Carlsberg (subtilisin A).

- Even under completely non-physiological conditions (substrate to buffer ratio of 1:1), the enzyme shows excellent chemical and enantiomeric purity (> 99 %).

- Optimase® M 440 is cheap; has high tolerance to high substrate concentrations, salt concentrations and organic solvents; is stable at high temperatures; shows high enantioselectivity (even at high temperatures) and has an acceptable reaction rate.

3) Flow scheme

Not published.

4) Process parameters

conversion:	50 %
yield:	43 %
selectivity:	high
ee:	> 99 %
chemical purity:	> 99 %
reactor type:	batch
reactor volume:	200 L
residence time:	69 h
space-time-yield:	$11 \ g \cdot L^{-1} \cdot d^{-1}$
down stream processing:	disc separator, crystallization
enzyme supplier:	Miles Kali (now Solvay Enzymes), Novo Nordisk
start-up date:	1992
closing date:	1992
production site:	Switzerland
company:	Hoffmann La-Roche, Switzerland

5) Product application

- (*S*)-**2** was prepared as an intermediate for the renin inhibitors ciprokiren and remikiren:

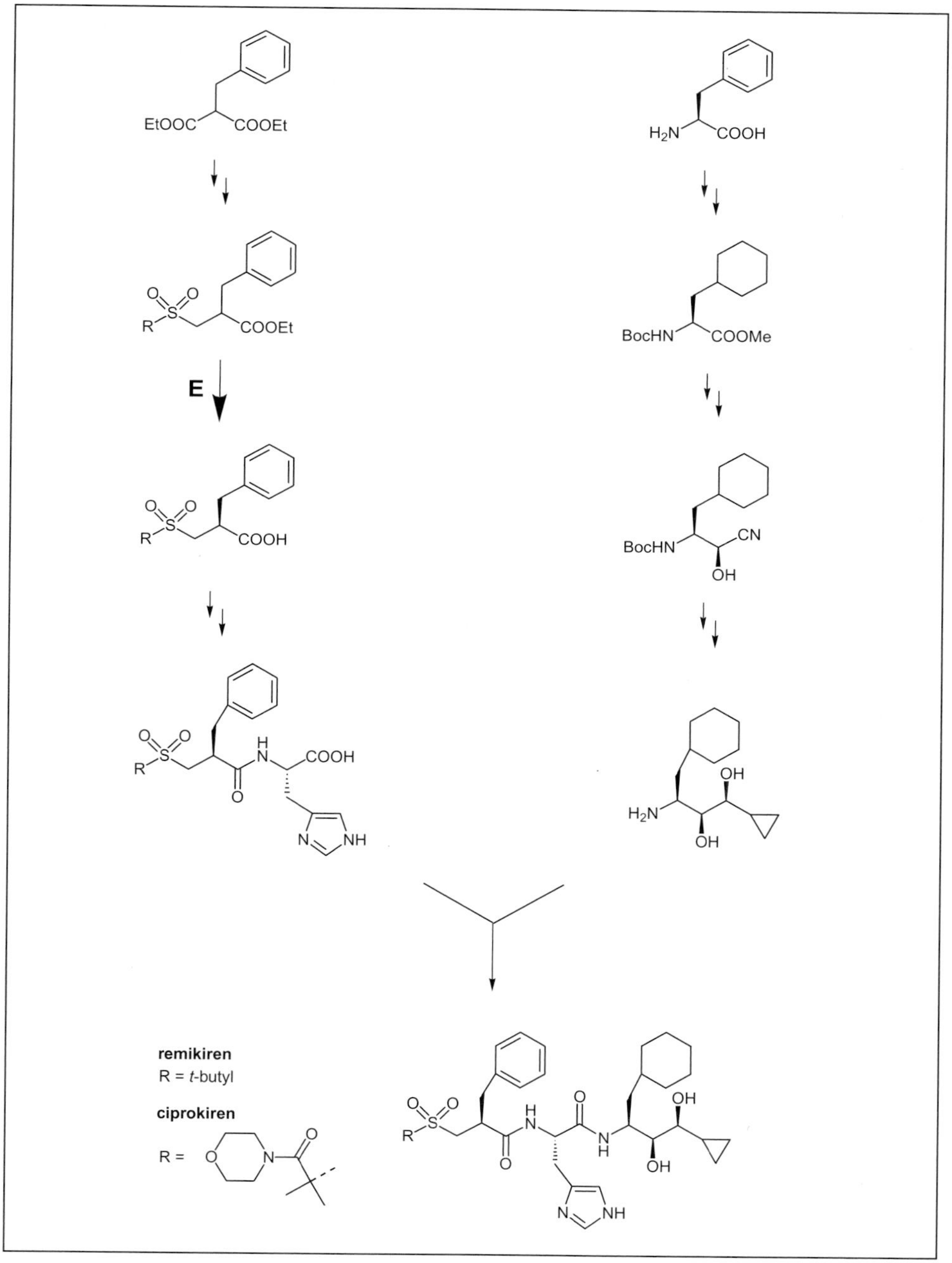

remikiren
R = *t*-butyl

ciprokiren

R =

Fig. 3.4.21.62 – 2

261

6) Literature

- Doswald, S., Estermann, H., Kupfer, E., Stadler, H., Walther, W., Weisbrod, T., Wirz, B., Wostl, W. (1994) Large scale preparation of chiral building blocks for the P_3 site of renin inhibitors, Bioorg. Med. Chem. **2**, 403–410

- Wirz, B., Weisbrod, T., Estermann, H. (1995) Enzymatic reactions in process research – The importance of parameter optimization and workup, Chimica Oggi. **14**, 37–41

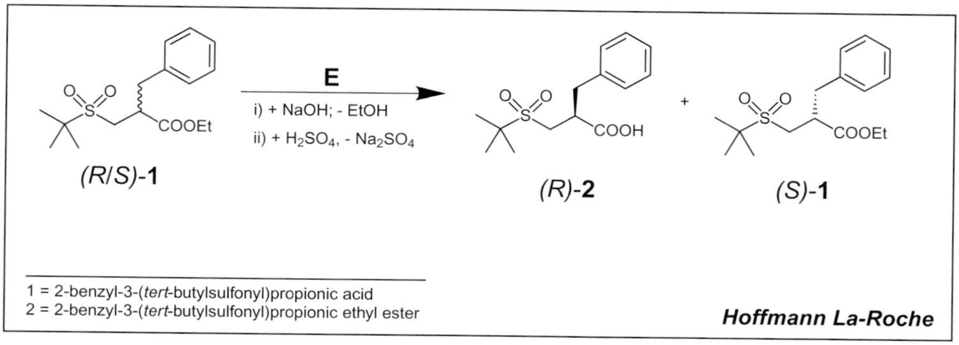
E
i) + NaOH; - EtOH
ii) + H2SO4, - Na2SO4
COOEt
(R/S)-1
COOH
(R)-2
COOEt
(S)-1
1 = 2-benzyl-3-(tert-butylsulfonyl)propionic acid
2 = 2-benzyl-3-(tert-butylsulfonyl)propionic ethyl ester
Hoffmann La-Roche

Fig. 3.4.21.62 – 1

1) Reaction conditions

[**1**]:	0.733 M, 228.25 g · L^{-1} [311.39 g · mol^{-1}]
pH:	7.5–8.5
T:	39–40 °C
medium:	two-phase: aqueous / organic
reaction type:	ester bond hydrolysis
catalyst:	solubilized enzyme
enzyme:	subtilisin Carlsberg (hydrolase)
strain:	*Bacillus licheniformis*
CAS (enzyme):	[9014-01–1]

2) Remarks

- The EC number was formerly 3.4.4.16 and 3.4.21.14.
- On a 150 kg scale, (*R,S*)-**1** is reacted under non-optimized conditions as a 22 % aqueous emulsion using solid Optimase® M 440 (5 % with respect to (*R,S*)-**1**, pH 7.5, 40 °C, 70 h) to yield 41 % of (*S*)-**2**.
- Under optimized conditions (*R,S*)-**1** is reacted as a 50 % emulsion in a 130 mM borax buffer using liquid Protease® L 660 (10 % with respect to (*R,S*)-**1**, pH 8.5, 40 °C, 42.3 h) to give 42 % of (*S*)-**2** (space-time yield of 550 g · L^{-1} · d^{-1}).
- Optimase® M 440 is a granular subtilisin preparation used in detergent formulations.
- Protease® L 660 is a stabilized, food-grade, water-miscible, liquid subtilisin preparation used in detergent formulations.
- Substilin Carlsberg is extremely tolerant to high concentrations of substrate, salt and organic solvent, to elevated temperature and to a combination of all these factors.
- Reduction of the reaction volume is crucial for lowering production costs. This can be achieved either by increasing the initial concentration of substrate (to about 20 %; higher concentrations lead to emulsification problems) or by using more concentrated sodium hydroxide solutions (2 M NaOH; higher concentrations affect hydrolysis rate) for pH control during reaction.

- Due to substrate precipitation at lower temperatures, it is necessary to keep the temperature above the melting point of the substrate (36 – 39 °C). The reaction temperature is maintained at 38 – 43 °C. The reaction rate decreases slightly above 45 °C.

- The poor substrate solubility can also be overcome by dissolving the substrate in a water-miscible cosolvent.

- Immobilization and reuse of the cheap enzyme are not commercially attractive in this case.

- A continuously operated disk separator reduces the number of extraction and filtration steps at neutral and acidic pH.

3) Flow scheme

Not published.

4) Process parameters

conversion:	50 %
yield:	41 – 46 %
selectivity:	high
ee:	> 99 %
chemical purity:	> 99 %
reactor type:	batch
capacity:	> 100 kg
residence time:	70 h
space-time-yield:	$20 \text{ g} \cdot \text{L}^{-1} \cdot \text{d}^{-1}$
down stream processing:	extraction, filtration, crystallization
enzyme supplier:	Miles Kali (now Solvay Enzymes)
start-up date:	1991
closing date:	1992
production site:	Switzerland
company:	Hoffmann La-Roche, Switzerland

5) Product application

- (*S*)-**1** was prepared as an intermediate for the renin inhibitor remikiren:

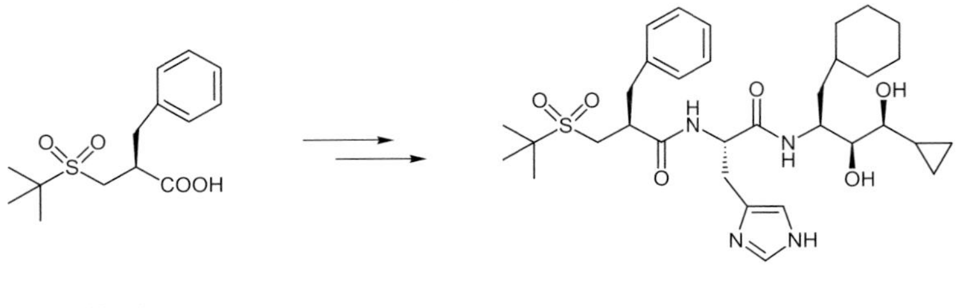

(*S*)-**1**

2

1 = 2-benzyl-3-(*tert*-butylsulfonyl)propionic acid
2 = remikiren

Fig. 3.4.21.62 – 2

6) Literature

- Doswald, S., Estermann, H., Kupfer, E., Stadler, H., Walther, W., Weisbrod, T., Wirz, B., Wostl, W. (1994) Large scale preparation of chiral building blocks for the P_3 site of renin inhibitors, Bioorg. Med. Chem. **2**, 403–410

- Wirz, B., Weisbrod, T., Estermann, H. (1995) Enzymatic reactions in process research – The importance of parameter optimization and workup, Chimica Oggi. **14**, 37–41

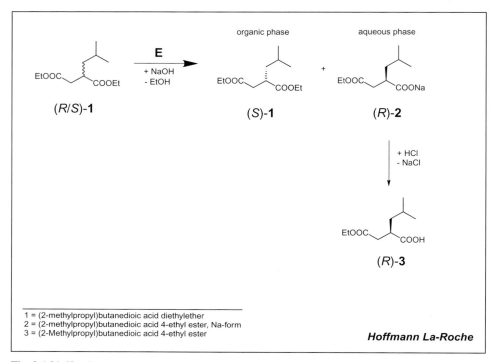

1 = (2-methylpropyl)butanedioic acid diethylether
2 = (2-methylpropyl)butanedioic acid 4-ethyl ester, Na-form
3 = (2-Methylpropyl)butanedioic acid 4-ethyl ester

Hoffmann La-Roche

Fig. 3.4.21.62 – 1

1) Reaction conditions

[**1**]:	0.84 M, 197.73 g·L^{-1} [235.39 g·mol^{-1}]
pH:	8.5
T:	22–25 °C
medium:	two-phase: aqueous / organic
reaction type:	ester bond hydrolysis
catalyst:	solubilized enzyme
enzyme:	subtilisin Carlsberg (hydrolase)
strain:	*Bacillus* sp.
CAS (enzyme):	[9014-01–1]

2) Remarks

- The EC number was formerly 3.4.4.16 and 3.4.21.14.

- The diester **1** is utilized as a 20 % emulsion in 30 mM aqueous NaHCO$_3$ using Protease® L 660 or Alcalase® 2.5 L (9 % each, with respect to **1**).

- The unconverted (*S*)-**1** can be extracted in a solvent such as toluene and racemized by heating the anhydrous extract with a catalytic amount of sodium ethoxide. The resulting racemic **1** can be recycled, thus improving the overall yield from 45 % up to 87 %.

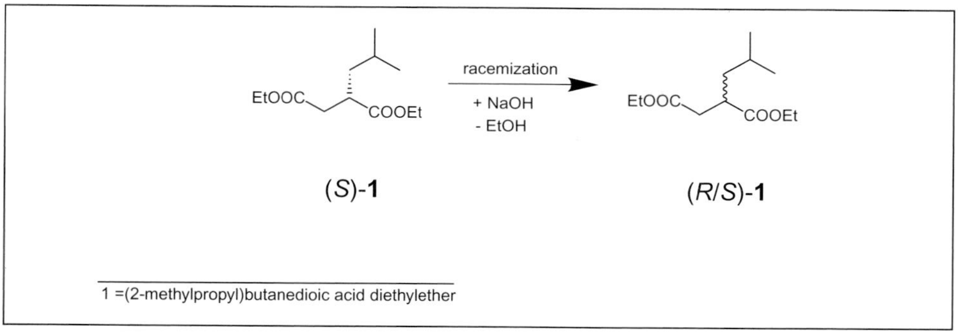

Fig. 3.4.21.62 – 2

- The reaction was repeatedly carried out on a 200 kg scale with respect to **1.**

- The enzyme is highly stereoselective even at high substrate concentrations (20 %).

- The existing chemical research synthesis for bulk amounts was replaced by the chemo-enzymatic route starting from the cheap bulk agents maleic anhydride and isobutylene:

Subtilisin
Bacillus sp.

EC 3.4.21.62

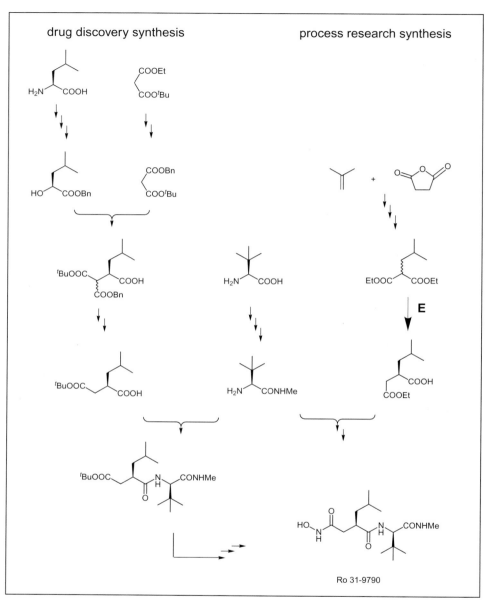

Fig. 3.4.21.62 – 3

3) Flow scheme

Not published.

268

4) Process parameters

conversion:	45 %
ee:	> 99 %
reactor type:	batch
residence time:	48 h
space-time-yield:	$32 \ g \cdot L^{-1} \cdot d^{-1}$
enzyme supplier:	Solvay (Protease® L-660), Novo Nordisk (Alcalase® 2.5 L)
start-up date:	1994
closing date:	1994
production site:	Switzerland
company:	Hoffmann La-Roche, Switzerland

5) Product application

- (R)-**3** is used as a building block for potential collagenase inhibitors (e.g. Ro 31–9790) in the treatment of osteoarthritis.

6) Literature

- Wirz, B., Weisbrod, T., Estermann, H. (1995) Enzymic reactions in process research – The importance of parameter optimization and work-up, Chimica Oggi. **14**, pp. 37–41

- Wirz, B., Weisbrod, T., Estermann, H. (1995) Enzymatic reactions in process research – The importance of parameter optimization and workup, Chimica Oggi. **14**, pp. 37–41

- Doswald, S., Estermann, H., Kupfer, E., Stadler, H., Walther, W., Weisbrod, T., Wirz, B., Wostl, W. (1994) Large scale preparation of chiral building blocks for the P3 site of renin inhibitors, Bioorg. Med. Chem. **2**, 403–410.

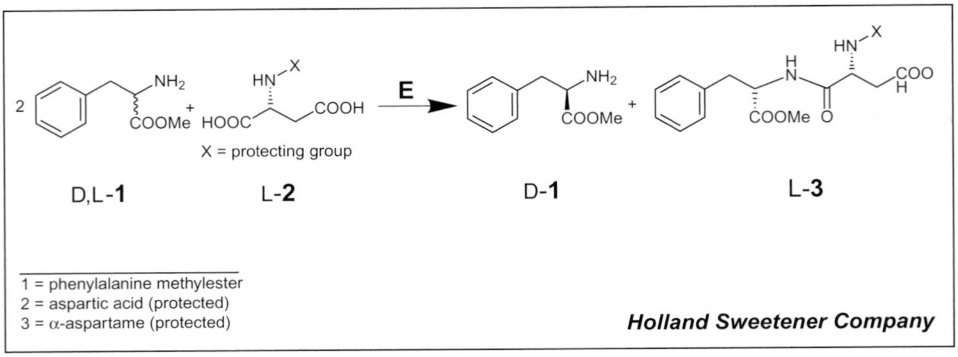

Thermolysin
Bacillus proteolicus

Fig. 3.4.24.27 – 1

1) Reaction conditions

pH: 7.0–7.5
T: 50 °C
medium: aqueous
reaction type: homogeneous
catalyst: solubilized enzyme
enzyme: thermolysin (thermoase)
strain: *Bacillus proteolicus / thermoproteolyticus*
CAS (enzyme):   [9073–78–3]

2) Remarks

- The main problem in chemical synthesis is the by-product formation of β-aspartame. This isomer is of bitter taste and has to be completely removed from the α-isomer.

- The advantages of the enzymatic route are:

  1) no β-isomer is produced,

  2) the enzyme is completely stereoselective, so that racemic mixtures of the substrate or the appropriate enantiomer of the amino acid can be used,

  3) no racemization occurs during synthesis, and

  4) the reaction takes place in aqueous media under mild conditions.

- The bacterial strain was found in the Rokko Hot Spring in central Japan. Consequently it is very stable up to temperatures of 60 °C.

- The enzyme contains a zinc, ion which is responsible for its activity and 4 calcium ions that play an important role in the stability.

- Since the reaction is limited by the equilibrium the products have to be removed from the reaction mixture to achieve high yields.

270

- If an excess of phenylalanine methylester (which is inert to the reaction) is added, the carboxylic anion of the protected aspartame forms a poorly soluble adduct which precipitates from the reaction mixture. The precipitate can be removed easily by filtration.

- Final steps of the process are the removal of protecting groups and racemization of the formed L-amino acid:

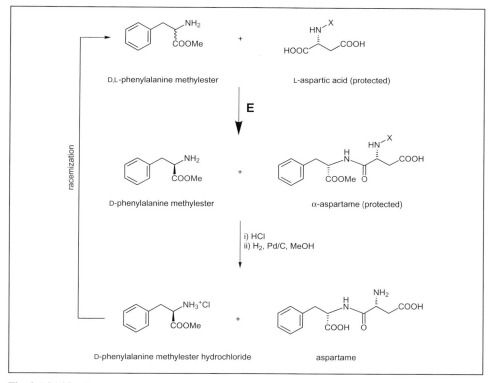

Fig. 3.4.24.27 – 2

3) Flow scheme

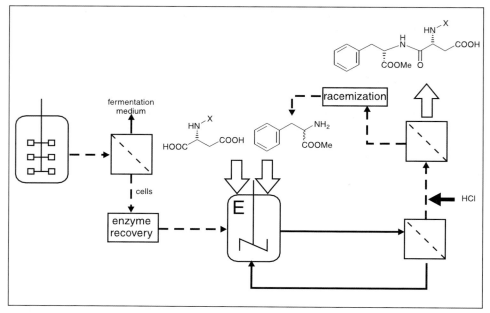

Fig. 3.4.24.27 – 3

4) Process parameters

selectivity:	> 99.9 %
ee:	> 99.9 %
reactor type:	batch
capacity:	> 2,000 t · a^{-1}
down stream processing:	filtration
start-up date:	1988
company:	Holland Sweetener Company, The Netherlands; joint venture between DSM, The Netherlands, and Toso, Japan; others

5) Product application

• Aspartame is used as a low-calorie sweeteener in food, including soft drinks, table-top sweeteners, dairy products, instant mixes, dressings, jams, confectionery, toppings and in pharmaceuticals.

• Aspartame is a sweetener which is 200 times sweeter than sucrose.

• It was first approved in France in 1979.

• The first trademark was Nutrasweet™ (by Monsanto).

6) Literature

- Oyama, K. (1992) The industrial production of aspartame. In: Chirality in Industry (Collins, A.N., Sheldrake, G.N., Crosby, J., eds.), pp. 237–247, John Wiley & Sons Ltd., New York

- Harada, T., Shinnanyo-Shi, Y., Irino, S., Kunisawa, Y., Oyama, K. (1996) Improved enzymatic coupling reaction of N-protected-L-aspartic acid and phenylalanine methyl ester, Holland Sweetener Company, The Netherlands, EP 0768384

- Kirchner, G., Salzbrenner, E., Werenka, C., Boesten, W. (1997) Verfahren zur Herstellung von Z-L-Asparaginsäure-Dinatriumsalz aus Fumarsäure, DSM Chemie Linz GmbH, Austria, Holland Sweetener Company, The Netherlands, EP 0832982

- Sheldon, R.A. (1993) Chirotechnology, Marcel Dekker, New York, 235

- Wiseman, A. (1995) Handbook of Enzyme and Biotechnology, Ellis Horwood, Chichester

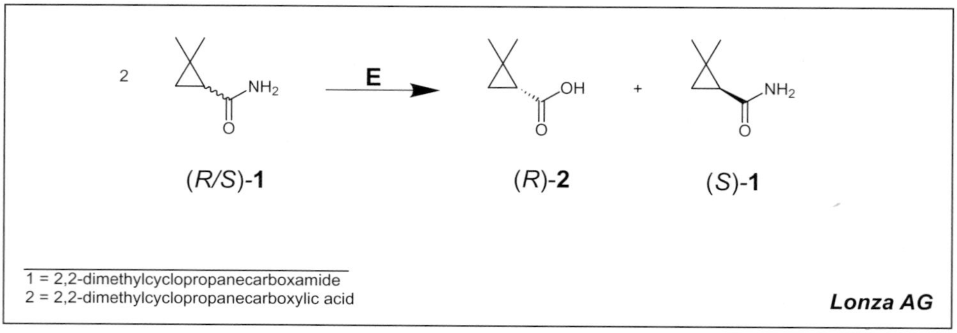

1 = 2,2-dimethylcyclopropanecarboxamide
2 = 2,2-dimethylcyclopropanecarboxylic acid

Lonza AG

Fig. 3.5.1.4 – 1

1) Reaction conditions

[**1**]:	0.247 M, 28 g · L^{-1} [113.17 g · mol^{-1}]
pH:	7.0
T:	37 °C
medium:	aqueous (10 mM sodium phosphate buffer)
reaction type:	amide hydrolysis
catalyst:	suspended whole cells
enzyme:	acylamide amidohydrolase (amidase)
strain:	*Comamonas acidovorans* (DSM: 6351), expressed in *Escherichia coli* XL1Blue/pCAR6
CAS (enzyme):	[9012–56-0]

2) Remarks

- The reaction rate and selectivity are increased by the addition of 5 % (v/v) ethanol.

- The product is purified by extraction, electrodialysis or drying.

- The isolation of the hydrolase enzyme is also possible. Prior to the application of the isolated enzyme in the biotransformation the hydrolase is immobilized on Eupergit C (1.5 U · g^{-1}$_{\text{wet weight Eupergit C}}$).

3) Flow scheme

Not published.

4) Process parameters

yield:	44 %
ee:	98.6 %
reactor type:	batch
capacity:	> 1 t · a^{-1}
residence time:	7 h
company:	Lonza AG, Switzerland

Amidase
Comamonas acidovorans

EC 3.5.1.4

5) Product application

- Intermediate for the synthesis of the dehydropeptidase-inhibitor cilastatin. This drug is administered in combination with penem-carbapenem antibiotics to prevent deactivation by renal dehydropeptidase in the kidney.

6) Literature

2222222222222222222222222222222

- Robins, K. (1993) Biotechnologisches Verfahren zur Herstellung von (S)-(+)-2,2-Dimethylcyclopropancarboxamid und (R)-(−)-2,2-Dimethylcyclopropancarbonsäure, Lonza AG, EP 0502525 A1
- Zimmermann, T., Robins, K., Birch, O.M., Böhlen, E. (1993) Gentechnologisches Verfahren zur Herstellung von (S)-(+)-2,2-Dimethylcyclopropancarboxamid mittels Mikroorganismen, Lonza AG, EP 0524604 A2

275

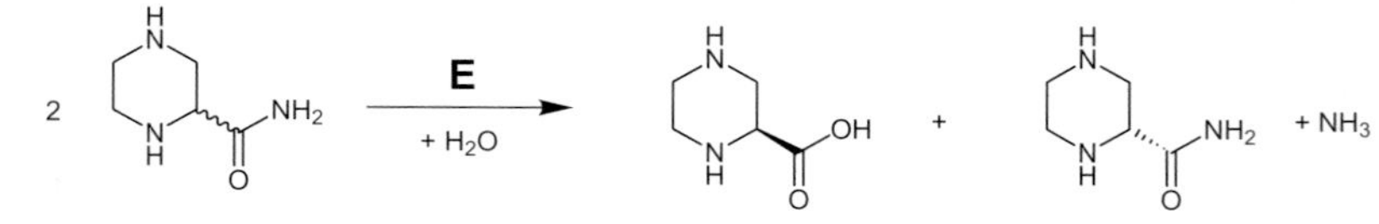

(*R*/*S*)-**1** (*S*)-**2** (*R*)-**1**

1 = piperazine-2-carboxamide
2 = piperazine-2-carboxylic acid

Lonza AG

Fig. 3.5.1.4 – 1

1) Reaction conditions

[**1**]:	0.15 M, 20 g · L^{-1} [129.16 g · mol^{-1}]
pH:	8.0
T:	47 °C
medium:	aqueous
reaction type:	carboxylic acid amide hydrolysis (kinetic resolution)
catalyst:	suspended whole cells
enzyme:	acylamide amidohydrolase (amidase)
strain:	*Klebsiella terrigena* DSM 9174
CAS (enzyme):	[9012–56-0]

2) Remarks

- Primary screening was done starting from soil samples using growth media that contained racemic carboxamides. The cell-free medium of the individual clones were then analysed on TLC plates, only strains with approximately 50 % of the racemic carboxamides were chosen.
- This kinetic resolution is attractive because the starting material can be easily prepared:

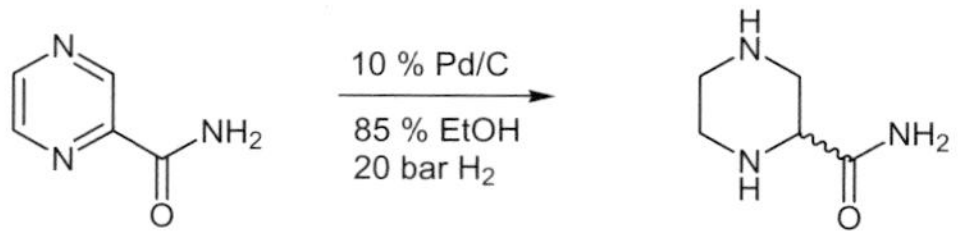

1 (*R*/*S*)-**2**

1 = pyrazine-2-carboxamide
2 = piperazine-2-carboxamide

Fig. 3.5.1.4 – 2

- The microorganism can be grown in fermenters on the racemic carboxamides at the same time as the biotransformations are taking place.

- After completition of conversion the cells are removed by centrifugation and the supernatant is concentrated 10-fold at 60 °C under reduced pressure. The product precipitates by acidifying with conc. HCl (pH 1).

- Using the strain *Burkholderia* DSM 9925 instead of *Klebsiella* DSM 9174 leads to (*R*)-piperazine-2-carboxylic acid (99 % ee, 22 % yield).

- (*S*)-Piperazine-2-carboxylic acid has also been prepared by kinetic resolution of racemic 4-(*tert*-butoxycarbonyl)piperazine-2-carboxamide with leucine aminopeptidase and of (*R,S*)-*n*-octyl-pipecolate with *Aspergillus* lipase. Both processes show practical disadvantages in starting material preparation and biocatalyst availability.

3) Flow scheme

Not published.

4) Process parameters

yield:	41.0 %
ee:	99.4 %
reactor type:	batch
capacity:	multi kg
residence time:	32 h – 76 h
down stream processing:	precipitation
company:	Lonza AG, Switzerland

5) Product application

- The product is used as an intermediate for pharmaceuticals, e.g. the orally active HIV protease inhibitor crixivan from Merck:

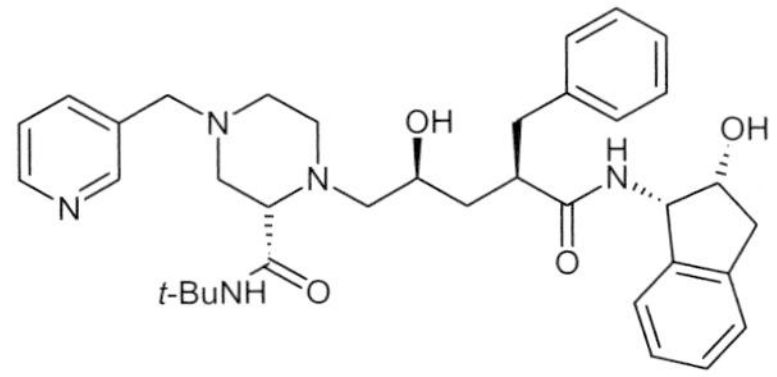

Fig. 3.5.1.4 – 3

- It is also a precursor of numerous bioactive compounds.

6) Literature

• Aebischer, B., Frey, P., Haerter, H.-P., Herrling, P. L., Mueller, W., Olverman, H. J., Watkins, J. C. (1989) Synthesis and NMDA antagonistic properties of the enantiomers of 4-(3-phosphonopropyl)piperazine-2-carboxylic acid (CPP) and of the unsaturated analogue (E)-4-(3-phosphonoprop-2-enyl)piperazine-2-carboxylic acid (CPP-ene), Helv. Chim. Acta **72**, 1043–1051

• Askin, D., Eng, K. K., Rossen, K., Purick, R. M., Wells, K. M., Volante, R. P., Reider, P. J. (1994) Highly diastereoselective reaction of a chiral, non-racemic amide enolate with (S)-glycidyl tosylate. Synthesis of the orally active HIV-1 protease inhibitor L-735,524, Tetrahedron Lett. **35**, 673–676

• Bigge, C. F., Johnson, G., Ortwine, D. F., Drummond, J. T., Retz, D. M., Brahce, L. J., Coughenour, L. L., Marcoux, F. W., Probert Jr., A. W. (1992) Exploration of N-phosphonoalkyl-, N-phosphoalkenyl-, and N-(phosphonoalkyl)phenyl-spaced α-amino acids as competitive N-methyl-D-aspartic acid antagonists, J. Med. Chem. **35**, 1371–1384

• Bruce, M. A., St Laurent, D. R., Poindexter, G. S., Monkovic, I., Hunag, S., Balasubramanian, N. (1995) Kinetic resolution of piperazine-2-carboxamide by leucine aminopeptidase. An application in the synthesis of the nucleoside transport blocker (Draflazine-), Synth. Commun. **25**, 2673–2684

• Eichhorn, E., Roduit, J.-P., Shaw, N., Heinzmann, K., Kiener A. (1997) Preparation of (S)-piperazine-2-carboxylic acid, (R)-piperazine-2-carboxylic acid, and (S)-pipecolic acid by kinetic resolution of the corresponding racemic carboxamides with stereoselective amidase in whole bacterial cells, Tetrahedron Asymm. **8(15)**, 2533–2536

• Kiener, A., Roduit, J.-P., Kohr, J., Shaw, N. (1995) Biotechnologisches Verfahren zur Herstellung von cyclischen (S)-α-Aminocarbonsäuren und (R)-α-Aminocarbonsäureamiden, Lonza AG, EP 06,86,698 A2

• Kiener, A., Roduit, J. P., Heinzmann, K. (1995) Biotechnical production process of piperazine (R)-α-carboxylic acids and piperazine (S)-α-carboxylic acid amide, Lonza AG, WO 96/35775

• Petersen,M., Kiener,A.(1999) Biocatalysis. Preparation and functionalization of N-heterocycles, Green Chem. **2**, 99–106

• Shiraiwa, T., Shinjo, K., Kurokawa, H. (1991) Asymmetric transformations of proline and 2-piperidinecarboxylic acid via formation of salts with optically active tartaric acid, Bull. Chem. Soc. Jpn. **64**, 3251–3255

Urease
Lactobacillus fermentum

$$H_2N-\overset{\displaystyle O}{\underset{}{C}}-NH_2 \quad + \quad H_2O \quad \xrightarrow{\ E\ } \quad 2\ NH_3 \quad + \quad CO_2$$

1   **2**

1 = urea
2 = ammonia

Asahi Chemical Industry
Toyo Jozo

Fig. 3.5.1.5. – 1

1) Reaction conditions

[**1**]:	12–37 ppm
pH:	4–4.5 (sake or other alcoholic beverages)
T:	10–15 °C
medium:	aqueous
reaction type:	carboxylic acid amide hydrolysis
catalyst:	immobilized enzyme
enzyme:	urea amidohydrolase (urease, acid urease)
strain:	*Lactobacillus fermentum*
CAS (enzyme):	[9002–13–5]

2) Remarks

- Ethyl carbamate is a natural component of sake or other fermentated beverages. It is known to be carcinogenic, teratogenic, and mutagenic.

- It is also formed in alcoholic beverages under heat:

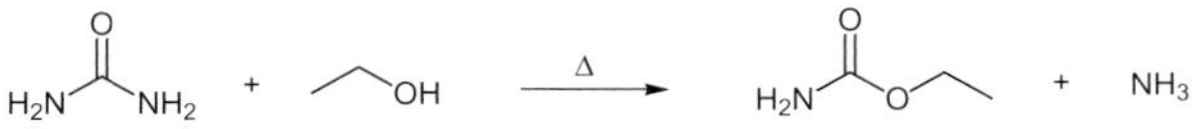

$$H_2N-\overset{\displaystyle O}{\underset{}{C}}-NH_2 \quad + \quad \diagup\!\!\diagdown\!OH \quad \xrightarrow{\ \Delta\ } \quad H_2N-\overset{\displaystyle O}{\underset{}{C}}-O-\diagup \quad + \quad NH_3$$

1   **2**   **3**   **4**

1 = urea
2 = ethanol
3 = ethyl carbamate
4 = ammonia

Fig. 3.5.1.5. – 2

- To prevent formation of ethyl carbamate the urea concentration has to be decreased to 3 ppm.

- The continuous process works in a stable mannner over more than 150 days.

- The isolated enzyme is immobilized on a polyacrylonitrile support. For high-speed treatment of urea the urease is immobilized on Chitopearl® (porous chitosan) beads.

- This method has been and is used for urea removal from sake by many companies in Japan.

- For table wine the limiting level of ethyl carbamate is 30 ppm. The removal of urea from red and white wine cannot be realized, if tannin being an inhibitor of urease is present in wine.

3) Flow scheme

Not published.

4) Process parameters

reactor type:	plug-flow reactor
reactor volume:	100 L
enzyme activity:	140 $U \cdot g^{-1}_{wet\ carrier}$
enzyme supplier:	Takeda Chemical Ind.
start-up date:	1988
production site:	Japan
company:	Asahi Chemical Industry, Japan and Toyo Jozo, Japan

5) Literature

- Matsumoto, K. (1993) Removal of urea from alcoholic beverages by immobilized acid urease, in: Industrial Application of Immobilized Biocatalysts (Tanaka, A., Tosa, T., Kobayashi, T., eds.), pp.255–273, Marcel Dekker Inc., New York

- Yoshizawa, K., Takahashi, K. (1988) Utilization of urease for decomposition of urea in sake, J. Brew. Soc. Jpn. **83** (2), 142–150

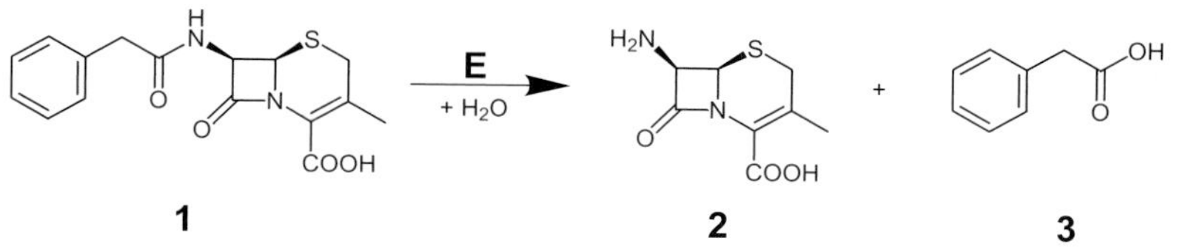

1 = cephalosporin G
2 = 7-amino deacetoxy cephalosporinic acid (7-ADCA)
3 = phenylacetic acid

Dr. Vig Medicaments

Fig. 3.5.1.11 – 1

1) Reaction conditions

[**1**]:	0.3 M, 100 g · L^{-1} [332.38 g · mol^{-1}]
pH:	8.0
T:	37 °C
medium:	aqueous
reaction type:	carboxylic acid amide hydrolysis
catalyst:	immobilized enzyme
enzyme:	penicillin amidohydrolase (penicillin acylase)
strain:	*E. coli*, optimized
CAS (enzyme):	[9014-06–6]

2) Remarks

- The reaction vessel is equipped with a filter sieve at the bottom to retain the immobilized penicillin acylase.

- By the same process 6-amino penicillanic acid (6-APA) is produced starting from penicillin G. In this case a yield of 87 % is achieved.

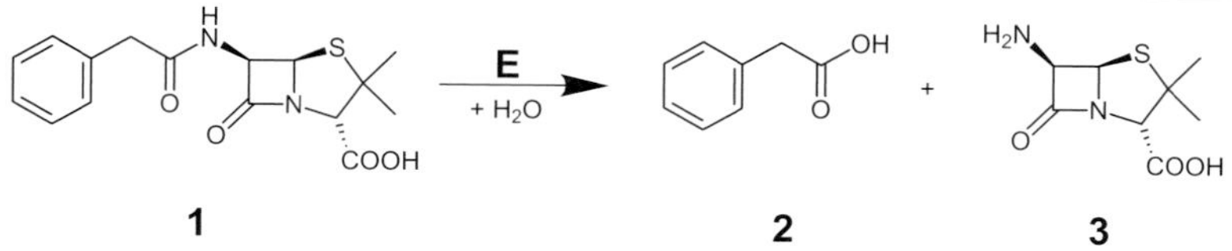

1 = penicillin G
2 = phenylacetic acid
3 = 6-amino penicillianic acid (6-APA)

Fig. 3.5.1.11 – 2

- In the case of 6-APA production the enzyme consumption is reduced by a factor of about 2 to 250 U · kg^{-1} in comparison to 7-ADCA production.

3) Flow scheme

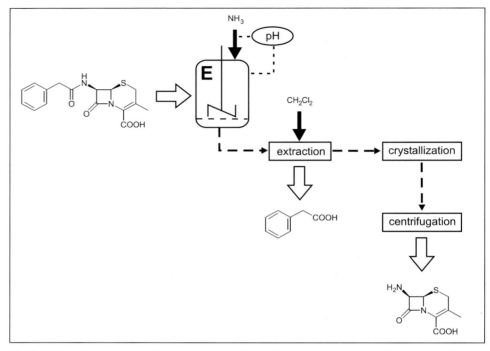

Fig. 3.5.1.11 – 3

4) Process parameters

conversion:	99 %
yield:	93 %
selectivity:	94 %
reactor type:	repetitive batch
reactor volume:	2,000 L
capacity:	$300 \ t \cdot a^{-1}$
residence time:	1.5 – 2 h
space-time-yield:	$30 \ g \cdot L^{-1}$
down stream processing:	crystallization
enzyme activity:	$1,000 \ U \cdot g^{-1}$
enzyme consumption:	$450 \ U \cdot kg^{-1}$
production site:	Baroda, India
company:	Dr. Vig Medicaments, India

5) Product application

- In the manufacture of semi-synthetic β-lactam antibiotics. The worldwide capacity is more than $20,000 \ t \cdot a^{-1}$.

6) Literature

- Vig, C. (1999) personal communication

282

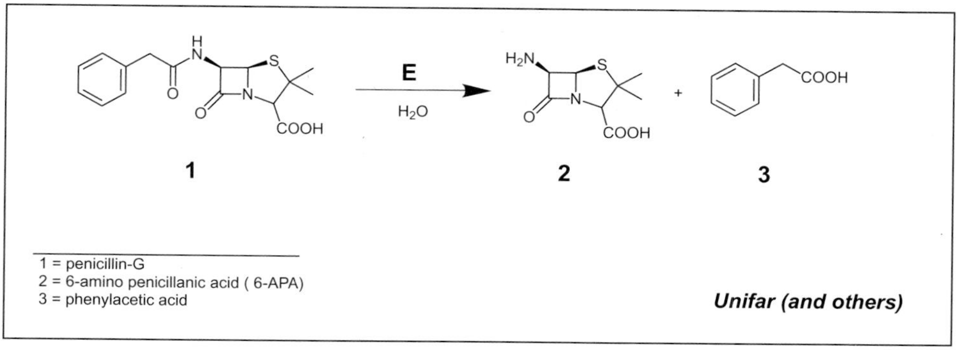

1 = penicillin-G
2 = 6-amino penicillanic acid (6-APA)
3 = phenylacetic acid

Unifar (and others)

Fig. 3.5.1.11 – 1

1) Reaction conditions

[**1**]:   0.24 M, 80 g · L^{-1} [334.39 g · mol^{-1}]
pH: 8.0
T: 30–35 °C
medium: aqueous
reaction type: carboxylic acid amide hydrolysis
catalyst: immobilized enzyme
enzyme: penicillin amidohydrolase (penicillin acylase, penicillin amidase)
strain: *Escherichia coli* and others (e.g. *Bacillus megaterium*)
CAS (enzyme):   [9014-06–6]

2) Remarks

- The enzyme is isolated and immobilized on Eupergit-C (Röhm, Germany).

- The production is carried out in a repetitive batch mode. The immobilized enzyme is retained by a sieve with a mesh size of 400.

- The time for filling and emptying the reactor is approximately 30 min.

- The residual activity of biocatalyst after 800 batch cycles, which is one production campaign, is about 50 % of the initial activity.

- The hydrolysis time after 800 batch cycles increases from the initial 60 min to 120 min.

- Phenylacetic acid is removed by extraction and 6-APA can be crystallized.

- The yield can be increased by concentrating the split-solution and/or the mother liquor of crystallization via vacuum evaporation or reverse osmosis.

- The production operates for 300 days per year with an average production of 12.8 batch cycles per day (production campaigns of 800 cycles per campaign).

- Several chemical steps are replaced by a single enzyme reaction. Organic solvents, the use of low temperature (–40 °C) and the need for absolutely anhydrous conditions, which made the process difficult and expensive, are no longer necessary in the enzymatic process:

283

Penicillin amidase
Escherichia coli

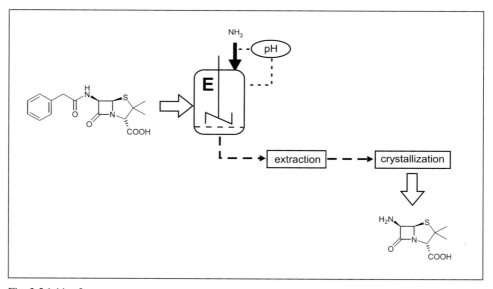

Fig. 3.5.1.11 – 2

1 = penicillin-G
2 = 6-amino penicillanic acid (6-APA)

3) Flow scheme

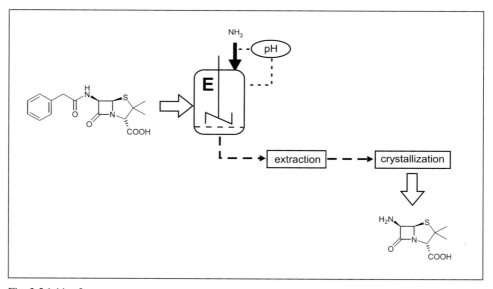

Fig. 3.5.1.11 – 3

4) Process parameters

conversion:	98 %
yield:	86 % (in reaction: 97 %)
selectivity:	> 99 %
chemical purity:	99 %
reactor type:	repetitive batch
reactor volume:	3,000 L
capacity:	300 t $\cdot$ a^{-1}
residence time:	1.5 h (average over 800 cycles; initial: 1 h)
space-time-yield:	445 g $\cdot$ L^{-1} $\cdot$ d^{-1} (which is the average for a production campaign of 800 batch cycles)
down stream processing:	extraction, crystallization (see remarks)
enzyme activity:	22 M U, corresponding to approx. 100 kg wet biocatalyst (27.5 kg of dry Eupergit-C)
enzyme consumption:	345 U $\cdot$ kg^{-1} (6-APA)
start-up date:	1973
production site:	Unifar, Turkey (and elsewhere)
company:	Unifar, Turkey (and others: Fujisawa Pharmaceutical Co., Japan, Gist-Brocades/DSM, The Netherlands, Novo-Nordisk, Denmark, Pfizer, USA)

5) Product application

- 6-APA is used as an intermediate for the manufacture of semi-synthetic penicillins.

6) Literature

- Cheetham, P. (1995) The application of enzymes in industry, in: Handbook of Enzyme Biotechnology (Wiseman, A. ed.), pp. 493–498, Ellis Harwood, London

- Krämer, D., Boller, C. (1998) personal communication.

- Matsumoto, K. (1993) Production of 6-APA, 7-ACA, and 7-ADCA by immobilized penicillin and cephalosporin amidases, in: Industrial Application of Immobilized Biocatalysts (Tanaka, A, Tosa, T., Kobayashi, T. eds.) pp. 67–88, Marcel Dekker Inc., New York

- Tramper, J. (1996) Chemical versus biochemical conversion: when and how to use biocatalysts, Biotechnol. Bioeng. **52**, 290–295

- Verweij, J, Vroom, E, (1993) Industrial transformations of penicillins and cephalosporins, Rec. Trav. Chim. Pays-Bas, **112** (2) 66–81

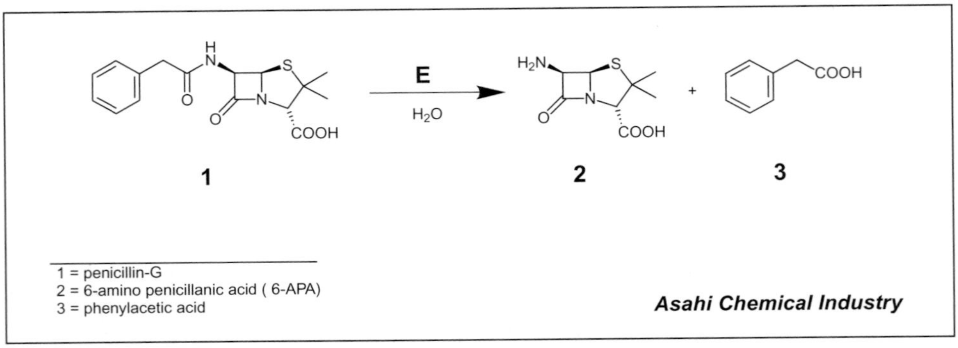

1 = penicillin-G
2 = 6-amino penicillanic acid (6-APA)
3 = phenylacetic acid

Asahi Chemical Industry

Fig. 3.5.1.11 – 1

1) Reaction conditions

[**1**]:	0.3 M, 100 g · L^{-1} [334.39 g · mol^{-1}]
pH:	8.4
T:	30–36 °C
medium:	aqueous
reaction type:	carboxylic acid amide hydrolysis
catalyst:	immobilized enzyme
enzyme:	penicillin amidohydrolase (penicillin acylase, penicillin amidase)
strain:	*Bacillus megaterium*
CAS (enzyme):	[9014-06–6]

2) Remarks

- The enzyme from *Bacillus megaterium* is an exoenzyme and immobilized using an aminated porous polyacrylonitrile (PAN) fiber as solid support.

- The production is carried out in a recirculation reactor consisting of 18 parallel columns with immobilized enzyme. Each column has a volume of 30 L. The circulation of the reaction solution is established with a flow rate off 6,000 L · h^{-1}. One cycle time takes 3 hours.

- The life time of each column is 360 cycles.

- Purification of 6-APA is done by isoelectric precipitation at pH 4.2 with subsequent filtration and washing with methanol.

286

3) Flow scheme

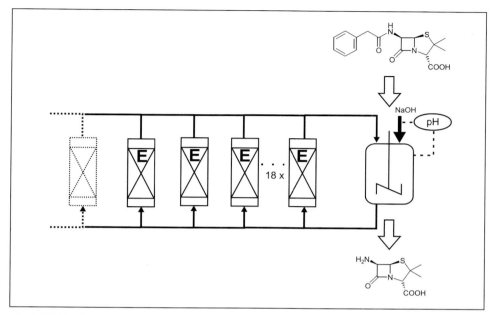

Fig. 3.5.1.11 – 2

4) Process parameters

yield:	86 %
chemical purity:	> 98 %
reactor type:	repetitive batch
reactor volume:	> 500 L
residence time:	3 h
space-time-yield:	$800 \text{ g} \cdot \text{L}^{-1} \cdot \text{d}^{-1}$
down stream processing:	isoelectric precipitation
enzyme activity:	$1,800 \text{ U} \cdot \text{g}^{-1}$(dry)
company:	Toyo Jozo Inc., Japan, Asahi Chemical Industry Co., Ltd., Japan

5) Product application

- 6-APA is used as an intermediate for manufacturing semi-synthetic penicillins.

6) Literature

- Matsumoto, K. (1993) Production of 6-APA, 7-ACA, and 7-ADCA by immobilized penicillin and cephalosporin amidases, in: Industrial Application of Immobilized Biocatalysts (Tanaka, A, Tosa, T., Kobayashi, T. eds.) pp. 67–88, Marcel Dekker Inc., New York

- Tramper, J. (1996) Chemical versus biochemical conversion: when and how to use biocatalysts, Biotechnol. Bioeng. **52**, 290–295

- Verweij, J., Vroom, E. (1993) Industrial transformations of penicillins and cephalosporins, Rec. Trav. Chim. Pays-Bas, **112** (2) 66–81

287

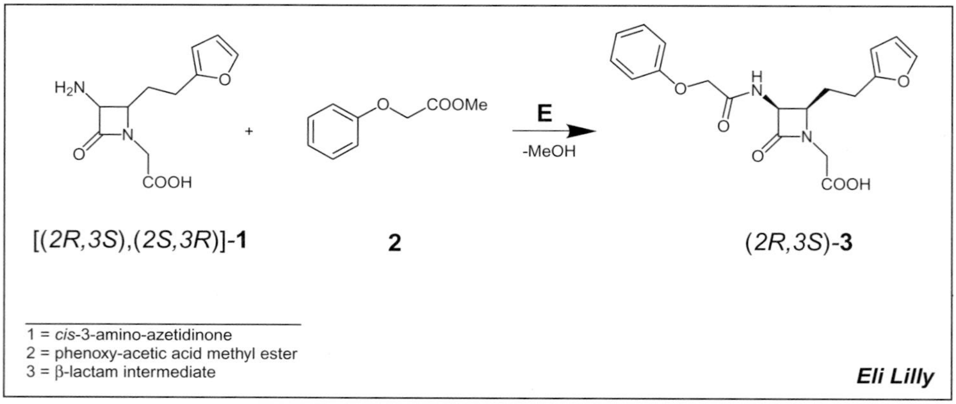

[(2R,3S),(2S,3R)]-**1**   **2**   (2R,3S)-**3**

1 = *cis*-3-amino-azetidinone
2 = phenoxy-acetic acid methyl ester
3 = β-lactam intermediate

Eli Lilly

Fig. 3.5.1.11 – 1

1) Reaction conditions

[**1**]:	0.042 M, 10 g·L^{-1} [238.24 g·mol^{-1}]
[**2**]:	0.063 M, 10.5 g·L^{-1} [166.17 g·mol^{-1}]
pH:	6.0
T:	28 °C
medium:	aqueous
reaction type:	carboxylic acid amide hydrolysis
catalyst:	immobilized enzyme
enzyme:	penicillin amidohydrolase (penicillin acylase, gen G amidase)
strain:	*Escherichia coli*
CAS (enzyme):	[9014-06-06]

2) Remarks

- The chemical resolution of the racemic azetidinone gives low yields.

- It was thought that the pen G amidase would exhibit only a limited substrate spectrum, since it does not hydrolyze the phenoxyacetyl side chain of penicillin V. Nevertheless the Lilly process shows that the pen G amidase acylates the 3-amino function with the methyl ester of phenoxyacetic acid.

- The acylation occurs using methyl phenylacetate (MPA) or methyl phenoxyacetate (MPOA) as the acylating agents.

- The enzyme displays similiar enantioselectivity with MPA or MPOA. It is immobilized on Eupergit (Röhm GmbH, Germany).

3) Flow scheme

Not published.

4) Process parameters

yield:	45 %
ee:	> 99.9 %
reactor type:	batch
down stream processing:	filtration
company:	Eli Lilly, USA

5) Product application

- The (2*R*,3*S*)-azetidinone is a key intermediate in the synthesis of loracarbef a carbacephalosporin antibiotic:

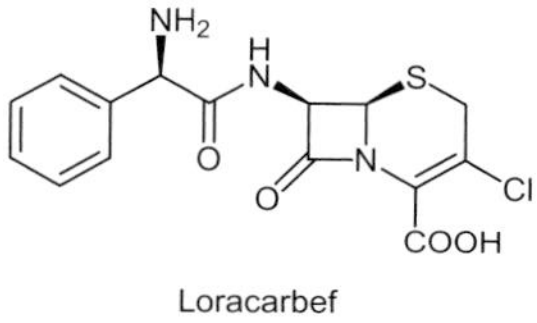

Loracarbef

Fig. 3.5.1.11 – 2

- Loracarbef is a stable analog of the antibiotic cefaclor.

6) Literature

- Zaks, A., Dodds, D.R. (1997) Application of biocatalysis and biotransformations to the synthesis of pharmaceuticals, Drug Discovery Today **2** (12), 513–530

- Zmijewski, M.J., Briggs, B.S., Thompson, A.R., Wright, I.G. (1991) Enantioselective acylation of a beta-lactam intermediate in the synthesis of Loracarbef using penicillin G amidase, Tetrahedron Lett. **32** (13), 1621–1622

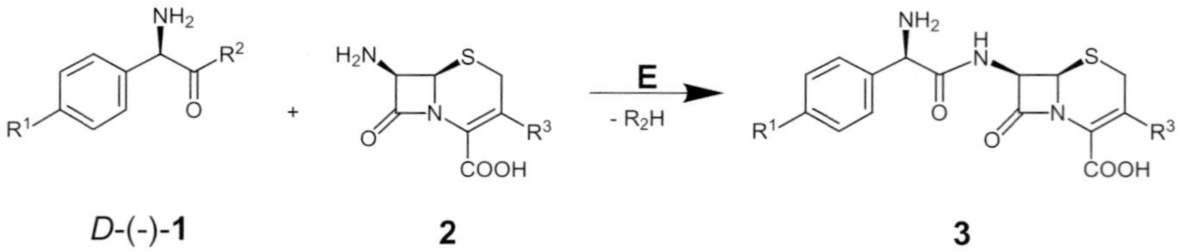

1a = phenylglycineamide (R^1=H, R^2=NH$_2$) = PGA
1b = phenylglycinemethylester (R^1=H, R^2=OMe) = PGM
1c = hydroxyphenylglycineamide (R^1=OH, R^2=NH$_2$) = HPGA
1d = hydroxyphenylglycinemethylester (R^1=OH, R^2=OMe) = HPGM
2a = 7-aminodeacetoxycephalosporanic acid (R^3=Me) = 7-ADCA
2b = 7-aminodeacetoxymethyl-3-chlorocephalosporanic acid (R^3=Cl) = 7-ACCA
3a = cefaclor (R^1=H, R^3=Cl)
3b = cephalexin (R^1=H, R^3=Me)
3c = cefadroxil (R^1=OH, R^3=Me)

Chemferm

Fig. 3.5.1.11 – 1

1) Reaction conditions

medium:	aqueous
reaction type:	carboxylic acid amide hydrolysis
catalyst:	immobilized whole cells or enzyme
enzyme:	penicillin amidase (penicillin amidase, α-acylamino-β-lactam acylhydrolase)
strain:	*Escherichia coli* and others
CAS (enzyme):	[9014-06-06]

2) Remarks

- The established chemical synthesis started from benzaldehyde and included fermentation of penicillin. The process consists of ten steps with a waste stream of 30–40 kg waste per kg product. The waste contained methylene chlorid, other solvents, silylating agents and many by-products from side chain protection and acylating promoters.

- In comparison the chemoenzymatic route needs only six steps including three biocatalytic ones.

- The following figure compares the chemical and chemo-enzymatic routes (Bruggink, 1996):

Penicillin acylase
Escherichia coli

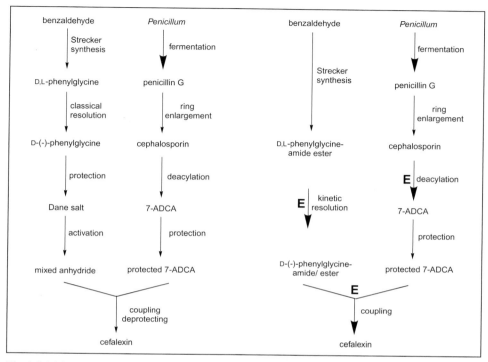

Fig. 3.5.1.11 – 2

- Beside *E. coli*, e.g., the strains *Klyveromyces citrophila* and *Bacillus megaterium* produce penicillin acylase.

- The penicillin acylases do not accept charged amino groups. Therefore phenylglycine itself cannot be used since at a pH value which the carboxyl function is necessarily uncharged the amino group will be charged.

- To reach non-equilibrium concentrations of the product, the substrate must be activated as an ester or amide. By this means the amino group can be partly uncharged at the optimal pH value of the enzyme. In biological systems the activation energy is delivered by ATP.

- The enzyme can be covalently attached on a gelatin-based carrier. Consequently the catalyst becomes water insoluble and can be easily separated from the reaction solution. Additionally the selectivity can be improved by choice of the right carrier composition. By-products resulting from hydrolysis of the educt can be avoided.

- To reach high conversions and high yields the educts and the by-products have to be added in a molar excess.

- Since the characteristics of the shown substances are different for each antibiotic, a special synthetic way had to be established.

- The production of **cefalexin** was the first successful application.

- If an excess of D-(–)PGA is used, surplus or non-converted D-(–)PGA has to be separated and recycled.

- The separation of D-(–)PGA can be done by addition of benzaldehyde and formation of the precipitating Schiff base which can be filtered off subsequent to separation of enzyme and solid products by filtration.

- Also the D-(–)-PGA is almost in-soluble in aqueous solution so that at the end of the reaction three solids (d-(–)-PGA, cefalexin, D-(–)-HPGA) have to be separated.

- One solution of this problem was the use of a special immobilized enzyme, which floats after stopping of the stirrer (Novo Nordisk, Denmark). The reaction solution can be removed from the bottom of the reactor containing the solid products.

- A better technique uses enzyme immobilizates with a defined diameter. At the end the reaction solution and solid substances can be removed from the reactor using a special sieve that is not permeable to the immobilized enzyme. This technique is shown in the flow scheme (Fig. 3.5.1.11–4).

- 7-ACCA (= 7-aminodeacetoxymethyl-3-chlorocephalosporanic acid) can be obtained by ozonolysis and chlorination of 3-methylene cephams. It is the precursor for the synthesis of cefaclor. Cefaclor is unstable at pH values above 6.5 while the solubility of 7-ACCA is very low at pH values under 6.5.

- One strategy, also established for cefalexin too, is to add a complexing agent (β-naphthol). The complex crystallizes and yields above 90 % are possible.

- Using this technique the concentration of product in the reaction mixture is very low, so that succeeding reactions can be suppressed and the mother liquor can be rejected. The disadvantage is the necessity of an organic solvent to yield a two phase system in which the decomplexation at low pH is possible.

- Using the same synthetic pathway alternatively to 7-ADCA and 7-ACCA also 6-APA derivatives can be synthesized:

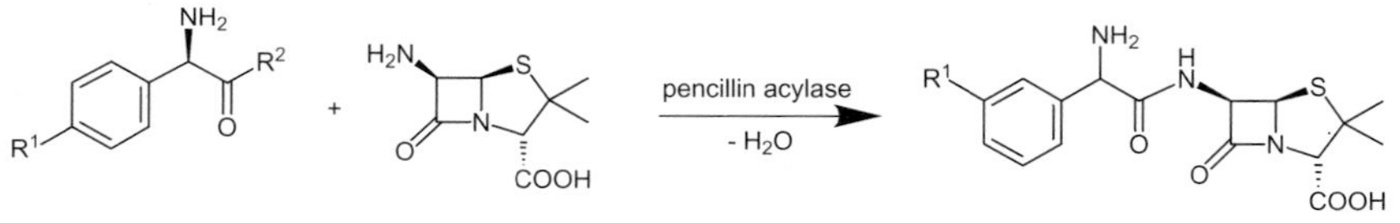

$$D\text{-}(\text{-})\text{-}1 \qquad\qquad 2 \qquad\qquad 3$$

1a = phenylglycineamide (R^1=H, R^2=NH$_2$) = PGA
1b = phenylglycinmethyl ester (R^1=H, R^2=OMe) = PGM
1c = hydroxyphenylglycineamide (R^1=OH, R^2=NH$_2$) = HPGA
1d = hydroxyphenylglycinemethyl ester (R^1=OH, R^2=OMe) = HPGM
2 = 6-APA
3a = ampicillin (R^1=H)
3b = amoxicillin (R^1=OH)

Fig. 3.5.1.11 – 3

- In contrast to cefalexin **ampicillin** has a better solubility, so that by using the recovery strategy of cefalexin too much product would be lost.

- Since the penicillanic acid derivatives are more sensitive towards degradation than cephalosporanic acid derivatives at almost all pH values, the conversion of 6-APA has to be complete and the product has to be recovered rapidly by crystallization (Fig. 3.5.1.11–5).

- The biocatalyst can be retained in the reactor by the sieve method analogous to the cephalexin procedure. Solubilized and precipitated product and D-(–)-PGA crystals are dissolved at acidic pH. Ampicillin is precipitated by adjusting the pH to its isoelectric point.

- In a similiar manner **amoxicillin** can be recovered. The advantage in this case is the low solubility of the product under reaction conditions so that hydrolysis of the product is suppressed since it precipitates at first. A semi-continuous reactor system with high substrate concentrations can be applied. This is shown in the last flow scheme (Fig. 3.5.1.11–6).

3) Flow scheme

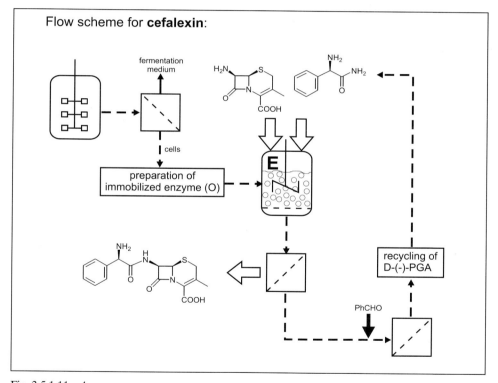

Flow scheme for **cefalexin**:

Fig. 3.5.1.11 – 4

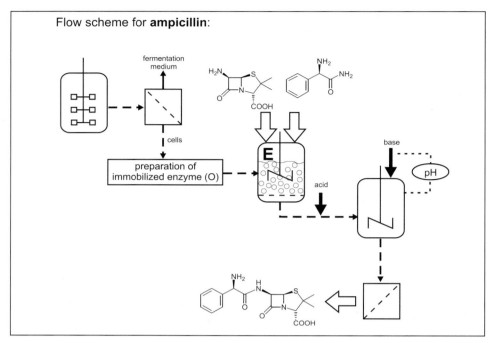

Fig. 3.5.1.11 – 5

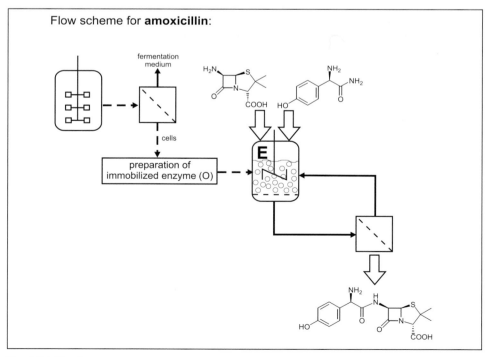

Fig. 3.5.1.11 – 6

4) Process parameters

yield:	> 90 %
selectivity:	> 95 %
ee:	> 99 %
reactor type:	repetitive batch (see flow schemes)
capacity:	2000 t · a^{-1} (worldwide)
down stream processing:	filtration
company:	Chemferm, The Netherlands (joint venture of Gist-Brocades and DSM Deretril, both The Netherlands); others

5) Product application

- The products are β-lactam antibiotics.

6) Literature

- Bruggink, A. (1996) Biocatalysis and process integration in the synthesis of semi-synthetic antibiotics, CHIMIA **50**, 431–432

- Bruggink, A., Roos, E.C., Vroom, E. de (1998) Penicillin acylase in the industrial production of β-lactam antibiotics, Org. Proc. Res. Dev. **2**, 128–133

- Clausen, K. (1995) Method for the preparation of certain β-lactam antibiotics, Gist-Brocades N. V., US 5,470,717

- Hernandez-Justiz, O., Fernandez-Lafuente, R., Terrini, Guisan, J. M. (1998) Use of aqueous two-phase systems for *in situ* extraction of water soluble antibiotics during their synthesis by enzymes immobilized on porous supports, Biotech. Bioeng. **59**, 1, 73–79

1 = *N*-acetyl-D,L-3-(4-thiazolyl)alanine
2 = 3-(4-thiazolyl)alanine
3 = *N*-acetyl-D-3-(4-thiazolyl)alanine

Chiroscience Ltd.

Fig. 3.5.1.14 – 1

1) Reaction conditions

pH:	7.0
medium:	aqueous
reaction type:	carboxylic acid amide hydrolysis
catalyst:	immobilized enzyme
enzyme:	*N*-acetyl-L-amino-acid amidohydrolase (aminoacylase)
strain:	*Aspergillus niger*
CAS (enzyme):	[9012–37–7]

2) Remarks

- The whole process consisting of the following reaction steps is performed in one vessel:

1 = 4-chloromethylthiazole hydrochloride
2 = *N*-acetyl-3-(4-thiazolyl)alanine
3 = amino-3-(4-thiazolyl)acid
4 = *t*-butoxycarbonyl-L-thiazoylalanine (BOC-protected thiazoylalanine)

Fig. 3.5.1.14 – 2

- The hydrolysis is performed by slow addition of hydroxide solution, maintaining the mixture at pH 9–10; at this pH the initial product undergoes decarboxylation; this causes a reduction in pH and, by careful control of the base addition, the resulting mixture can be kept approximately neutral.

- The neutral solution can be directly used for the biotransformation.

- The product is extracted directly from the aqueous reaction mixture with methyl-*tert*-butyl ether (MTBE) and is of high enantiomeric purity.

- The D-isomer can be recycled via an oxazolinone that tautomerizes to the enol:

1 = *N*-acetyl-3-(4-thiazolyl)alanine
2 = oxazolinone (azlactone)
3 = enol

Fig. 3.5.1.14 – 3

- Since the enzyme is immobilized and employed in a packed bed reactor it can be used several times resulting in a protein-free solution after down-stream processing.

- The process can be applied to other unnatural alanine derivatives, e.g.:

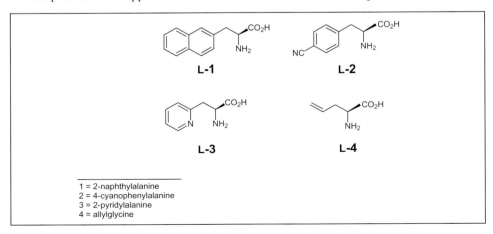

1 = 2-naphthylalanine
2 = 4-cyanophenylalanine
3 = 2-pyridylalanine
4 = allylglycine

Fig. 3.5.1.14 – 4

3) Flow scheme

Not published.

4) Process parameters

ee: >99 %
reactor type: plug-flow reactor
capacity: several kg
down stream processing: solvent extraction
company: Chiroscience Ltd., UK

5) Product application

- The product is used as a component of antihypertensive inhibitors of the enzyme renin, where it acts as a mimic of histidine.

- Two examples of L-thiazolylalanine containing inhibitors are:

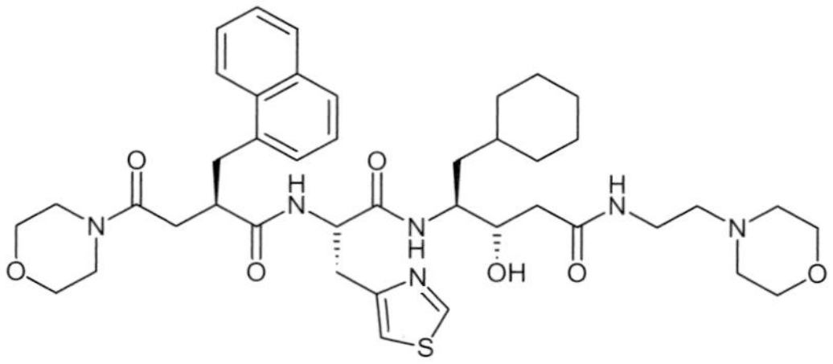

L-1

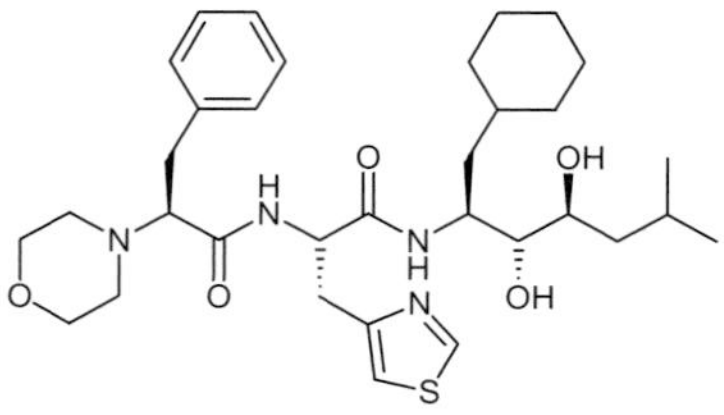

L-2

1 = thiazolalanine containing inhibitor (Sankyo)
2 = thiazolalanine containing inhibitor (Abbott)

Fig. 3.5.1.14 – 5

6) Literature

- Hsiao, C.-N., Leanna, M.R., Bhagavatula, L., Lara, E., Zydowsky, T.M. (1990) Synthesis of *N*-(*tert*-butoxycarbonyl)-3-(4-thiazolyl)-L-alanine, Synth. Commun. **20**, 3507–3517

- Miyazawa, T., Iwanaga, H., Ueji, S., Yamada, T., Kuwata, S. (1989) Porcine pancreatic lipase catalyzed enantioselective hydrolysis of esters of *N*-protected unusual amino acids, Chem. Lett. 2219–2222

- Nishi, T., Saito, F., Nagahori, H., Kataoka, M., Morisawa, Y. (1990) Syntheses and biological activities of renin inhibitors containing statine analogues, Chem. Pharm. Bull. **38**, 103–109

- Rosenberg, S.H., Spina, K.P., Woods, K.W., Polakowski, J., Martin, D.L. (1993) Studies directed toward the design of orally active renin inhibitors. 1. Some factors influencing the absorption of small peptides, J. Med. Chem. **36** (4), 449–459

- Taylor, S.J.C., Mc Cague, R. (1997) Dynamic resolution of an oxazolinone by lipase biocatalysis: Synthesis of (*S*)-*tert*-leucine, in: Chirality In Industry II (Collins, A. N., Sheldrake, G. N., Crosby, J., eds.), pp. 194–200, John Wiley & Sons, New York

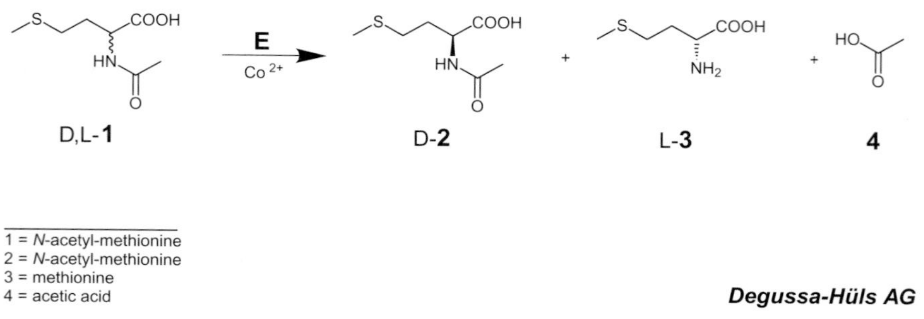

Aminoacylase
Aspergillus oryzae

1 = *N*-acetyl-methionine
2 = *N*-acetyl-methionine
3 = methionine
4 = acetic acid

Degussa-Hüls AG

Fig. 3.5.1.14 – 1

1) Reaction conditions

[**1**]:	0.6 M, 97.96 g · L^{-1} [163.27 g · mol^{-1}]
[**Co^{2+}**]:	0.5 · 10^{-3} M, 0.029 g · L^{-1} [58.93 g · mol^{-1}] (activator)
pH:	7.0
T:	37 °C
medium:	aqueous
reaction type:	hydrolysis
catalyst:	solubilized enzyme
enzyme:	*N*-acyl-L-amino-acid amidohydrolase (aminoacylase, acylase 1)
strain:	*Aspergillus oryzae*
CAS (enzyme):	[9012–37–7]

2) Remarks

- The *N*-acetyl-D,L-amino acid precursors are conveniently accessible through acetylation of D,L-amino acids with acetyl chloride or acetic anhydride under alkine conditions in a Schotten-Baumann reaction.

- As effector Co^{2+} is added to increase the operational stability of the acylase.

- The unconverted acetyl-D-methionine is racemized by acetic anhydride under alkaline conditions and the racemic acetyl-D,L-methionine is recycled.

- The racemization can also be carried out in a molten bath or by racemase.

- Product recovery of L-methionine is achieved by crystallization, because L-methionine is much less soluble than the substrate.

- A polyamide ultrafiltration membrane with a cutoff of 10,000 dalton is used.

- Several proteinogenic and non-proteinogenic amino acids are produced in the same way by Degussa-Hüls:

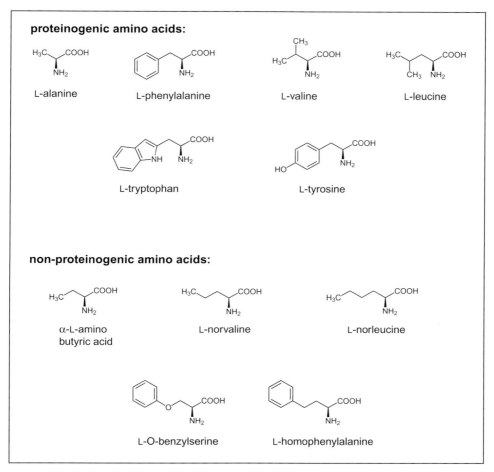

proteinogenic amino acids:

L-alanine

L-phenylalanine

L-valine

L-leucine

L-tryptophan

L-tyrosine

non-proteinogenic amino acids:

α-L-amino
butyric acid

L-norvaline

L-norleucine

L-O-benzylserine

L-homophenylalanine

Fig. 3.5.1.14 – 2

3) Flow scheme

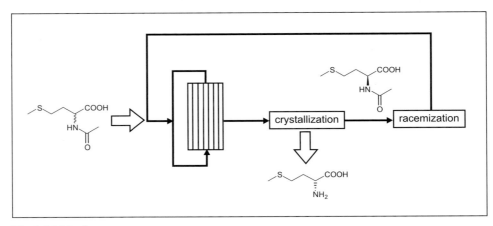

Fig. 3.5.1.14 – 3

4) Process parameters

yield:	80 %
ee:	99.5 %
reactor type:	cstr, UF-membrane reactor
capacity:	$200 \; t \cdot a^{-1}$
residence time:	2.9 h
space-time-yield:	$592 \; g \cdot L^{-1} \cdot d^{-1}$
down stream processing:	crystallization
enzyme activity:	$1,067 \; U \cdot g^{-1}_{protein}$
enzyme consumption:	$2,067 \; U \cdot kg^{-1}$
enzyme supplier:	Amano Corp., Nagoya, Japan
company:	Degussa-Hüls AG, Germany

5) Product application

- L-Amino acids are used for parenteral nutrition (infusion solutions), feed and food additives, cosmetics, pesticides and as intermediates for pharmaceuticals as well as chiral synthons for organic synthesis.

6) Literature

- Bommarius, A.S., Drauz, K., Klenk, H., Wandrey, C. (1992) Operational stability of enzymes – acylase-catalyzed resolution of *N*-acetyl amino acids to enantiomerically pure L-amino acids, Ann. N. Y. Acad. Sci. **672**, 126–136

- Chenault, H.K., Dahmer, J., Whitesides, G.M. (1989) Kinetic resolution of unnatural and rarely occuring amino acids: enantioselective hydrolysis of *N*-acyl amino acids catalyzed by acylase, J. Am. Chem. Soc. **111**, 6354–6364

- Leuchtenberger, W., Karrenbauer, M., Plöcker, U. (1984) Scale-up of an enzyme membrane reactor process for the manufacture of L-enantiomeric compounds, Enzyme Engineering 7, Ann. N. Y. Acad. Sci. **434**, 78

- Takahashi, T., Izumi, O., Hatano, K. (1989) Acetylamino acid racemase, production and use thereof, Takeda Chemical Industries, Ltd., EP 0 304 021 A2

- Wandrey, C., Flaschel, E. (1979) Process development and economic aspects in enzyme engineering. Acylase L-methionine system. In: Advances in Biochemical Engineering 12 (Ghose, T.K., Fiechter A., Blakebrough, N., eds.),pp. 147–218, Springer-Verlag, Berlin

- Wandrey, C., Wichmann, R., Leuchtenberger, W., Kula, M.R. (1981) Process for the continuous enzymatic change of water soluble α-ketocarboxylic acids into the corresponding amino acids, Degussa AG, US 4,304,858

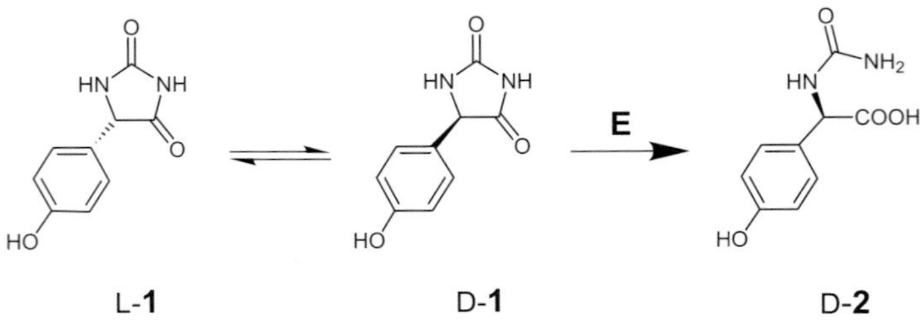

L-**1**   D-**1**   D-**2**

1 = 5-(p-hydroxybenzyl)-hydantoin
2 = D-*N*-carbamoyl amino acid

Kanegafuchi

Fig. 3.5.2.2 – 1

1) Reaction conditions

pH:	8.0
medium:	aqueous
reaction type:	carboxylic acid amide hydrolysis
catalyst:	immobilized whole cells
enzyme:	5,6-dihydropyridine amidohydrolase (dihydropyrimidinase, hydantoin peptidase, hydantoinase)
strain:	*Bacillus brevis*
CAS (enzyme):	[9030–74–4]

2) Remarks

- The hydantoinase is D-specific.

- The unreacted L-hydantoins are readily racemized under the conditions of enzymatic hydrolysis.

- Quantitative conversion is achieved because of the *in situ* racemization.

- L-Specific hydantoinases are also known.

- This process enables the stereospecific preparation of various amino acids, such as D-tryptophan, D-phenylalanine, D-valine, D-alanine and D-methionine.

- The carbamoyl group can be removed by use of a carbamoylase (EC 3.5.1.77, see page 308) or alternatively by chemical treatment with sodium nitrite:

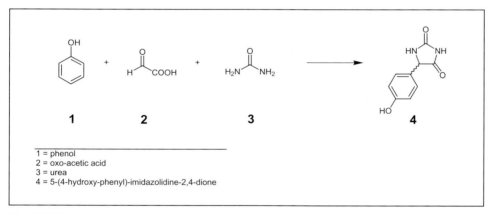

Fig. 3.5.2.2 – 2

- Racemic hydantoins are synthesized starting from phenol derivatives, glyoxylic acid and urea via the Mannich condensation:

1 = phenol
2 = oxo-acetic acid
3 = urea
4 = 5-(4-hydroxy-phenyl)-imidazolidine-2,4-dione

Fig. 3.5.2.2 – 3

- Re-use of cells by immobilization cells is possible.
- Several other companies have developed patented processes leading to D-hydroxyphenyl glycine (Ajinomoto, DSM, Snamprogetti, Ricordati and others, for example see page 308).

3) Flow scheme

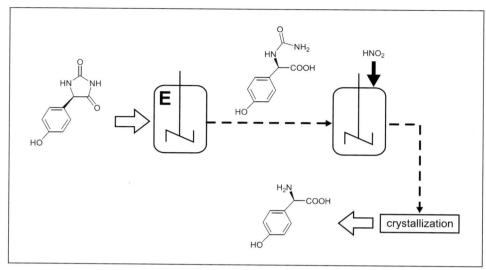

Fig. 3.5.2.2 – 4

4) Process parameters

conversion: 100 %
capacity: $200 \, t \cdot a^{-1}$
enzyme activity: $17.14 \, U \cdot g^{-1}$
start-up date: 1983
production site: Jurong Industrial Estate, Singapore
company: Kanegafuchi, Japan

5) Product application

- D-*p*-Hydroxyphenyl glycine is a key raw material for the semisynthetic penicillins, ampicillin and amoxycillin.

- It is also used in photographic developers.

6) Literature

- Cheetham, P.S.J. (1994) Case studies in applied biocatalysis, in: Applied Biocatalysis (Cabral, J.M.S., Best, D., Boross, L., Tramper, J.; eds.) pp. 68–70, Harwood Academic Publishers, Chur, Switzerland

- Crosby, J. (1991) Synthesis of optically active compounds: a large scale perspective, Tetrahedron **47**, 4789–4846

306

- Ikemi, M. (1994) Industrial chemicals: enzymic transformation by recombinant microbes, Bioprocess Technology **19**, pp. 797–813

- Schmidt-Kastner, G., Egerer, P. (1984) Amino acids and peptides. In: Biotechnology, Vol. 6a (Kieslich, K., ed.), pp. 387–419, Verlag Chemie, Weinheim

- Sheldon, R.A. (1993) Chirotechnology, Marcel Dekker Inc., New York

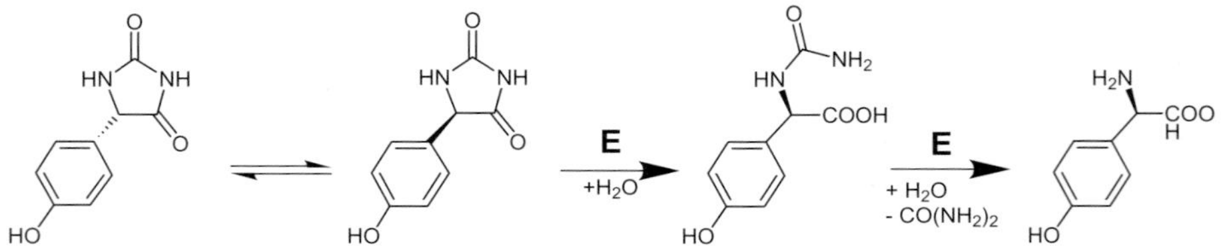

Fig. 3.5.2.4 / 3.5.1.77 – 1

1) Reaction conditions

[**1**]:	>0.21 M, 40 g · L^{-1} [192.17 g · mol^{-1}]
pH:	8.0
T:	38 °C
medium:	aqueous
reaction type:	carboxylic acid amide hydrolysis
catalyst:	whole cells
enzyme:	L-5-carboxymethylhydantoin amidohydrolase / *N*-carbamoyl-D-amino acid hydrolase
strain:	*Pseudomonas* sp.
CAS (enzyme):	[9025–14–3] / –

2) Remarks

- The strain contains both enzymes (hydantoinase and carbamoylase).

- The cell biomass is used directly in the process. Alternatively the enzymes may be extracted and immobilized.

- Instead of hydroxyphenylglycine the non-hydroxylated phenylglycine can be produced in the same way.

- Kanegafuchi also uses the hydantoinase to hydrolyze D,L-5-(*p*-hydroxyphenyl)hydantoin (see page 304). The second step, the hydrolysis of the D-*N*-carbamoyl-D-hydroxyphenyl glycine is performed either chemically with HNO$_2$ or enzymatically with carbamoylase.

3) Flow scheme

Not published.

Hydantoinase / Carbamoylase
Pseudomonas sp.

EC 3.5.2.4 / 3.5.1.77

4) Process parameters

conversion:	95 %
yield:	80 %
selectivity:	84 %
chemical purity:	98.5 %
reactor type:	batch
reactor volume:	15,000 L
capacity:	pilot scale
residence time:	15 – 20 h
space-time-yield:	$57.6 \, g \cdot L^{-1} \cdot h^{-1}$
down stream processing:	concentration and crystallization
enzyme supplier:	Captive Production, India
production site:	Baroda, India
company:	Dr. Vig Medicaments, India

5) Product application

- The product is used as a chemical intermediate in the synthesis of amoxycillin/cefadroxil through 'dane'-salt formation and of ampicillin/cefalexin through acid chloride formation with 6-APA/7-ADCA respectively.

6) Literature

- Vig, C. (1997) personal communication

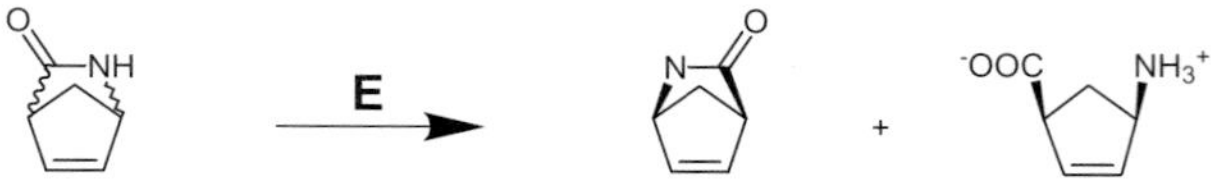

1

(+)-1

(-)-2

1 = 2-azabicylo[2.2.1]hept-5-en-3-one (γ-lactam)
2 = 4-amino-cyclopent-2-enecarboxylic acid

Chiroscience Ltd.

Fig. 3.5.2.6 – 1

1) Reaction conditions

[1]:	1.83 M, 200 g · L^{-1} [109.05 g · mol^{-1}]
T:	70 °C
medium:	aqueous
reaction type:	carboxylic acid amide hydrolysis
catalyst:	immobilized enzyme
enzyme:	β-lactamhydrolase (β-lactamase)
strain:	*Aureobacterium* sp.
CAS (enzyme):	[9001–74–5]

2) Remarks

- The enzyme is purified (ammonium sulphate fractionation and anion-exchange chromatography) and immobilized on a glutaraldehyde-activated solid support.

- The biotransformation is operated in a batch reaction wherein an aqueous solution of the racemic lactam is cycled through the fixed bed of immobilized enzyme.

- The reaction is complete when the (–)-enantiomer is hydrolyzed completely (E-value > 7,000).

- The stability of the enzyme is improved by immobilization so that a nearly steady-state production can be achieved for more than 6 months.

- To separate the amino acid from the reaction mixtur the solution has to be slurried with acetone. The (+)-lactam stays in solution while the amino acid crystallizes and can be removed by filtration.

- In comparison to the chemical resolution in which the lactam is converted to the salt of the amino acid, the biotransformation produces the neutral, zwitter-ionic form of the acid.

3) Flow scheme

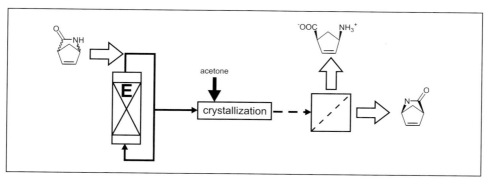

Fig. 3.5.2.6 – 2

4) Process parameters

reactor type: plug-flow reactor with circulation (batch)
down stream processing: extraction, filtration
company: Chiroscience Ltd., UK

5) Product application

- 4-Amino-cyclopent-2-enecarboxylic acid is used as building block in the synthesis of carbocyclic nucleosides of the natural configuration.

6) Literature

- Taylor, S.C. (1997) Development of an immobilised lactamase resolution process for the (+)-γ-lactam and the ring-opened (–)-amino acid, in: Chirality In Industry II (Collins, A.N., Sheldrake, G.N., Crosby, J., eds.), pp. 187–188, John Wiley & Sons, New York

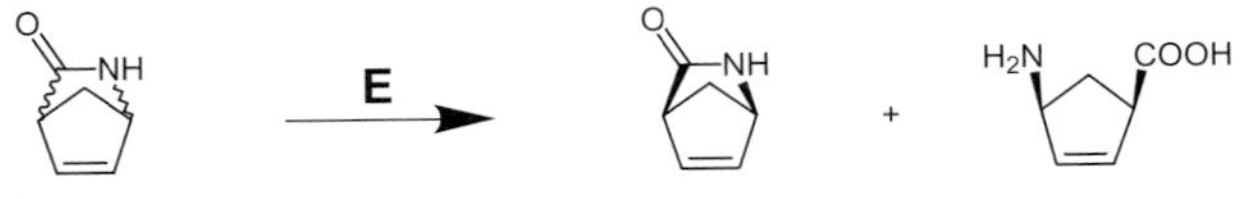

β-Lactamase
Pseudomonas solanacearum

1 = 2-azabicylo[2.2.1]hept-5-en-3-one (γ-lactam)
2 = 4-amino-cyclopent-2-enecarboxylic acid

Chiroscience Ltd.

Fig. 3.5.2.6 – 1

1) Reaction conditions

[1]:	0.92 M, 100 g·L⁻¹ [109.13 g·mol⁻¹]

[1]: 0.92 M, 100 g$\cdot$L^{-1} [109.13 g$\cdot$mol^{-1}]
pH: 7.0
T: 25 °C
medium: aqueous
reaction type: carboxylic acid amide hydrolysis
catalyst: suspended whole cells
enzyme: β-lactamhydrolase (β-lactamase)
strain: *Pseudomonas solanacearum*
CAS (enzyme):   [9001–74–5]

2) Remarks

- The cells are grown by fermentation and the resulting cell mass is frozen and stored.

- The frozen cell mass is added in its crude form to the aqueous reaction solution. The advantage is that the biotransformation can be performed at different sites.

- The reaction rate is increased by some cell lysis, which is caused by the freeze-thaw process, liberating the enzyme.

- The whole cells are used since the isolated microbial enzyme is only of limited stability.

- Control of pH is not required, since the product 2-amino-cyclopent-2-ene carboxylic acid acts as its own buffer.

- The lactam is extracted with dichloromethane.

- The amino acid can be recovered from the aqueous phase as the hydrochloride by acidification (HCl) and evaporation of the water.

3) Flow scheme

Not published.

4) Process parameters

conversion:	55 %
yield:	45 %
selectivity:	82%
ee:	> 98 %
reactor type:	batch
capacity:	ton scale
residence time:	3 h
down stream processing:	harvesting cells, extraction of product with dichloromethane
production site:	Cambridge, UK
company:	Chiroscience Ltd., UK

5) Product application

- The lactam can be converted to carbovir:

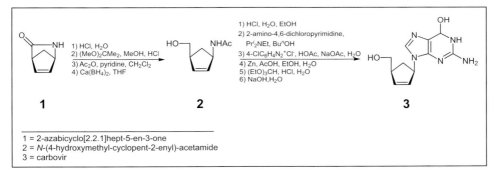

1) HCl, H₂O, EtOH
2) 2-amino-4,6-dichloropyrimidine, Pr$^i_2$NEt, Bu^{n}OH
3) 4-ClC₆H₄N₂$^+$Cl$^-$, HOAc, NaOAc, H₂O
4) Zn, AcOH, EtOH, H₂O
5) (EtO)₃CH, HCl, H₂O
6) NaOH, H₂O

1) HCl, H₂O
2) (MeO)₂CMe₂, MeOH, HCl
3) Ac₂O, pyridine, CH₂Cl₂
4) Ca(BH₄)₂, THF

1   **2**   **3**

1 = 2-azabicyclo[2.2.1]hept-5-en-3-one
2 = *N*-(4-hydroxymethyl-cyclopent-2-enyl)-acetamide
3 = carbovir

Fig. 3.5.2.6 – 2

- Carbovir is a potent and selective inhibitor of HIV-1.

- Its ability to inhibit infectivity and replication of the virus in human T-cell lines at concentrations 200–400 fold below toxic levels makes carbovir a promising candidate for development as a potential antiretroviral agent.

6) Literature

- Taylor, S.J.C., Sutherland, A.G., Lee, C., Wisdom, R., Thomas, S., Roberts, S. M., Evans, C. (1990) Chemoenzymatic synthesis of (–)-Carbovir utilizing a whole cell catalyzed resolution of 2-aza-bicyclo[2.2.1]hept-5-en-3-one, J. Chem. Soc., Chem. Commun., 1120–1121

- Taylor, S.J.C., Mc Cague, R. (1997) Resolution of the carbocyclic nucleoside synthon 2-aza-bicyclo[2.2.1]hept-5-en-3-one with lactamses, in: Chirality In Industry II (Collins, A.N., Sheldrake, G.N. and Crosby, J., eds.), pp. 184–190, John Wiley & Sons, New York

313

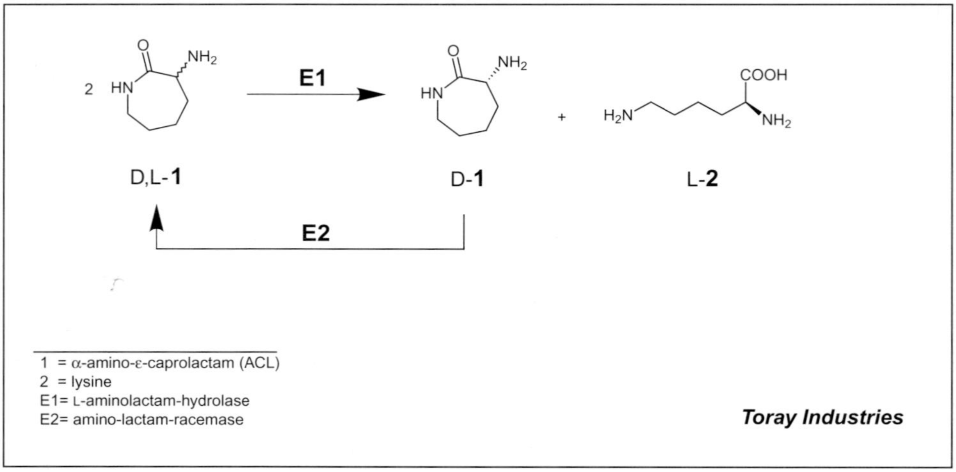

1 = α-amino-ε-caprolactam (ACL)
2 = lysine
E1= L-aminolactam-hydrolase
E2= amino-lactam-racemase

Toray Industries

Fig. 3.5.2.11 / 5.1.1.15 – 1

1) Reaction conditions

[**1**]:	0.78 M, 100 g · L^{-1} [128.09 g · mol^{-1}]
pH:	8.0–9.0
T:	40 °C
medium:	aqueous
reaction type:	carboxylic acid amide hydrolysis / racemization
catalyst:	suspended whole cells
enzyme:	**E1** L-lysine-1,6-lactam hydrolase (L-lysine lactamase)/
	E2 2-aminohexano-6-lactam racemase (α-amino-ε-caprolactam racemase)
strain:	*Cryptococcus laurentii / Achromobacter obae*

2) Remarks

● The lactamase and racemase are fortunately active at the same pH, so that they can be used in one reactor.

● The combination of cells from *Candida humicola* and *Alcaligenes faecalis* is used alternatively.

● The reaction starts from cyclohexene leading to the oxin as intermediate. The Beckmann rearrangement gives the caprolactam for the enzymatic steps.

3) Flow scheme

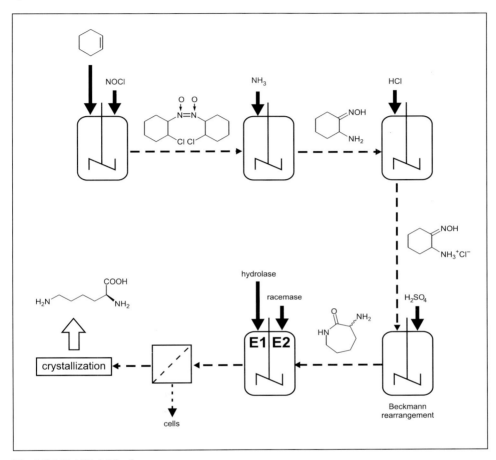

Fig. 3.5.2.11 / 5.1.1.15 – 2

4) Process parameters

ee:	> 99.5 %
reactor type:	batch
capacity:	4,000 t · a^{-1}
residence time:	25 h
down stream processing:	crystallization
start-up date:	1970
production site:	Japan
company:	Toray Industries, Japan

5) Product application

• As nutrient and food supplement.

6) Literature

- Crosby, J. (1991) Synthesis of optically active compounds: a large scale perspective, Tetrahedron **47**, 4789–4846

- Sheldon, R.A. (1993) Chirotechnology, Marcel Dekker Inc., New York

- Atkinson, B., Mavituna, F. (1991) Biochemical Engineering and Biotechnology Handbook, Stockton Press, New York

- Schmidt-Kastner, G., Egerer, P. (1984) Amino acids and peptides. In: Biotechnology, Vol. 6a, (Kieslich, K., ed.) pp. 387–419, Verlag Chemie, Weinheim

- Wiseman, A. (1995) Handbook of Enzyme and Biotechnology, Ellis Horwood, Chichester

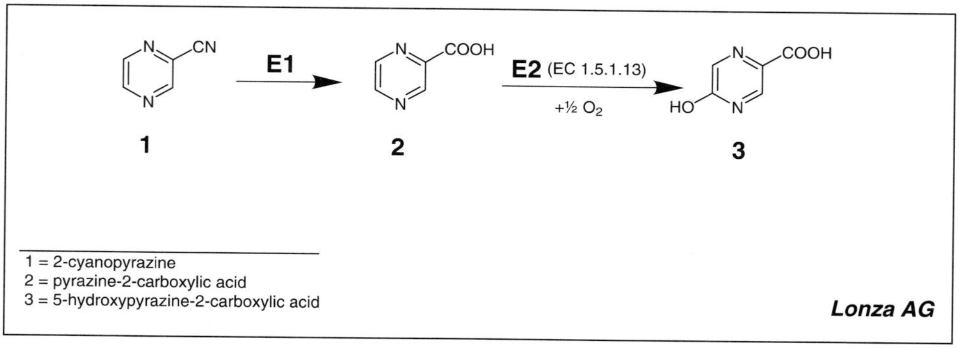

Nitrilase / Hydroxylase
Agrobacterium sp.

1 = 2-cyanopyrazine
2 = pyrazine-2-carboxylic acid
3 = 5-hydroxypyrazine-2-carboxylic acid

Lonza AG

Fig. 3.5.5.1 / 1.5.1.13 – 1

1) Reaction conditions

[1]:	0.29 M, 30 g·L^{-1} [105.1 g·mol^{-1}]
pH:	6.0–8.0
T:	15 °C – 45 °C
medium:	aqueous
reaction type:	*N*-bond hydrolysis (nitrile hydrolysis)
catalyst:	suspended, living whole cells
enzyme:	**E1**: nitrilase (nitrile aminohydrolase) and **E2**: hydroxylase (nicotinate dehydrogenase) (EC 1.5.1.13)
strain:	*Agrobacterium* sp. DSM 6336 (**E1** + **E2**)
CAS (enzyme):	**E1**: [9024–90–2] / **E2**: [9059-03–4]

2) Remarks

- In contrast to the biotransformation the chemical synthesis of 5-substituted pyrazine-2-carboxylic acid leads to a mixture of 5- and 6-substituted pyrazinecarboxylic acids and requires multiple steps.

- Although the reaction sequence of growth and biotransformation are pretty similar, the cells grown on 2-cyanopyridine are much more active due to an optimized expression of the hydroxylase. Additonally the degradation pathway in case of 2-cyanopyrazine can be stopped.

317

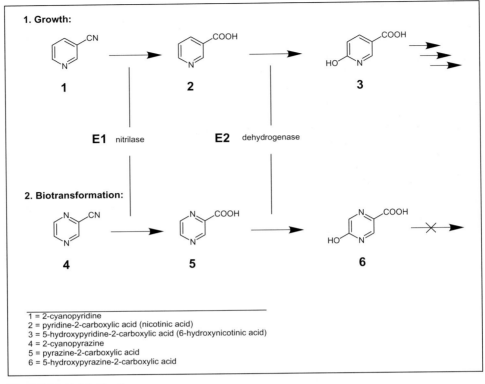

1. Growth:

1   2   3

E1 nitrilase   **E2** dehydrogenase

2. Biotransformation:

4   5   6

1 = 2-cyanopyridine
2 = pyridine-2-carboxylic acid (nicotinic acid)
3 = 5-hydroxypyridine-2-carboxylic acid (6-hydroxynicotinic acid)
4 = 2-cyanopyrazine
5 = pyrazine-2-carboxylic acid
6 = 5-hydroxypyrazine-2-carboxylic acid

Fig. 3.5.5.1 / 1.5.1.13 – 2

- Since a high pO_2 induces cell death, the oxygen partial pressure is limited to 90 mbar during cell growth. During the transformation it is reduced to 50 mbar.

- High oxygen partial pressure does also introduce the autooxidative dimerisation to di- and trihydroxylated pyridines.

- The biomass is separated by ultrafiltration (cutoff 10 kDa) after the biotransformation.

- 5-Hydroxypyrazine-2-carboxylic acid is precipitated from the permeate by acidification with sulfuric acid to pH 2.5.

- The lowered practical yield of 80 % in comparison to the analytical yield of 95 % is due to repeated precipitation during product isolation.

3) Flow scheme

Not published.

4) Process parameters

yield:	95 % (analytical) / 80 % (isolated)
selectivity:	high
chemical purity:	> 99 % (isolated)
reactor type:	batch
reactor volume:	20 L
capacity:	multi kg
space-time-yield:	$36 \; g \cdot L^{-1} \cdot d^{-1}$
down stream processing:	ultrafiltration and precipitation
company:	Lonza AG, Switzerland

5) Product application

- Versatile building block for the synthesis of new antitubercular agents, e.g. 5-chloro-pyrazine-2-carboxylic acid esters:

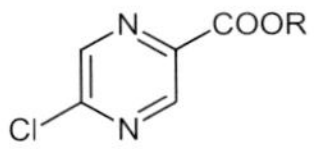

6) Literature

- Kiener, A. (1994) Mikrobiologisches Verfahren zur Herstellung von 5-Hydroxy-2-pyrazincarbonsäure, Lonza AG, EP 578137 A1

- Kiener, A., Roduit, J.-P., Tschech, A., Tinschert, A. Heinzmann, K. (1994) Regiospecific enzymatic hydroxylations of pyrazinecarboxylic acid and a practical synthesis of 5-chloropyrazine-2-carboxylic acid, Synlett **10**, 814–816

- Roduit, J.-P. (1993) Verfahren zur Herstellung von Carbonsäurechloriden aromatischer Stickstoff-Heterocyclen, EP 0561421 A1

- Wieser, M., Heinzmann, K., Kiener A. (1997) Bioconversion of 2-cyanopyrazine to 5-hydroxypyrazine-2-carboxylic acid with *Agrobacterium sp.* DSM 6336, Appl. Microbiol. Biotechnol. **48**, 174–180

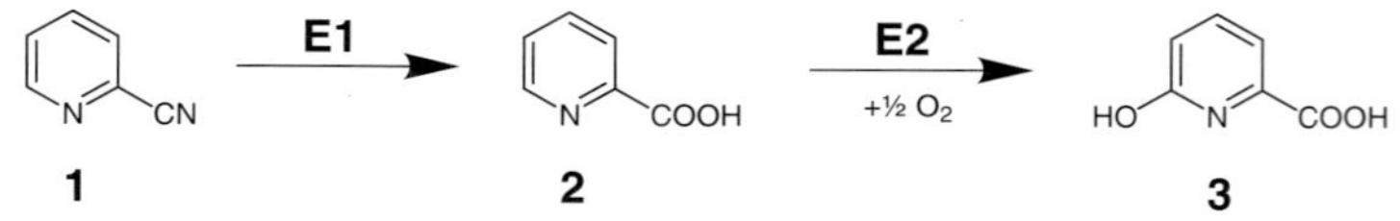

1 = 2-cyanopyridine
2 = picolinic acid
3 = 6-hydroxypicolinic acid

Lonza AG

Fig. 3.5.5.1 / 1.5.1.13 – 1

1) Reaction conditions

[2]:	> 0.016 M, > 2 g·L^{-1} [123.11 g·mol^{-1}]
[3]:	0.543 M, 75 g·L^{-1} [138 g·mol^{-1}]
pH:	7.0
T:	30 °C
medium:	aqueous
reaction type:	N-bond hydrolysis (nitrile hydrolysis)
catalyst:	suspended whole cells
enzyme:	**E1**: nitrile aminohydrolase (nitrilase), **E2**: hydroxylase (EC 1.5.1.13)
strain:	*Alcaligenes faecalis* DSM 6335 (**E1** + **E2**)
CAS (enzyme):	[9024–90–2]

2) Remarks

- The cells are grown on sodium fumarate (32 g·L^{-1}) as C-source up to an OD$_{650}$ of 16.

- Resting cells are employed as biocatalyst under aerobic conditions.

- Since 2-cyanopyridine is a solid at room temperature, the substrate solution is heated to 50 °C in order to add a liquid substrate.

- Since the intermediate picolinic acid is inhibiting the reaction to 6-hydroxypicolinic acid, the educt 2-cyanopyridine has to be maintained at a low concentration level. Therefore, the educt 2-cyanopyridine is continuously fed to the reaction solution. The feed rate is controlled by the on-line-analysis concentration of the intermediate picolinic acid.

- The reaction rate for the first step (**E1**) is 2.5 times faster than for the second step (**E2**).

- At the end of the biotransformation no intermediate can be found.

- To precipitate the product, the reaction solution is filtrated to remove the cells and the pH is adjusted to 2.5 using sulfuric acid at a temperature of 60 °C.

- Under strictly anaerobic conditions the hydroxylase activity is suppressed enabling the production of picolinic acid using similar conditions with a space-time yield of 138 g·L^{-1}·d^{-1} (chemical purity: 86 %). To prevent hydroxylation caused by oxygen, the reactor is aerated with nitrogen during the whole biotransformation.

- This enzyme acts on a wide range of aromatic nitriles and also on some aliphatic nitriles.

3) Flow scheme

Not published.

4) Process parameters

conversion:	100 %
yield:	87 %
reactor type:	fed-batch
capacity:	$1 \, t \cdot a^{-1}$
residence time:	31 h
space-time-yield:	$58 \, g \cdot L^{-1} \cdot d^{-1}$
down stream processing:	precipitation
company:	Lonza AG, Switzerland

5) Product application

- The product is used as an intermediate for pharmaceuticals, e.g. 2-oxypyrimidin, and herbicides.

6) Literature

- Fischer, E., Heß, K., Stahlschmidt, A. (1912) Verwandlung der Dihydrofurandicarbonsäure in Oxypyridincarbonsäure, Ber. Dtsch. Chem. Ges. **45**, 2456–2467

- Foster, C.J., Gilkerson, T., Stocker R. (1991) Herbical carboxamide derivatives, Shell International Research Maatschappij B.V., EP 0447004 A2

- Glöckler, R., Roduit, J.-P.(1996) Industrial bioprocesses for the production of substituted aromatic heterocycles, Chimia **50**, 413–415

- Kiener, A.(1992) Mikrobiologisches Verfahren zur Herstellung von 6-Hydroxypicolinsäure, Lonza AG, EP 0504818 A2

- Kiener, A. Glöckler, R., Heinzmann, K. (1993) Preparation of 6-O-oxo-1,6-dihydropyridine-2-carboxylic acid by microbial hydroxylation of pyridine-2-carboxylic acid, J. Chem. Soc. Perkin Trans. I **11**, 1201–1202

- Kiener, A., Roduit, J.-P., Glöckner, R. (1996) Mikrobiologisches Verfahren zur Herstellung von heteroaromatischen Carbonsäuren mittels Mikroorganismen der Gattung *Alcaligenes*, Lonza AG, EP 0747486 A1

- Petersen, M., Kiener, A. (1999) Biocatalysis – Preparation and functionalization of *N*-heterocycles, Green Chem. **2**, 99–106

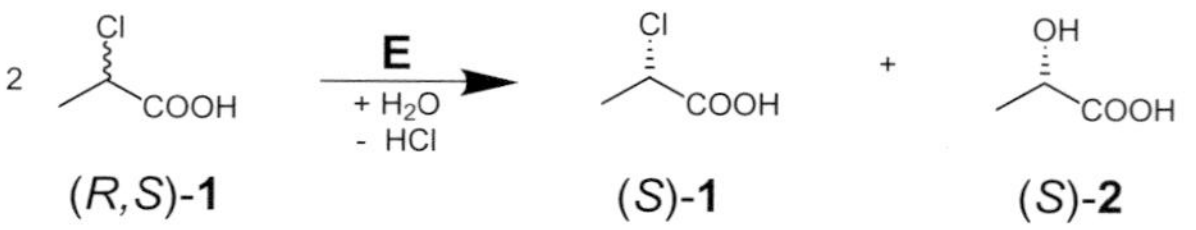

Dehalogenase
Pseudomonas putida

$$2 \quad \overset{\underset{\displaystyle\mathrm{Cl}}{|}}{\diagup}\text{COOH} \quad \xrightarrow[\substack{-\ \mathrm{HCl}}]{\substack{\mathbf{E}\\ +\ H_2O}} \quad \overset{\underset{\displaystyle\mathrm{Cl}}{\vdots}}{\diagup}\text{COOH} \quad + \quad \overset{\underset{\displaystyle\mathrm{OH}}{\vdots}}{\diagup}\text{COOH}$$

(R,S)-**1** $\qquad\qquad\qquad$ (S)-**1** $\qquad\qquad$ (S)-**2**

1 = 2-chloropropionic acid
2 = lactic acid

Zeneca Life Science Molecules

Fig. 3.8.1.2 – 1

1) Reaction conditions

[**1**]:	> 0.1 M, 10.9 g·L^{-1} [108.52 g·mol^{-1}]
pH:	7.4
T:	30 °C
medium:	aqueous
reaction type:	C-halide hydrolysis
catalyst:	suspended whole cells
enzyme:	2-haloacid halidohydrolase (2-haloacid dehalogenase)
strain:	*Pseudomonas putida*
CAS (enzyme):	[37289–39–7]

2) Remarks

- *Pseudomonas putida* is a robust organism that grows rapidly on cheap carbon sources even in presence of high concentrations.

- Since at alkaline pH 2-chloropropionic acid racemizes rapidly it is important that the dehalogenase retains specific activity at neutral pH.

- The K_M-value is very low so that the reaction rate remains high with typical end concentrations of the substrate of 1 g·L^{-1}.

- Since the stability of the cells is poor, the steps of fermentation and biotransformation are carried out separately.

- Usually an immobilization method would be used but in this case a special cell drying method is applied, where only 10 % enzyme activity is lost and the solid biocatalyst can be stored for over 12 month without deactivation.

- The main steps for the production of (S)-2-chloropropionic acid are:
  1. Continuous fermentation
  2. Biocatalyst preparation and drying
  3. Biocatalyst storage
  4. Fed-batch biotransformation
  5. Biocatalyst separation
  6. Solvent extraction of product

- The gene responsible for the intracellular enzyme production could be determined, cloned and overexpressed in an *E. coli* strain. Several manipulations of the strain increased the production level by about 20 times.

3) Flow scheme

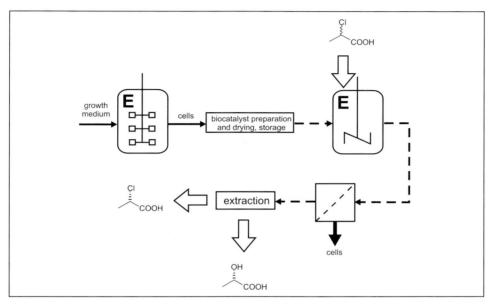

Fig. 3.8.1.2 – 2

4) Process parameters

yield:	> 42 %
ee:	99 %
chemical purity:	99 %
reactor type:	fed-batch
capacity:	2,000 t · a^{-1}
down stream processing:	solvent extraction / distillation
enzyme activity:	22 U · mg^{-1} (Protein)
enzyme supplier:	in-house manufacture
start-up date:	1988
production site:	Billingham and Huddersfield, England
company:	Zeneca Life Science Molecules, England

5) Product application

- Applications include the synthesis of other optically active compounds such as pharmaceuticals.

- The main application of (S)-2-chloropropionic acid is in the synthesis of optically active phenoxypropionic acid herbicides:

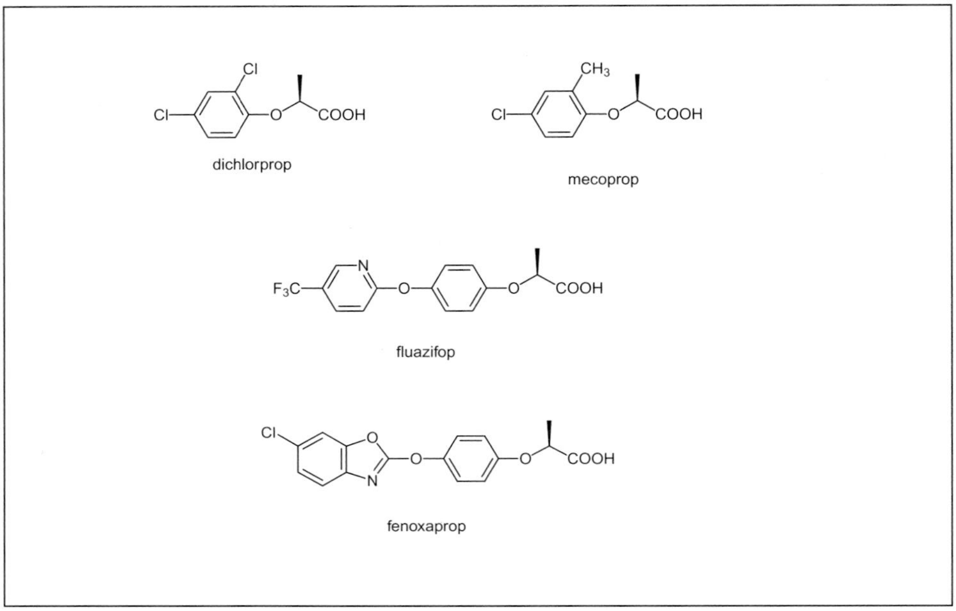

Fig. 3.8.1.2 – 3

- Many of these herbicides were sold as racemates but because of environmental considerations, the switch to the active enantiomer is preferred, thereby doubling capacities in existing plants and leading to cost reduction through raw material savings.

- Other chiral products synthesized by this dehalogenase technique are shown in the following figure:

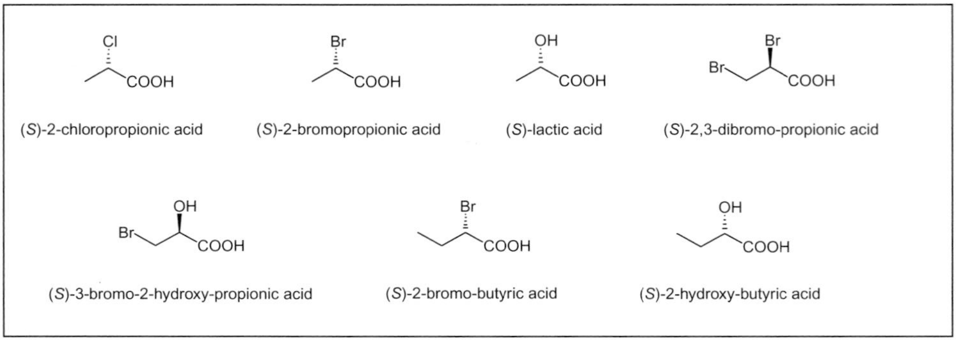

Fig. 3.8.1.2 – 4

324

6) Literature

- Taylor, S.C. (1988) D-2-Haloalkanoic acid halidohydrolase, Zeneca Limited, Imperial Chemical Industries PLC, US 4758518

- Liddell, J.M., Greer, W. (1995) Biocatalysts, Zeneca Limited, EP 366303

- Smith, J.M., Harrison, K., Colby, J. (1990) Purification and characterization of D-2-haloacid dehalogenase from *Pseudomonas putida* strain AJ1/23, J. Gen. Microbiol. **136**, 881–886

- Barth, P.T., Bolton, L., Thomson, J. C. (1992) Cloning and partial sequencing of an operon encoding two *Pseudomonas putida* haloalkanoate dehalogenases of opposite stereospecificity, J. Bacteriol. **174**, 2612–2619

- Taylor, S.C. (1997) (*S*)-2-Chloropropanoic acid: developments in its industrial manufacture. In: Chirality In Industry II (Collins, A.N., Sheldrake, G.N., Crosby, J., eds.) pp. 207–223, John Wiley & Sons

- Cheetham, P.S.J. (1994) Case studies in applied biocatalysis, in: Applied Biocatalysis (Cabral, J.M.S., Best, D., Boross, L., Tramper, J.; eds.) p. 70, Chur, Switzerland, Harwood Academic Publishers

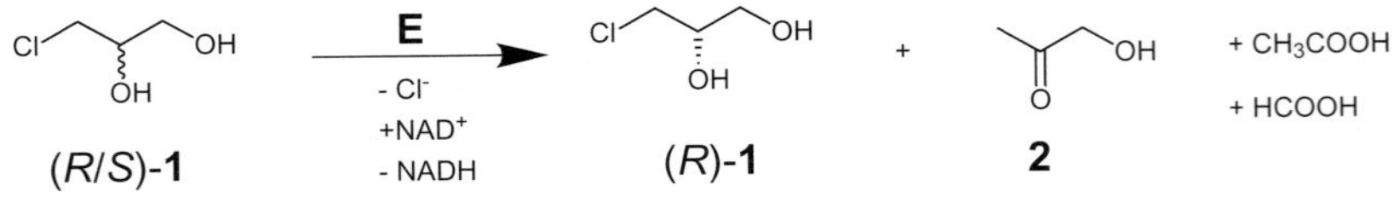

1 = 3-chloropropane-1,2-diol
2 = 2-oxo-propionaldehyde

Daiso Co. Ltd

Fig. 3.8.1.5 – 1

1) Reaction conditions

[**1**]:	< 0.9 M, < 100 g $\cdot$ L^{-1} [110.55 g $\cdot$ mol^{-1}]
pH:	6.8
T:	30 °C
medium:	aqueous
reaction type:	hydrolysis of halide bonds
catalyst:	immobilized whole cells
enzyme:	1-haloalkane halidohydrolasehalohydrin (haloalkane dehalogenase, HDDase,)
strain:	*Alcaligenes* sp. or *Pseudomonas* sp. depending on substrate
CAS (enzyme):	[95990–29–7]

2) Remarks

- The racemic starting materials are economically produced from propylene in the petrochemical industry. Therefore the limited yield of 50 % due to the kinetic resolution is economically tolerable.

- This microbial resolution can be carried out in an inorganic medium using bacteria that can assimilate (*R*)- or (*S*)- 2,3-dichloro-1-propanol and (*R*)- or (*S*)-3-chloro-1,2-propanediol as the sole source of carbon.

- Some related optically active halohydrins can also be resolved.

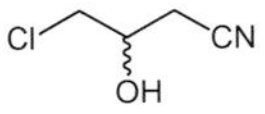

4-chloro-3-hydroxy-butyronitrile 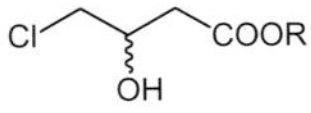   4-chloro-3-hydroxy-butyroester 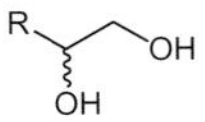   1,2-diol

- The enzyme shows a broad substrate specificity for alcohols, but not for acids.

326

- The whole cells are immobilized in calcium alginate.

- The fermenter is aerated with air at 20 L · min⁻¹.

- Glycidol can be easily synthesized from 3-chloropropane-1,2-diol:

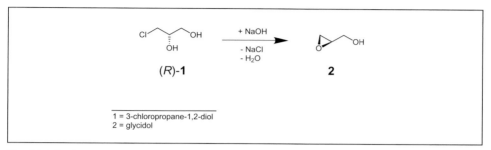

1 = 3-chloropropane-1,2-diol
2 = glycidol

Fig. 3.8.1.5 – 2

3) Flow scheme

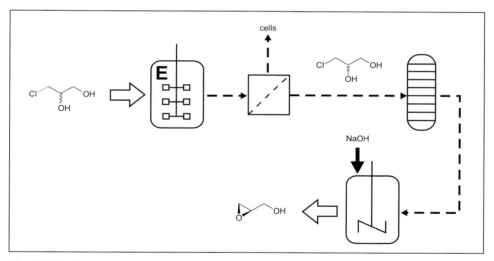

Fig. 3.8.1.5 – 3

4) Process parameters

conversion:	50 %
yield:	max. 50 %
ee:	99.5–100 %
chemical purity:	> 98 %
reactor type:	batch
reactor volume:	25,000 L
residence time:	48 h – 60 h
down stream processing:	extraction, distillation
enzyme activity:	HDDase: (in the presence of NAD^+ and O_2) 3.5 $U \cdot g^{-1}$ (*Alcanigenes* sp.) and 4.8 $U \cdot g^{-1}$ (*Pseudomonas* sp.)
start-up date:	1994
production site:	Amagasaki City, Matsuyama, Japan
company:	Daiso Co. Ltd, Japan

5) Product application

- The products are chiral synthons for various chiral pharmaceuticals, agrochemicals and ferro-electroliquid crystals.

- Other possible products:

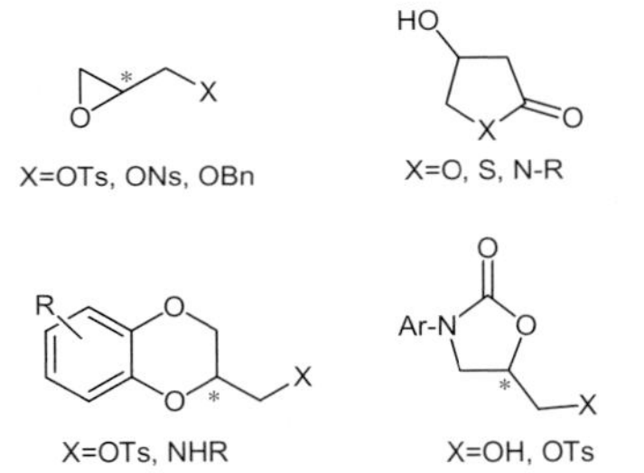

X=OTs, ONs, OBn

X=O, S, N-R

X=OTs, NHR

X=OH, OTs

Fig. 3.8.1.5 – 4

6) Literature

- Crosby, J., (1991) Synthesis of optically active compounds: A large scale perspective, Tetrahedron **47**, 4789–4846

- Kasai, N., Suzuki, T., Furukawa, Y. (1998) Chiral epoxides and halohydrins: Their preparation and synthetic application, J. Mol. Catal. B: Enzymatic **4**, 237–252

- Kasai, N., Tsujimura, K., Unoura, K., Suzuki, T. (1992) Preparation of (*S*)-2,3-dichloro-1-propanol by *Pseudomonas* species and its use in the synthesis of (*R*)-epichlorohydrin, J. Indust. Microbiol. **9**, 97–101

- Suzuki, T., Kasai, N. (1991) A novel method for the generation of (*R*)-and (*S*)-3-chloro-1,2-propanediol by stereospecific dehalogenating bacteria and their use in the preparation of (*R*)-and (*S*)-glycidol, Bioorg. Med. Chem. Lett. **1**, 343

- Suzuki, T., Kasai, N., Yamamoto, R., Minamiura, N. (1992) Isolation of a bacterium assimilating (*R*)-3-chloro-1,2-propanediol and production of (*S*)-3-chloro-1,2-propanediol using microbial resolution, J. Ferment. Bioeng. **73**, 443–448

- Suzuki, T., Kasai, N., Yamamoto, R., Minamiura, N. (1993) Production of highly optically active (*R*)-3-chloro-1,2-propanediol using a bacterium assimilating the (*S*)-isomer, Appl. Microbiol. Biotechnol. **40**, 273–278

- Suzuki, T., Kasai, N., Yamamoto, R., Minamiura, N. (1994) A novel enzymatic dehalogenation of (*R*)-3-chloro-1,2-propanediol in *Alicaligenes sp.* DS-S-7G, Appl. Microbiol. Biotechnol. **42**, 270–279

- Suzuki, T., Kasai, N., Minamiura, N. (1994) A novel generation of optically active 1,2-diols from the racemates by using halohydrin dehydro-dehalogenase, Tetrahedron: Asymmetry **5**, 239–246

- Suzuki, T., Kasai, N., Yamamoto, R., Minamiura, N. (1994) Microbial production of optically active 1,2-diols using resting cells of *Alcaligenes sp.* DS-S-7G, J. Ferment. Bioeng. **78**, 194–196

- Suzuki, T., Idogaki, H., Kasai, N. (1996) A novel generation of optically active ethyl-4-chloro-3-hydroxybutyrate as a C4 chiral building unit using microbial dechlorination, Tetrahedron: Asymmetry **7**, 3109–3112

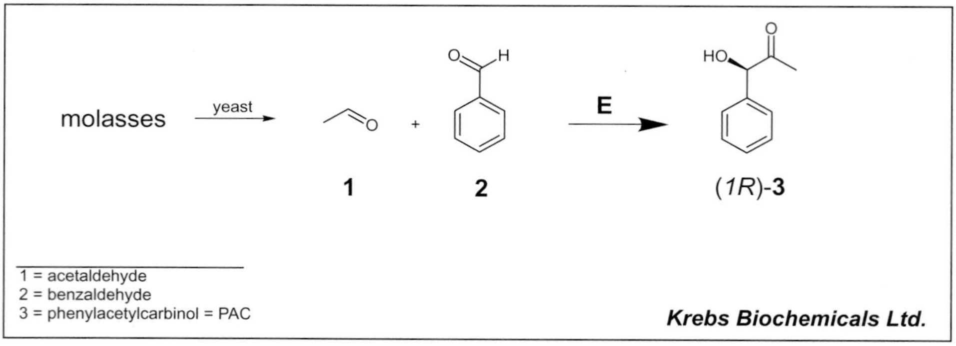

1 = acetaldehyde
2 = benzaldehyde
3 = phenylacetylcarbinol = PAC

Krebs Biochemicals Ltd.

Fig. 4.1.1.1 – 1

1) Reaction conditions

[1]:	0.022 M, 3.3 g $\cdot$ L^{-1} [150.17 g $\cdot$ mol^{-1}]
medium:	aqueous
reaction type:	carboligation
catalyst:	whole cells
enzyme:	2-oxo-acid carboxy-lyase (α-ketoacid carboxylase, pyruvate decarboxylase)
strain:	*Saccharomyces cerevisiae*
CAS (enzyme):	[9001-04-01]

2) Remarks

- Phenylacetylcarbinol production from benzaldehyde by yeast is also operated on a large scale by Knoll (BASF, Germany) and Malladi Drugs (India).

- Phenylacetylcarbinol is chemically converted in a two-step process to D-pseudoephedrine:

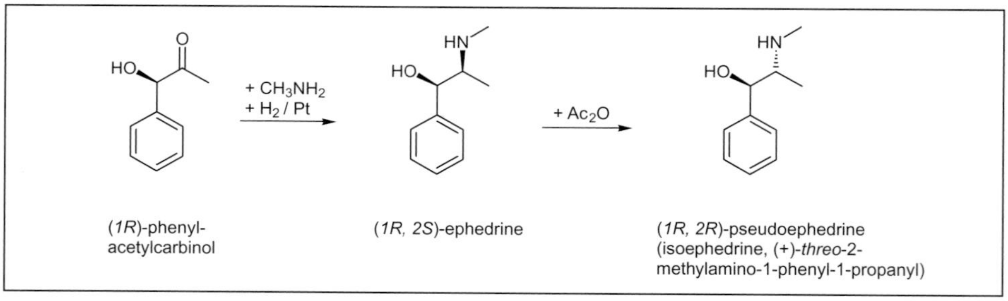

(1R)-phenyl-
acetylcarbinol

(1R, 2S)-ephedrine

(1R, 2R)-pseudoephedrine
(isoephedrine, (+)-*threo*-2-
methylamino-1-phenyl-1-propanyl)

Fig. 4.1.1.1 – 2

3) Flow scheme

Not published.

4) Process parameters

reactor type:	batch
reactor volume:	30,000 L
capacity:	120 t $\cdot$ a^{-1}
enzyme consumption:	6 g$_{yeast}$ $\cdot$ kg$_{PAC}$
production site:	Hyderabad, India
company:	Krebs Biochemicals Ltd., India

5) Product application

- Ephedrine and pseudoephedrine are used for the treatment of asthma, hay fever and as a bronchodilating agent and decongestant.

- Ephedrine is also an important building block for the side chain of taxol that is used against breast cancer:

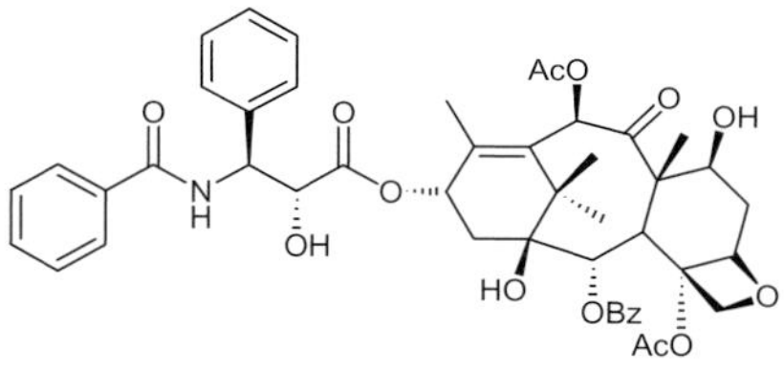

taxol

Fig. 4.1.1.1 – 3

6) Literature

- Dr. Ravi R.T., Krebs Biochemicals Ltd., personal communication

- Cheetham, P.S.J. (1994) Case studies in applied biocatalysis – from ideas to products, in: Applied Biocatalysis (Cabral, J. M. S., Best, D., Boross, L., Tramper, J., eds.), pp. 87–89, Harwood Academic Publishers, USA

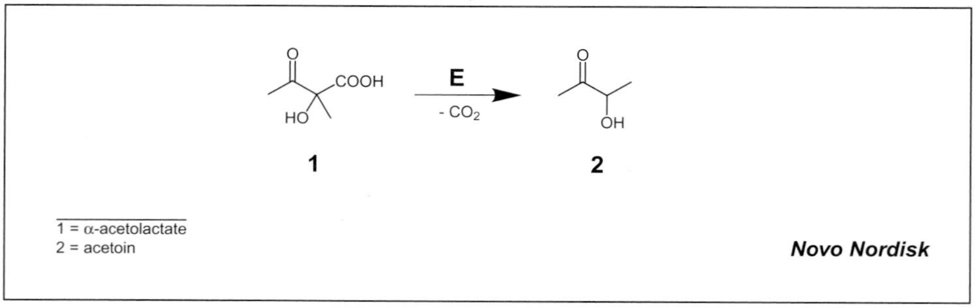

Acetolactate decarboxylase
Bacillus brevis

1 = α-acetolactate
2 = acetoin

Novo Nordisk

Fig. 4.1.1.5 – 1

1) Reaction conditions

[2]:	$< 1.13\ \mu M$, $< 0.1\ mg \cdot L^{-1}$ $[88.11\ g \cdot mol^{-1}]$
pH:	< 4
T:	13°C
medium:	aqueous
reaction type:	decarboxylation
catalyst:	solubilized enzyme
enzyme:	(*S*)-2-hydroxy-2-methyl-3-oxobutanoate carboxy-lyase (α-acetolactate decarboxylase)
strain:	*Bacillus brevis*
CAS (enzyme):	[9025-02–9]

2) Remarks

- During beer fermentation diacetyl is formed by a non-enzymatic oxidative decarboxylation of α-acetolactate.

- Diacetyl has a very low flavor threshhold.

- The problem is the slow reaction rate for the conversion of α-acetolactate to diacetyl with subsequent conversion of diacetyl to acetoin. The addition of acetolactate decarboxylase allows the bypassing of the slow oxidation step:

332

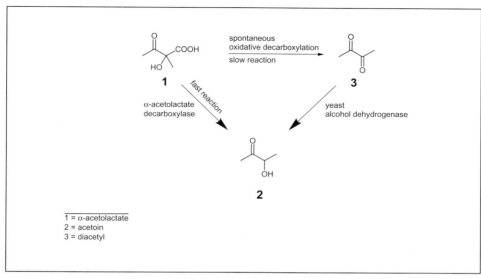

Fig. 4.1.1.5 – 2

1 = α-acetolactate
2 = acetoin
3 = diacetyl

- The enzyme can be activated at low pH values by addition of glutaraldehyde (about 0.05 %) which intermolecularly cross-links the active dimer, which otherwise dissociates under acidic conditions.

3) Flow scheme

Not published.

4) Process parameters

residence time: > 6 d
company: Novo Nordisk, Denmark (and others)

5) Product application

- The process can be applied for beer fermentation procedures, in which diacetyl is produced as a by-product and has to be eliminated.

6) Literature

- Pedersen, S., Lange, N.K., Nissen, A.M. (1995) Novel industrial enzyme applications, Ann. N. Y. Acad. Sci. **750**, 376–390

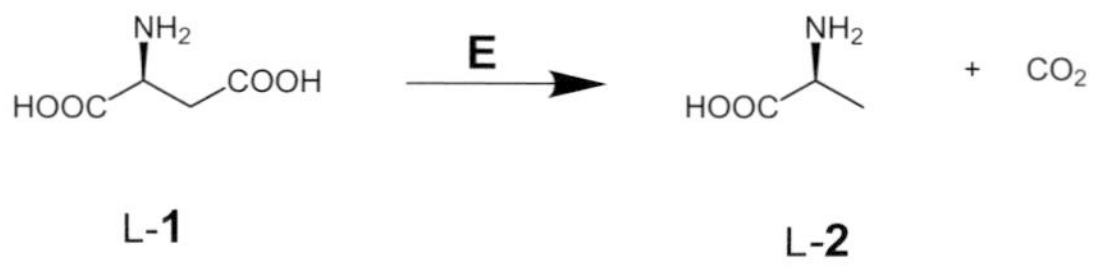

Aspartate β-decarboxylase
Pseudomonas dacunhae

L-1　　　　　　　　　**L-2**

1 = aspartic acid
2 = alanine

Tanabe Seiyaku Co., Ltd.

Fig. 4.1.1.12 – 1

1) Reaction conditions

[1]:	2.5 M, 332.75 g · L^{-1} [133.1 g · mol^{-1}]
pH:	6.2
T:	37 °C
medium:	aqueous
reaction type:	decarboxylation
catalyst:	immobilized whole cells
enzyme:	L-aspartate 4-decarboxylase (L-aspartate β-decarboxylase)
strain:	*Pseudomonas dacunhae*
CAS (enzyme):	[9024–57–1]

2) Remarks

- L-Alanine is produced industrially by Tanabe Seiyaku, Japan, since 1965 via a batch process with L-aspartate β-decarboxylase from *Pseudomonas dacunhae.*

- To improve the productivity a continuous production was established in 1982. Here the formation of carbon dioxide was the main problem in comparison to the catalyst stability and the microbial enzyme activity. The production of CO_2 occurs stoichiometricaly (nearly 50 L of CO_2 for each liter of reaction mixture with 2 M aspartate). The consequences are difficulties in obtaining a plug-flow condition in fixed bed reactors and the pH shift that takes places due to formation of CO_2. Therefore a pressurized fixed bed reactor with 10 bar was designed.

- The enzyme stability is not affected by the elevated pressure.

- The main side reaction, the formation of L-malic acid, can be completely avoided.

- To improve the yield of L-alanine the alanine racemase and fumarase activities can be destroyed by acid treatment of the microorganisms (pH 4.75, 30 °C). The L-aspartate β-decarboxylase activity is stabilized by the addition of pyruvate and pyridoxal phosphate.

- The process is often combined with the aspartase catalyzed synthesis of L-aspartic acid from fumarate (see page 381) in a two step biotransformation (Fig. 4.1.1.12 – 4). The main reason for the separation in two reactors is the difference in pH optimum for the two enzymes (aspartase from *E. coli*: pH 8.5, L-aspartate β-decarboxylase: pH 6.0). This is the first commercialized system of a sequential enzyme reaction using two kinds of immobilized microbial cells:

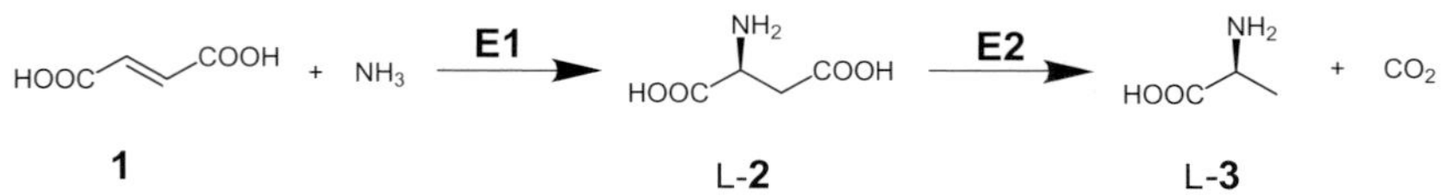

1 = fumaric acid
2 = aspartic acid
3 = alanine
E1 = L-aspartase ammonia-lyase (fumaric aminase)
E2 = aspartate §-decarboxylase (L-aspartate 4-decarboxylase)

Fig. 4.1.1.12 – 2

- In this combination L-alanine can efficiently be produced by co-immobilization of *E. coli* and *Pseudomonas dacunhae* cells.
- If D,L-aspartic acid is used as a substrate for the reaction, L-aspartic acid is converted to L-alanine and D-aspartic acid remains unchanged in one resolution step. Both products can be separated after crystallization by addition of sulfuric acid. The continuous variant of the L-alanine and D-aspartic acid production is commercially in operation since 1988 (Fig. 4.1.1.12 – 5).

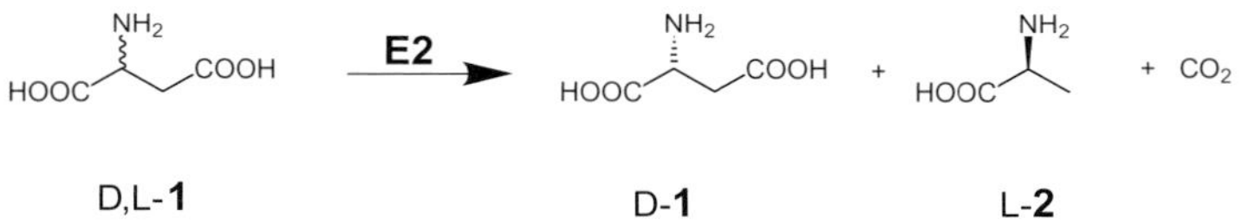

1 = aspartic acid
2 = alanine

Fig. 4.1.1.12 – 3

Aspartate β-decarboxylase
Pseudomonas dacunhae

3) Flow scheme

- Production of L-aspartic acid from fumarate in a two step biotransformation:

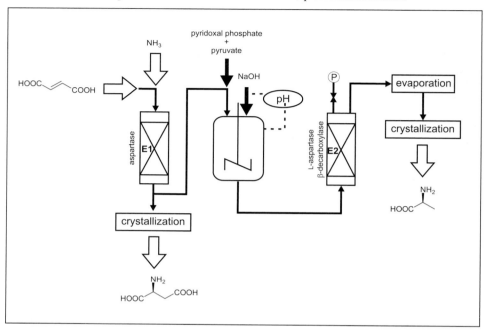

Fig. 4.1.1.12 – 4

- Production of L-alanine and D-aspartic acid:

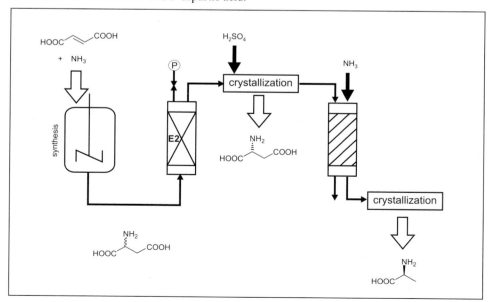

Fig. 4.1.1.12 – 5

4) Process parameters

conversion:	99 %
yield:	86 % after down stream processing
reactor type:	plug flow reactor
reactor volume:	1,000 L (each fixed bed column)
capacity:	114 t $\cdot$ a^{-1} D-aspartic acid; 61 t $\cdot$ a^{-1} L-alanine
residence time:	11 h
space-time-yield:	317 g $\cdot$ L^{-1} $\cdot$ d^{-1} D-aspartic acid; 170 g $\cdot$ L^{-1} $\cdot$ d^{-1} L-alanine
down stream processing:	crystallization
start-up date:	1982
production site:	Japan
company:	Tanabe Seiyaku Co., Ltd., Japan

5) Product application

- The products are used in infusion solutions and as food additives.

- D-Aspartic acid is an intermediate for the synthetic penicillin apoxycillin:

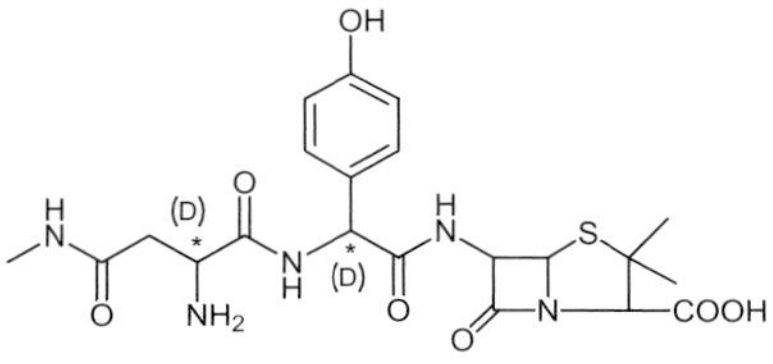

apoxicillin

Fig. 4.1.1.12 – 6

6) Literature

- Chibata, L., Tosa, T., Shibatani, T. (1992) The industrial production of optically active compounds by immobilized biocatalysts, in: Chirality in Industry (Collins, A.N., Sheldrake, G.N., Crosby, J., eds.) pp. 351–370, John Wiley & Sons Ltd, New York

- Furui, M., Yamashita, K. (1983) Pressurized reaction method for continuous production of L-Alanine by immobilized *Pseudomonas dacunhae* cells, J. Ferment. Technol. **61**, 587–591

- Schmidt-Kastner, G., Egerer, P. (1984) Amino acids and peptides, in: Biotechnology, Vol. 6a, (Kieslich, K., ed.) pp. 387–419, Verlag Chemie, Weinheim

- Takamatsu, S., Umemura, J., Yamamoto, K., Sato, T., Tosa, T., Chibata, I. (1982) Production of L-alanine from ammonium fumarate using two immobilized microorganisms, Eur. J. Appl. Biotechnol. **15**, 147–152

- Tanaka, A., Tosa, T., Kobayashi, T. (1993) Industrial Application of Immobilized Biocatalysts, Marcel Dekker Inc., New York

N-Acetyl-D-neuraminic acid aldolase
Escherichia coli

EC 4.1.3.3

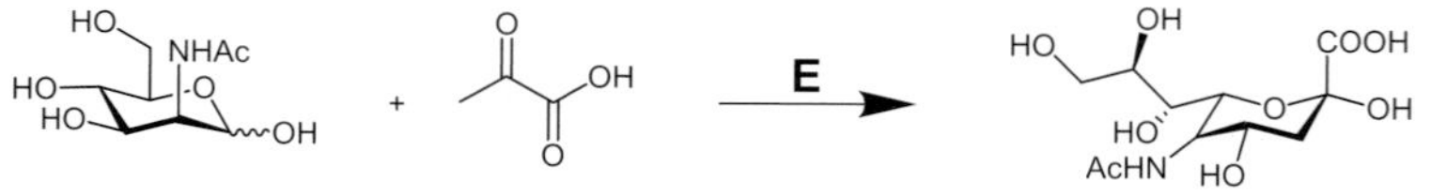

1   2   3

1 = N-acetyl-D-mannosamine
2 = pyruvic acid
3 = N-acetyl-D-neuraminic acid

Glaxo

Fig. 4.1.3.3 – 1

1) Reaction conditions

[**1**]:	0.9 M, 200 g · L^{-1} [221.21 g · mol^{-1}]
pH:	7.5
T:	20°C
medium:	aqueous
reaction type:	C-C bond cleavage
catalyst:	immobilized enzyme
enzyme:	*N*-acetylneuraminate pyruvate-lyase (sialic aldolase)
strain:	*Escherichia coli*
CAS (enzyme):	[9027–60–5]

2) Remarks

- The chemical synthesis of Neu5Ac is costly since it requires complex protection and deprotection steps.
- The enzyme is covalently immobilized on Eupergit-C.
- Since *N*-acetyl-D-mannosamine is very expensive, it is synthesized from *N*-acetyl-D-glucosamine by epimerization at C_2. The equilibrium of the epimerization is on the side of *N*-acetyl-D-glucosamine (GlcNAc:ManNAc = 4:1). After neutralization and addition of isopropanol GlcNAc precipitates. In the remaining solution a ratio of GlcNAc:ManNAc = 1:1 is reached. After evaporation to dryness and extraction with methanol a ratio of GlcNAc:-ManNAc = 1:4 is reached. In contrast to this chemical epimerization enzymatic epimerization using an epimerase is applied by another company, see pages 340 and 385.
- *N*-Acetyl-D-glucosamine is not a substrate for the aldolase, but it is an inhibitor and limits by this way the applied maximal concentration of ManNAc.
- Non-converted GlcNAc can be recycled after down stream processing by epimerization to ManNAc.
- The natural direction of the aldolase-catalyzed reaction is the cleavage of Neu5Ac to pyruvate and ManNAc. The K_M for ManNAc is 700 mM. Therefore a very high ManNAc concentration of up to 20 % w/v is used. By this means ManNAc itself drives the equilibrium. Pyruvate is used in a 1.5 molar ratio.

- Neu5Ac can be crystallized directly from the reaction mixture simply by the addition of acetic acid.

- In the repetitive batch mode the immobilized enzyme can be reused for at least nine cycles without any significant loss in activity.

3) Flow scheme

Not published.

4) Process parameters

yield:	60 % based on ManAc; 27 % based on GlcNAc
reactor type:	repetitive batch
capacity:	multi t
residence time:	5 h
down stream processing:	crystallization
enzyme supplier:	Röhm GmbH, Germany
company:	Glaxo Wellcome, UK

5) Product application

- Neu5Ac is the major representative of amino sugars (sialic acids) that are incorporated at the terminal positions of glycoprotein, glycolipid and play an important role in a wide range of biological recognition processes.

- The synthesis of Neu5Ac analogues and Neu5Ac-containing oligosaccharides is of interest in studies towards inhibitors of neuraminidase, hemagglutinin and selectin-mediated leucocyte adhesion.

6) Literature

- Dawson, M., Noble, D., Mahmoudian, D.S. (1994) Process for the preparation of *N*-acetyl-ᴅ-neuraminic acid, PCT WO 9429476

- Mahmoudian, M., Noble, D., Drake, C.S., Middleton, R.F., Montgomery, D.S., Piercey, J.E., Ramlakhan, D., Todd, M., Dawson, M.J. (1997) An efficient process for production of *N*-acetylneuraminic acid using *N*-acetylneuraminic acid aldolase, Enzyme Microb. Tech. **45**, 393–400

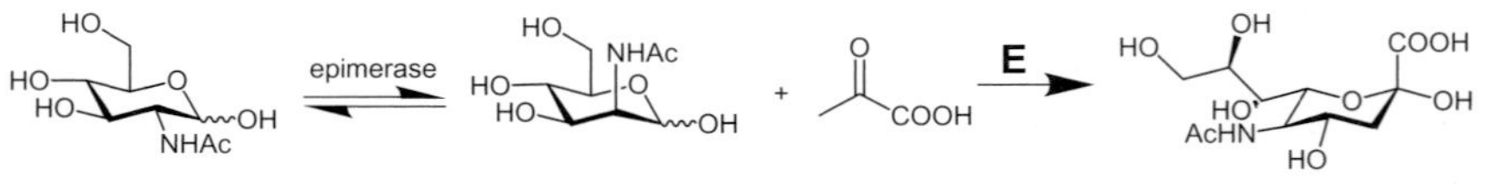

N-Acetyl-D-neuraminic acid aldolase
Escherichia coli

1 = N-acetyl-D-glucosamine (GlcNAc)
2 = N-acetyl-D-mannosamine (ManNAc)
3 = pyruvic acid
3 = N-acetyl-D-neuraminic acid (Neu5Ac)

Marukin Shoyu
Research Center Jülich

Fig. 4.1.3.3 – 1

1) Reaction conditions

[1]:   0.8 M, 177 g $\cdot$ L^{-1} [221.21 g $\cdot$ mol^{-1}]
pH: 7.2
T: 30 °C
medium: aqueous
reaction type: C-C bond cleavage
catalyst: solubilized enzyme
enzyme: *N*-acetylneuraminate pyruvate-lyase (sialic aldolase)
strain: *Escherichia coli*
CAS (enzyme):   [9027–60–5]

2) Remarks

- *N*-Acetyl-D-neuraminic acid aldolase from *E. coli* K-12 has been cloned and overexpressed in *E. coli*.

- The enzyme catalyzed aldol condensation to *N*-acetylneuraminic acid is combined in a one-vessel synthesis with the enzyme-catalyzed epimerization of *N*-acetyl-glucosamine (GlcNAc), see page 385.

- The production of Neu5Ac on a multi ton scale is performed by Glaxo utilizing chemical epimerization (see page 338). In contrast to this synthesis here the native, non-immobilized enzyme is applied.

- Both enzymes (epimerase and aldolase) can be used in a pH range of 7.0 to 8.0. For the biotransformation pH 7.2 was chosen.

- Since excess amounts of pyruvate (educt for the aldolase) inhibit the epimerase, a fed batch in regard to pyruvate is performed. After the start of the reaction with a ratio of pyruvate to GlcNAc of 1:0.6, two times pyruvate is added twice up to a total amount of 251 mol (ratio of pyruvate to GlcNAc at start: 1:0.6; after first addition: 1:1.5; after second addition 1:2).

- Before the product is purified by crystallization (initiated by the addition of 5 volumes of glacial acetic acid), the enzymes are denatured by heating to 80 °C for 5 minutes, afterwards the reaction solution is filtered.

- Another process layout is realized by the Research Center Jülich, Germany, that already established in 1991 the one-pot synthesis of Neu5Ac with the combined use of epimerase and aldolase. Here a continuously operated membrane reactor is used, where the enzymes are retained by an ultrafiltration membrane. By this technology kg quantities of N-acetylneuraminic acid were produced (GlcNAc = 300 mM; pyruvate = 600 mM; pH = 7.5; T = 25 °C; conversion = 78 %; residence time = 4 h; reactor volume = 0.44 L; space-time-yield = 655 g · L^{-1} · d^{-1}; enzyme consumption = 10,000 U · kg^{-1}). The advantage of this approach is a simplified downstream processing, since the catalysts are already separated. The product is therefore pyrogen free.

3) Flow scheme

Not published.

4) Process parameters

conversion:	60 %
reactor type:	fed batch
reactor volume:	200 L
capacity:	multi kg
down stream processing:	crystallization
enzyme supplier:	Marukin Shoyu Co., Ltd., Japan, and others
company:	Marukin Shoyu Co., Ltd., Japan and Research Center Jülich GmbH, Germany

5) Product application

- Please see page 338.

6) Literature

- Ghisalba, O., Gygax, D., Kragl, U., Wandrey, C. (1991) Enzymatic method for N-acetylneuramin acid production, Novartis, EP 0428947

- Kragl, U., Gygax, D., Ghisalba, O., Wandrey, C. (1991) Enzymatic process for prepaing N-acetylneuraminic acid, Angew. Chem. Int. Ed. Engl. **30**, 827–828

- Kragl, U., Kittelmann, M., Ghisalba, O., Wandrey, C. (1995) N-Acetylneuraminic acid: from a rare chemical isolated from natural sources to a multi-kilogram enzymatic synthesis for industrial application, Ann. N. Y. Acad. Sci. **750**, 300–305

- Maru, I., Ohnishi, J., Ohta, Y., Tsukada, Y. (1998) Simple and large-scale production of N-acetylneuraminic acid from N-acetyl-D-glucosamine and pyruvate using N-acyl-D-glucosamine 2-epimerase and N-acetylneuraminate lyase, Carbohydrate Res. **306**, 575–578

- Ohta, Y., Tsukada, Y., Sugimori, T., Murata, K., Kimura, A. (1989) Isolation of a constitutive N-acetylneuraminate lyase-producing mutant of *Escherichia coli* and its use for NPL production, Agric. Biol. Chem. **53**, 477–481

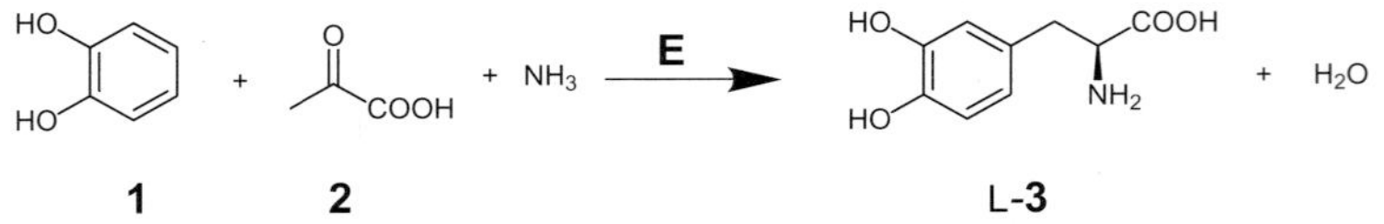

1 = catechol
2 = pyruvic acid
3 = dopa

Ajinomoto Co., Ltd.

Fig. 4.1.99.2 – 1

1) Reaction conditions

[**3**]:	0.558 M, 110 g · L^{-1} [197.19 g · mol^{-1}]<pH>
medium:	aqueous
reaction type:	α,β-elimination (reversed)
catalyst:	suspended whole cells
enzyme:	L-tyrosine phenol lyase (deaminating) (β-tyrosinase)
strain:	*Erwinia herbicola*
CAS (enzyme):	[9059–31–8]

2) Remarks

- Monsanto has successfully scaled up the chemical synthesis of L-dopa:

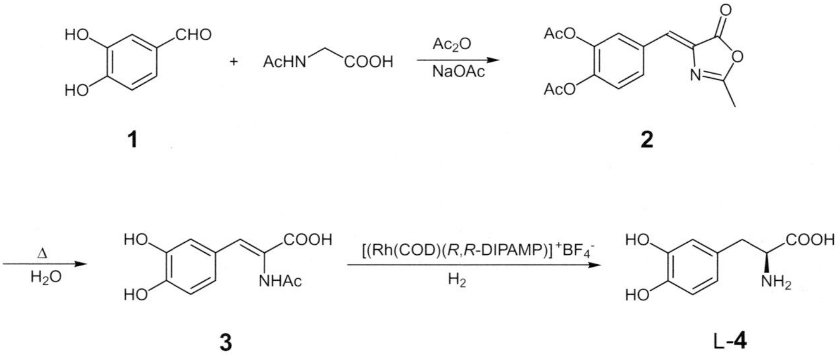

Fig. 4.1.99.2 – 2

3) Flow scheme

Not published.

Tyrosine phenol lyase
Erwinia herbicola

4) Process parameters

reactor type:	fed batch
reactor volume:	60,000 L
capacity:	$250 \, t \cdot a^{-1}$
production site:	Kawasaki, Kanagaua Prefecture, Japan
company:	Ajinomoto Co., Ltd., Japan

5) Product application

- The product is applied for the treatment of Parkinsonism. Parkinsonism is caused by a lack of L-dopamine and its receptors in the brain. L-Dopamine is synthesized in organisms by decarboxylation of L-dopa. Since L-dopamine cannot pass the blood-brain barrier L-dopa is applied in combination with dopadecarboxylase-inhibitors to avoid formation of L-dopamine outside the brain:

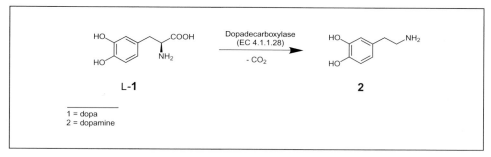

L-**1**   2

1 = dopa
2 = dopamine

Fig. 4.1.99.2 – 3

6) Literature

- Ager, D.J. (1999) Handbook of Chiral Chemicals, Marcel Dekker Inc., New York

- Tsuchida, T., Nishimoto, Y., Kotani, T., Iiizumi, K. (1993) Production of L-3,4-dihydroxyphenylalanine, Ajinomoto Co., Ltd., JP 5123177A

- Yamada, H. (1998) Screening of novel enzymes for the production of useful compounds, in: New Frontiers in Screening for Microbial Biocatalysis (Kieslich, K., van der Beek, C.P., de Bont, J.A.M., van den Tweel, W.J.J., eds.) pp. 13–17, Studies in Organic Chemistry **53**, Elsevier, Amsterdam

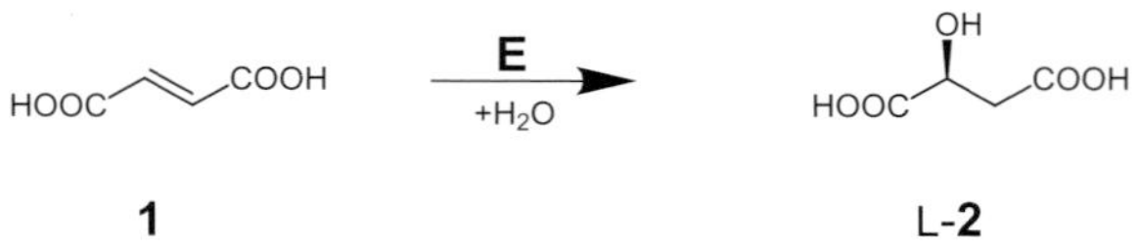

1 = fumaric acid
2 = malic acid

Amino GmbH

Fig. 4.2.1.2 – 1

1) Reaction conditions

[**1**]:	0.97 M, 150 g · L^{-1} [154.14 g · mol^{-1}] (slurry of calcium salt)
[**2**]:	0.87 M, 150 g · L^{-1} [172.16 g · mol^{-1}] (slurry of calcium salt)
pH:	8.0
T:	25 °C
medium:	aqueous
reaction type:	C-O bond formation
catalyst:	suspended whole cells
enzyme:	(*S*)-malate hydrolyase (fumarase, fumarate hydratase)
strain:	*Corynebacterium glutamicum*
CAS (enzyme):	[9032–88–6]

2) Remarks

- Only L-malate is produced, D-malate is not detectable.

- Microbial fumarases lead to a mixture of 85 % malate and 15 % fumarate.

- According to German drug regulations the fumaric acid content of malic acid has to be less than 0.15 %.

- Fumaric acid separation is circumvented by forcing a quantitative transformation in a slurry reaction (solubility of calcium malate and calcium fumarate is approx. 1 %).

- The reaction is carried out in a slurry of crystalline calcium fumarate and crystalline calcium malate.

- The precipitation of the product shifts the equilibrium towards calcium malate (figure 4.2.1.2 – 2):

344

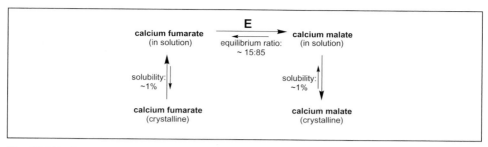

Fig. 4.2.1.2 – 2

- The biotransformation is carried out under non-sterile conditions, resulting in the addition of preservatives to prevent microbial growth.

- In this case p-hydroxybenzoic acid esters are used as preservatives because of their low effective concentrations and their biodegradability. They are permitted as food preservatives and therefore no special precautions need to be taken for their technical application.

- Enzyme stabilization is achieved by addition of soy bean protein or bovine serum albumin. The surplus of foreign protein can occupy interfaces at walls, stirrers and liquid surfaces and protect the enzyme from interface denaturation.

- Protein addition, temperature adjustment, pH adjustment and imidazole supplementation result in 25-fold increase of the fumarate hydratase's productivity.

- Separation of the biocatalyst is performed by filtration of the slurry.

- Down-stream processing: Acidification with H_2SO_4 yields L-malic acid and gypsum. The latter is separated by filtration. L-Malic acid is purified by ion exchange chromatography.

- The same reaction is carried out by Tanabe Seiyaku Co., Japan, with an immobilized fumurase as catalyst (see page 347).

3) Flow scheme

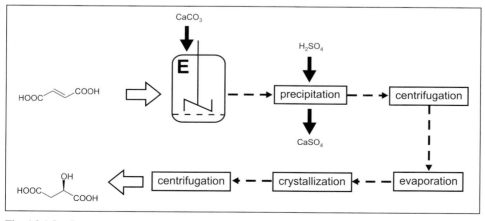

Fig. 4.2.1.2 – 3

4) Process parameters

yield: 85 %
chemical purity: > 99 %
reactor type: batch
capacity: $2,000 \text{ t} \cdot \text{a}^{-1}$
down stream processing: crystallization
enzyme consumption: $3,500 \text{ U} \cdot \text{kg}^{-1}$
start-up date: 1988
production site: Frellstedt, Germany
company: AMINO GmbH, Germany

5) Product application

• About 40,000 t of malic acid are used worldwide annually.

• L-Malic acid is used in food, cosmetic and pharmaceutical industries, e.g., as a component of amino acid infusions for parenteral nutrition.

6) Literature

• Daneel, H.J., Faurie, R., (1994) Verfahren zur Herstellung von L-Äpfelsäure aus Fumarsäure, AMINO GmbH, DE 4424664 C1

• Daneel, H.J., Faurie, R. (1994) Verfahren zur Abtrennung von Fumarsäure, Maleinsäure und/oder Bernsteinsäure von einem Hauptbestandteil Äpfelsäure, AMINO GmbH, DE 4430010 C1

• Daneel, H.J., Busse, M., Faurie, R. (1995) Pharmaceutical grade L-malic acid from fumaric acid – development of an integrated biotransformation and product purification process; Med. Fac. Landbouww. Univ. Gent **60 / 4a**, 2093–2096

• Daneel, H.J., Busse, M., Faurie, R. (1996) Fumarate hydratase from *Corynebacterium glutamicum* – process related optimization of enzyme productivity for biotechnical L-malic acid synthesis, Med. Fac. Landbouww. Univ. Gent **61 / 4a**, 1333–1340

• Mattey, M. (1988) The production of organic acids, Crit. Rev. Biotechnol. **12**, 87–132

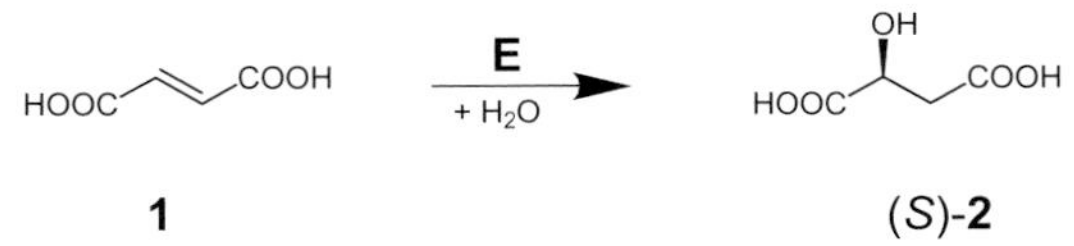

Fumarase
Brevibacterium flavum

HOOC—COOH $\xrightarrow[\text{+ H}_2\text{O}]{\text{E}}$ structure with OH, HOOC, COOH

1 (*S*)-**2**

1 = fumaric acid
2 = malic acid

Tanabe Seiyaku Co., Ltd.

Fig. 4.2.1.2 – 1

1) Reaction conditions

[**1**]:	1.0 M, 116.1 g·L^{-1} [116.1 g·mol^{-1}]
pH:	6.5–8.0
T:	37 °C
medium:	aqueous
reaction type:	C-O bond cleavage (elimination of H$_2$O)
catalyst:	immobilized whole cells
enzyme:	(*S*)-malate hydro-lyase (fumarate hydratase)
strain:	*Brevibacterium flavum*
CAS (enzyme):	[9032–88–6]

2) Remarks

- The cells are immobilized on κ-carrageenan gel (160 kg wet cells in 1,000 L of 3.5 % gel).

- The side reaction (formation of succinic acid) can be eliminated by treatment of immobilized cells with bile extracts. Additionally, the activity and stability can be improved by immobilization in κ-carregeenan in the presence of Chinese gallotannin.

- The operational temperature of the immobilized cells is 10 °C higher than that of native cells.

- First the strain *Brevibacterium ammoniagenes* was used for the process. During optimization *Brevibacterium flavum* was discovered. The productivity with *B. flavum* is more than 9 times higher than with *B. ammoniagenes*.

- The cultural age of the cells also had a marked effect on the enzyme activity and the operational stability of immobilized cells.

- The same process is also employed by Amino GmbH, Germany, with the difference that they use the non-immobilized, native fumarase (see page 344).

3) Flow scheme

Not published.

4) Process parameters

conversion:	80 % (equilibrium conversion)
yield:	> 70 %
reactor type:	plug-flow reactor
reactor volume:	1,000 L
capacity:	$468 \, t \cdot a^{-1}$
enzyme activity:	$17 \, U \cdot mL(gel)^{-1}$ (37 °C); $28 \, U \cdot mL(gel)^{-1}$ (50 °C)
enzyme consumption:	$t_{1/2} = 243$ d (37 °C); $t_{1/2} = 128$ d (50 °C)
start-up date:	1974
company:	Tanabe Seiyaku Co., Ltd., Japan

5) Product application

- The product is used as an acidulant in fruit and vegetable juices, carbonated soft drinks, jams and candies, in amino acid infusions and for the treatment of hepatic malfunctioning.

6) Literature

- Tosa, T., Shibatani, T. (1995) Industrial applications of immobilized biocatalysts in Japan, Ann. N. Y. Acad. Sci. **750**, 364–375

- Tanaka, A., Tosa, T., Kobayashi, T. (1993) Industrial Application of Immobilized Biocatalysts, Marcel Dekker Inc., New York

- Lilly, M.D. (1994) Advances in biotransformation processes. Eighth P. V. Danckwerts memorial lecture presented at Glaziers' Hall, London, U.K. 13 May 1993, Chem. Eng. Sci. **49**, 151–159

- Wiseman, A. (1995) Handbook of Enzyme and Biotechnology, Ellis Horwood, Chichester

- Sheldon, R.A. (1993) Chirotechnology, Marcel Dekker Inc., New York

- Crosby, J. (1991) Synthesis of optically active compounds: a large scale perspective, Tetrahedron **47**, 4789–4846

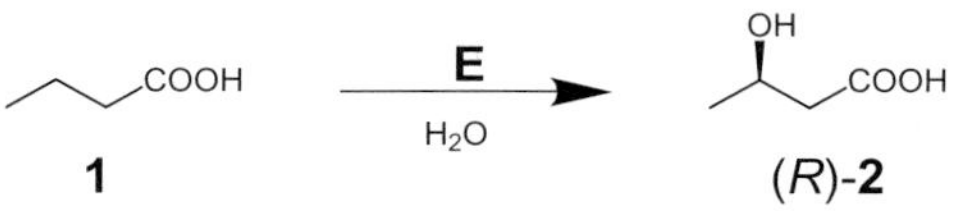

Enoyl-CoA hydratase
Candida rugosa

1 = butyric acid
2 = β-hydroxy-n-butyric acid

Kanegafuchi

Fig. 4.2.1.17 – 1

1) Reaction conditions

pH:	7.2–7.5
T:	30 °C – 33 °C
medium:	aqueous
reaction type:	hydration (addition of H_2O to carbon-carbon double bond)
catalyst:	whole cells
enzyme:	(3S)-3-hydroxyacyl-CoA hydro-lyase (β-hydroxyacid dehydrase, acyl-CoA dehydrase, enoyl-CoA hydratase)
strain:	*Candida rugosa* IFO 0750 M
CAS (enzyme):	[9027–13–8]

2) Remarks

- A three step reaction takes place. Initially the aliphatic acid is dehydrogenated to the α,β-unsaturated acid and in a subsequent step enantioselectively hydrated:

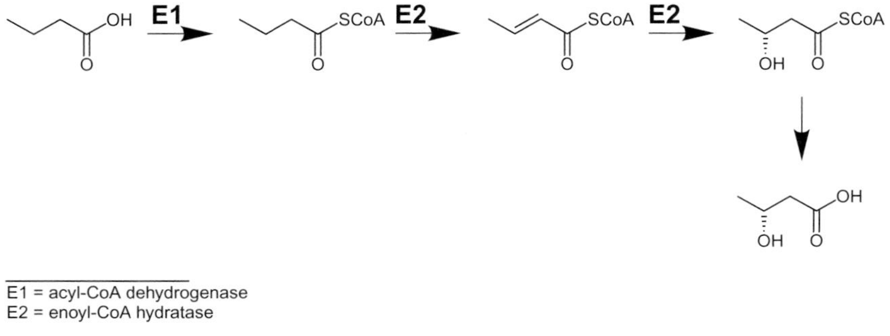

E1 = acyl-CoA dehydrogenase
E2 = enoyl-CoA hydratase

Fig. 4.2.1.17 – 2

3) Flow scheme

Not published.

4) Process parameters

ee:	> 98 %
space-time-yield:	$5–10 \; g \cdot L^{-1} \cdot d^{-1}$
company:	Kanegafuchi, Japan

5) Product application

- (*R*)-β-Hydroxy-*n*-butyric acid is used for the synthesis of a carbapenem intermediate:

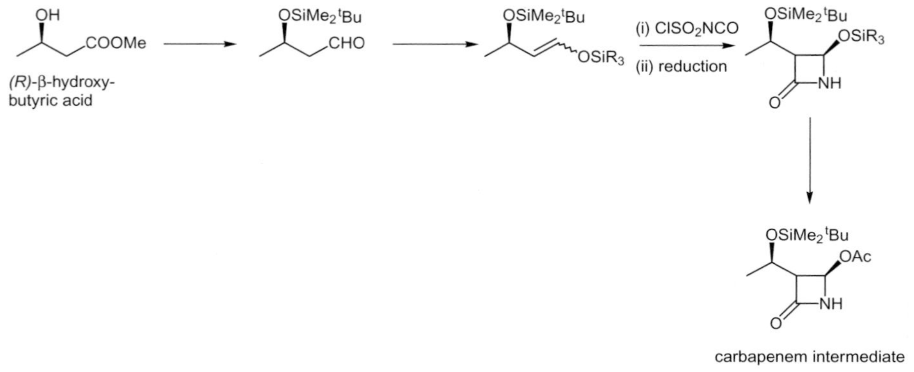

Fig. 4.2.1.17 – 3

6) Literature

- Sheldon, R.A. (1993) Chirotechnology, Marcel Dekker Inc., New York

- Kieslich, K., (1991), Biotransformations of industrial use, 5th Leipziger Biotechnologiesymposium 1990, Acta Biotechnol. **11** (6), 559–570

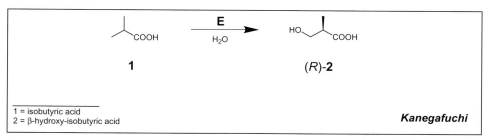

1 = isobutyric acid
2 = β-hydroxy-isobutyric acid

Kanegafuchi

Fig. 4.2.1.17 – 1

1) Reaction conditions

pH: 7.2–7.5
T: 30 °C – 33 °C
medium: aqueous
reaction type: hydration (addition of H_2O to carbon-carbon double bond)
catalyst: whole cells
enzyme: (3*S*)-3-hydroxyacyl-CoA hydro-lyase (β-hydroxyacid dehydrase,
                  acyl-CoA dehydrase, enoyl-CoA hydratase)
strain: *Candida rugosa* IFO 0750 M
CAS (enzyme): [9027–13–8]

2) Remarks

- A three step reaction takes place. Initially the aliphatic acid is dehydrogenated to the α,β-unsaturated acid and in a subsequent step enantioselectively hydrated:

E1 = acyl-CoA dehydrogenase
E2 = enoyl-CoA hydratase

Fig. 4.2.1.17 – 2

3) Flow scheme

Not published.

Enoyl-CoA hydratase
Candida rugosa

EC 4.2.1.17

4) Process parameters

yield: 98 %
ee: 97 %
space-time-yield: $5\text{–}10 \text{ g} \cdot \text{L}^{-1} \cdot \text{d}^{-1}$
company: Kanegafuchi, Japan

5) Product application

- (*R*)-β-Hydroxy-isobutyric acid is used as a chiral synthon in the synthesis of captopril, an ACE-inhibitor (ACE = angiotensin converting enzyme):

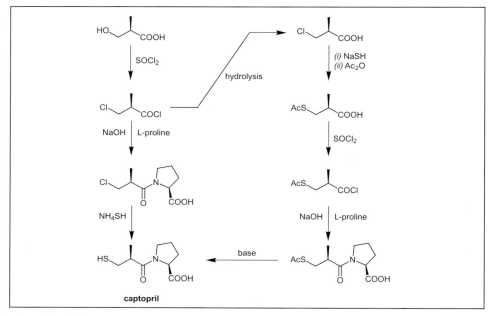

Fig. 4.2.1.17 – 3

6) Literature

- Crosby, J. (1991) Synthesis of optically active compounds: a large scale perspective, Tetrahedron **47**, 4789–4846

- Hasegawa, J., Ogura, M., Kanema, H., Noda, N., Kawaharada, H., Watanabe, K. (1982) Production of D-β-hydroxyisobutyric acid from isobutyric acid by *Candida rugosa* and its mutant, J. Ferment. Technol. **60**, 501–508

- Kieslich, K., (1991), Biotransformations of industrial use, 5th Leipziger Biotechnologiesymposium 1990, Acta Biotechnol. **11** (6), 559–570

- Sheldon, R.A. (1993) Chirotechnology, Marcel Dekker Inc., New York

Tryptophan synthase
Escherichia coli

EC 4.2.1.20

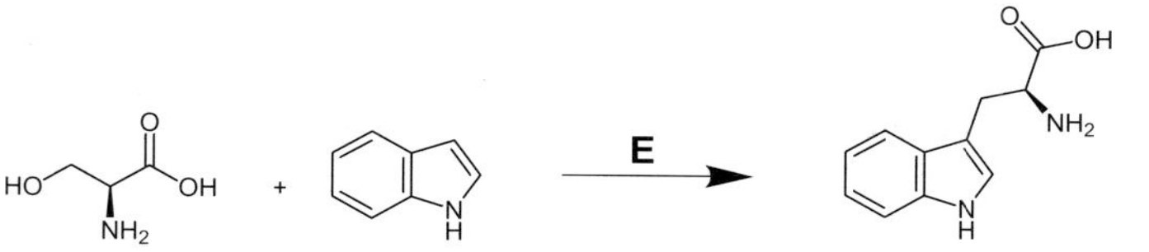

1 **2** **3**

1 = L-serine
2 = indole
3 = L-tryptophan

Amino GmbH

Fig. 4.2.1.20 – 1

1) Reaction conditions

[**1**]:	0.1 M, 10.41 $g \cdot L^{-1}$ [104.1 $g \cdot mol^{-1}$] (initial)
[**2**]:	0.01 M, 1.17 $g \cdot L^{-1}$ [117.16 $g \cdot mol^{-1}$] (steady state)
[pyridoxal phosphate]:	$> 5 \cdot 10^{-5}$ M (cofactor)
[**3**]:	0.06 M, 12.25 $g \cdot L^{-1}$ [204.23 $g \cdot mol^{-1}$](saturated solution)
pH:	8.0 – 9.0
T:	40 °C
medium:	aqueous
reaction type:	C-O bond cleavage
catalyst:	suspended whole cells
enzyme:	L-tryptophan synthase (L-serine hydrolyase)
strain:	*Escherichia coli*
CAS (enzyme):	[9014–52–2]

2) Remarks

- Pyridoxal phosphate is needed as cofactor.
- The enzyme works enantiospecifically for α-L-amino acid substrates of the type:

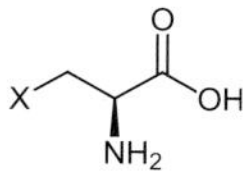

where X indicates a small nucleophilic substituent like -OH, -Cl.

- The established process is dedicated to the production of L-tryptophan as a pharmaceutically active ingredient.
- The educt L-serine is separated from molasses. The best separation is performed with ion exchange chromatography (polystyrene resin) close to the isoelectric point of serine, pH 5.7. By concentration of the serine fraction to 35 % dry mass the main fraction of D-serine can be separated by filtration, leaving a L-serine stock solution.
- The fed batch is pH regulated and the indole dosage is directed via on-line HPLC analysis of the product/educt ratio.

- L-Tryptophan is produced in such high concentrations that it crystallizes instantly and it is isolated together with the cells at the end of the fed batch.

- The crude L-tryptophan is solubilized in hot water and the cells are separated after addition of charcoal.

3) Flow scheme

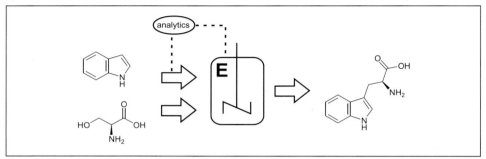

Fig. 4.2.1.20 – 2

4) Process parameters

yield:	> 95 % (based on indole)
reactor type:	fed batch
capacity:	$30 \ t \cdot a^{-1}$
residence time:	6 h / batch
space-time-yield:	$75 \ g \cdot L^{-1} \cdot d^{-1}$
down stream processing:	multiple fractional crystallizations from aqueous solution
enzyme activity:	$250 \ U \cdot L^{-1}$
enzyme consumption:	$4,500 \ U \cdot kg^{-1}$
enzyme supplier:	AMINO GmbH, Germany
start-up date:	1988/89
production site:	Frellstedt, Germany
company:	AMINO GmbH, Germany

5) Product application

- L-Tryptophan is used in parenteral nutrition (infusion solution) and as a pharmaceutical active ingredient in sedative, neuroleptica, antidepressiva and food additives.

- Intermediate for production of other pharmaceutical compounds.

6) Literature

- Bang, W., Lang, S., Sahm, H., Wagner, F. (1983) Production of L-tryptophan by *Escherichia coli* cells, Biotech. Bioeng. **25**, 999–1011

- Plischke, H., Steinmetzer, W., (1988) Verfahren zur Herstellung von L-Tryptophan und D,L-Serin, AMINO GmbH, DE 3630878 C1

- Wagner, F., Klein, J., Bang, W., Lang, S., Sahm,H. (1980) Verfahren zur mikrobiellen Herstellung von L-Tryptophan, DE 2841642 C2

Malease
Pseudomonas pseudoalcaligenes

EC 4.2.1.31

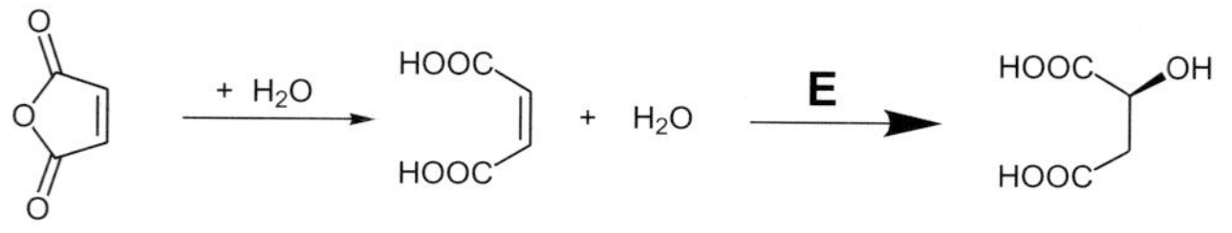

1 **2** **(*R*)-3**

1 = maleic anhydride
2 = maleic acid
3 = malic acid

DSM

Fig. 4.2.1.31 – 1

1) Reaction conditions

[**2**]:	0.8 M, 92.9 $g \cdot L^{-1}$ [116.07 $g \cdot mol^{-1}$]
pH:	7.0
T:	25 °C
medium:	aqueous
reaction type:	addition of H_2O
catalyst:	immobilized whole cells
enzyme:	(*R*)-malate hydro-lyase (maleate hydratase, malease)
strain:	*Pseudomonas pseudoalcaligenes*
CAS (enzyme):	[37290–71–4]

2) Remarks

- The cheaper maleic anhydride can be used instead of maleic acid which hydrolyses *in situ* to maleate.
- Two degradation pathways have been described for maleate as carbon and energy source:

Malease
Pseudomonas pseudoalcaligenes

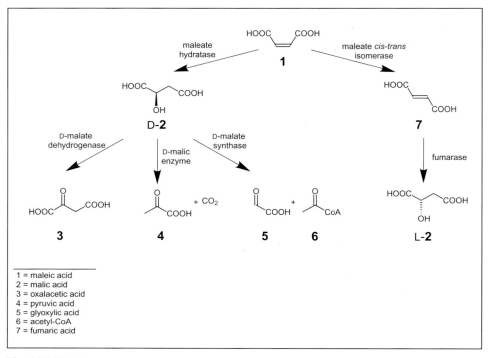

Fig. 4.2.1.31 – 2

- Fumarate accumulation during cell incubation with maleate under anaerobic conditions allows discrimination between the two degradation pathways and was used as selection criteria.

- 315 strains were screened.

- The chosen strain *Pseudomonas pseudoalcaligenes* is not able to grow on maleate as the sole source of carbon and energy. It is probably not capable of synthesizing a transport mechanism for maleate.

- To overcome transport problems of substrate and product across the cell membrane, the cells are permeabilized with Triton X-100.

- Cells have to be harvested before the substrate for growth is completely consumed, because otherwise malease activity drops rapidly. During growth the activity is constant in logarithmic phase. The presence of maleate does not influence the growth rate.

- The enzyme malease needs no cofactor.

- Although D-malate is a competitive inhibitor, it stabilizes the enzyme.

- The K_M-value is about 0.35 M.

Malease
Pseudomonas pseudoalcaligenes

- Malease can also catalyze the hydration of citraconate. The activity is only 56 % of the maleate hydration acitvity and the K_M-value is about 0.2 M:

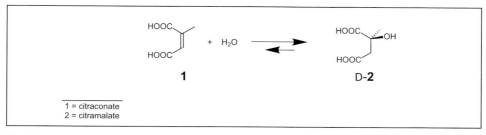

1 = citraconate
2 = citramalate

Fig. 4.2.1.31 – 3

3) Flow scheme

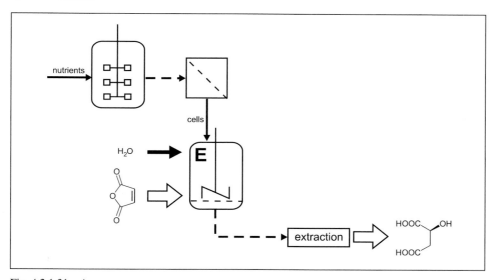

Fig. 4.2.1.31 – 4

4) Process parameters

yield:	> 99 %
ee:	> 99.99 %
enzyme activity:	2,000 U · g⁻¹_Protein (permeabilized cells grown on 3-hydroxy-benzoate)
company:	DSM, The Netherlands

yield: > 99 %
ee: > 99.99 %
enzyme activity: $2{,}000 \ U \cdot g^{-1}_{Protein}$ (permeabilized cells grown on 3-hydroxy-benzoate)
company: DSM, The Netherlands

5) Product application

- D-Malate can be used as chiral synthon in organic chemistry or as resolving agent in resolution of racemic compounds.

6) Literature

- Chibata, I., Tosa, T., Shibatani, T. (1992) The industrial production of optically active compounds by immobilized biocatalysts, in: Chirality in Industry (Collins, A.N., Sheldrake, G., Crosby, J., eds.) pp. 351–370, John Wiley & Sons Ltd., New York

- Submarmarian, S.S., Raghavendra Rao, M.R. (1968) Purification and properties of citraconase, J. Biol. Chem. **243**, 2367–2372

- van der Werf. M.J., van den Tweel, W.J.J., Hartmans, S. (1992) Screening for microorganisms producing D-malate from maleate, Appl. Environ. Microbiol. **58**, 2854–2860

- van der Werf, M.J., van den Tweel, W.J.J., Kamphuis, J., Hartmans, S., de Bont J.A.M. (1993) D-Malate and D-citramalate production with maleate hydro-lyase from *Pseudomonas pseudoalcaligenes*, Proc. 6th Eur. Congr. Biotechnol., 471–474, Florence

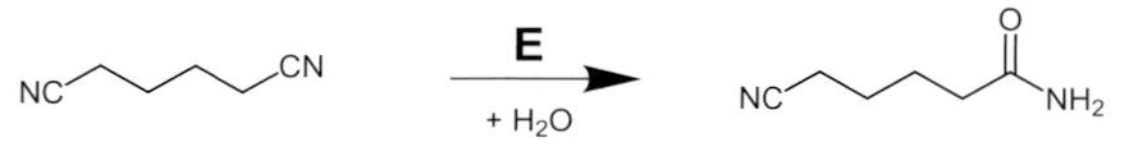

Nitrile hydratase
Pseudomonas chlororaphis

1 = adiponitrile
2 = 5-cyano-valeramide

DuPont

Fig. 4.2.1.84 – 1

1) Reaction conditions

[1]:	2.015 mol (organic phase), 218 kg [108.14 g·mol^{-1}]
pH:	7.0
T:	5 °C
medium:	two-phase: aqueous/organic
reaction type:	C–O bond cleavage by elimination of water
catalyst:	immobilized whole cells
enzyme:	nitrile hydro-lyase (nitrile hydratase, acrylonitrile hydratase, NHase, L-NHase, H-NHase)
strain:	*Pseudomonas chlororaphis* B23
CAS (enzyme):	[82391–37–5]

2) Remarks

- The cells are immobilized in calcium alginate beads.

- For strain selection it is important that the cells do not show an amidase activity that would further hydrolyze the amide to the carboxylic acid.

- By this method 13.6 t are produced in 58 repetitive batch cycles.

- As by-product adipodiamide is formed.

- This biotransformation is chosen over the chemical transformation due to a higher conversion and selectivity, production of more product per weight (3,150 kg·kg^{-1} (dry cell weight)), and less waste:

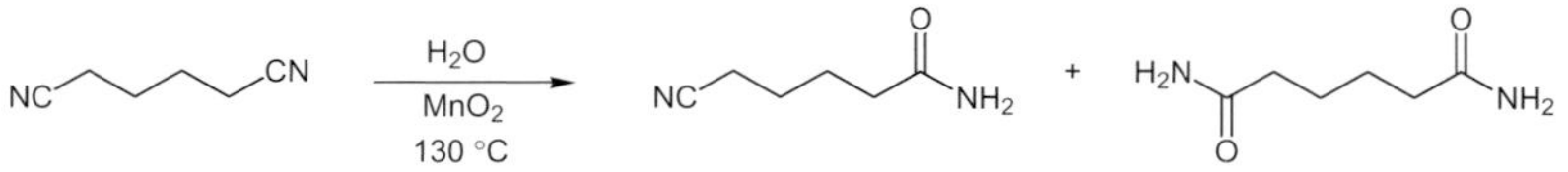

1 = adiponitrile
2 = 5-cyano-valeramide
3 = adipodiamide

Fig. 4.2.1.84 – 2

- Reactions with product concentrations higher than 0.45 M form two-phase systems.

- As reaction temperature 5 °C is chosen, since the solubility of the by-product adipodiamide is only 37–42 mM in 1 to 1.5 M 5-cyanovaleramide.

359

- A batch reactor is preferred over a fixed-bed packed column reactor, because of the lower selectivity to 5-cyanovaleramide that is observed and the possibility of precipitation of adipamide and plugging of the column.

- After completion of the reaction 90 % of the product mixture is decanted and the reactor is recharged with 1,007 L of reaction buffer (23 mM sodium butyrate and 5 mM calcium chlorid) and 218 kg (2,015 mol) adiponitrile.

- The catalyst productivity is 3,150 kg 5-cyanovaleramide per kg of dry cell weight.

- The catalyst consumption is 0.006 kg per kg product.

- Excess water is removed at the end of the reaction by distillation. The by-product adipamide as well as calcium and butyrate salts are precipitated by dissolution of the resulting oil in methanol at > 65 °C. The raw product solution is directly transferred to the herbicide synthesis.

3) Flow scheme

Not published.

4) Process parameters

conversion:	97 %
yield:	93 %
selectivity:	96 %
reactor type:	repetitive batch
reactor volume:	2,300 L
capacity:	several $t \cdot a^{-1}$
residence time:	4 h
down stream processing:	distillation
company:	DuPont, USA

5) Product application

- 5-Cyanovaleramide is used as an intermediate for the synthesis of herbicides.

6) Literature

- Hann, E.C., Eisenberg, A., Fager, S.K., Perkins, N.E., Gallagher, F.G., Cooper, S.M., Gavagan, J.E., Stieglitz, B., Hennesey, S.M., DiCosimo, R (1999) 5-Cyanovaleramide production using immobilized *Pseudomonas chlororaphis* B23, Bioorg. Med. Chem. **7**, 2239–2245

- Yamada, H., Ryuno, K., Nagasawa, T., Enomoto, K., Watanabe, I. (1986), Optimum culture conditions for production of *Pseudomonas chlororaphis* B23 of uitril hydratase, Agric. Biol. Chem. **50**, 2859–2865

Nitrile hydratase
Rhodococcus rhodochrous

EC 4.2.1.84

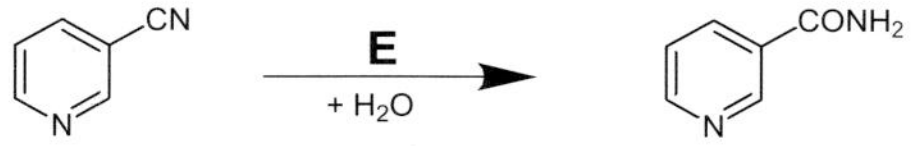

1 **2**

1 = 3-cyanopyridine
2 = nicotinamide = vitamin B3

Lonza

Fig. 4.2.1.84 – 1

1) Reaction conditions

[**1**]:	[104.11 g · mol^{-1}]
medium:	aqueous
reaction type:	C-O bond cleavage by elimination of water
catalyst:	immobilized whole cells
enzyme:	nitrile hydro-lyase (nitrile hydratase, acrylonitrile hydratase, NHase, L-NHase, H-NHase)
strain:	*Rhodococcus rhodochrous* J1
CAS (enzyme):	[82391–37–5]

2) Remarks

- In contrast to the chemical alkaline hydrolysis of 3-cyanopyridine with 4 % by-product of nicotinic acid (96 % yield) the biotransformation works with absolute selectivity.
- The same strain is used in the industrial production of acrylamide.

3) Flow scheme

Not published.

4) Process parameters

yield:	100 %
selectivity:	100 %
capacity:	3,000 t · a^{-1}
production site:	China
company:	Lonza AG, Switzerland

5) Product application

- Vitamin supplement for food and animal feed.

6) Literature

- Petersen, M., Kiener, A. (1999) Biocatalysis – Preparation and functionalization of *N*-heterocycles, Green Chem. **4**, 99–106

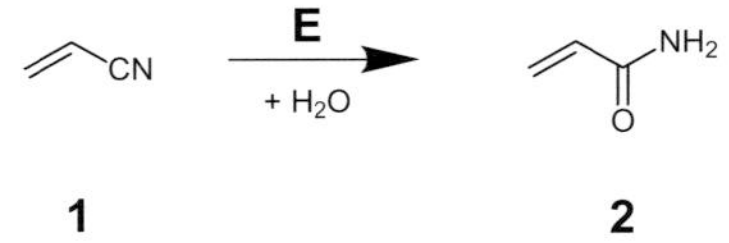

Nitto Chemical Industry

1 = acrylonitrile
2 = acrylamide

Fig. 4.2.1.84 – 1

1) Reaction conditions

[**1**]:	0.11 M, 6 g·L^{-1} [53.06 g·mol^{-1}] (fed batch)
[**2**]:	5.6 M, 400 g·L^{-1} [71.08 g·mol^{-1}]
pH:	7.0
T:	5 °C
medium:	aqueous
reaction type:	C-O bond cleavage by elimination of water
catalyst:	immobilized whole cells
enzyme:	nitrile hydro-lyase (nitrile hydratase, acrylonitrile hydratase, NHase, L-NHase, H-NHase)
strain:	*Rhodococcus rhodochrous* J1
CAS (enzyme):	[82391–37–5]

2) Remarks

- The chemical synthesis uses copper salt as catalyst for the hydration of acrylonitrile and has several disadvantages:

    1) The rate of acrylamide formation is lower than the acrylic acid formation,

    2) the double bond of educts and products causes by-product formations such as ethylene, cyanohydrin and nitrilotrispropionamide and

    3) at the double bonds polymerization occurs.

- The biotransformation has the advantages that recovering the unreacted nitrile is not necessary since the conversion is 100 % and that no copper catalyst removal is needed.

- This biotransformation is the first example of an application in the petrochemical industry and the successful enzymatic manufacture of a bulk chemical.

- Although nitriles are generally toxic some microorganism can use nitriles as carbon / nitrogen source for growth.

- Since acrylonitrile is the most poisonous one among the nitriles, screening for microorganisms was conducted with low-molecular mass nitriles instead.

- More than 1,000 microbial strains were examined.

- Two degradation ways of nitriles are known:

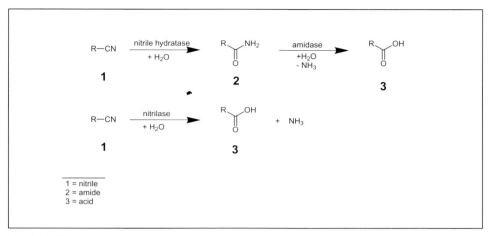

Fig. 4.2.1.84 – 2

- Microorganisms that produce amidases beside the nitrile hydratase are not suitable for the production of acrylamide without adding an amidase inhibitor.

- In the course of improvement of the biocatalyst for the production of acrylamide three main strains were used:

  1) *Rhodococcus* sp. N774

  2) *Pseudomonas chlororaphis*

  3) *Rhodococcus rhodochrous*

- The *Rhodococcus* sp. N774 strain was used for three years before the better *Pseudomonas chlororaphis* strain was found.

- The *Pseudomonas* strain cannot grow on acrylonitrile but grows on isobutyronitrile.

- The optimization of the *Pseudomonas* strain reveals that methacrylamide causes the greatest induction of nitrile hydratase. The addition of ferrous or ferric ions to the culture medium increases enzyme formation, no other ionic addition shows improvements, indicating that the nitrile hydratase contain Fe^{2+} ions as a cofactor.

- The growth medium can be optimized resulting in an amount of nitrile hydratase of 40 % of the total soluble protein formed in the cells.

- A problem during growth of *Pseudomonas chlororaphis* strain in the first optimized sucrose containing medium is the production of mucilaginous polysaccharides. These causes a high viscosity, resulting in difficulties during cell harvest.

- Using chemical mutagenesis methods (*N*-methyl-*N′*-nitro-*N*-nitrosoguanidine = MNNG) mucilage polysaccharide-non-producing mutants could be isolated. The following table shows the improvements (total activity increases 3,000-times) by optimizing the fermentation medium and by mutagenesis:

strain	specific activity (U/mg of dry cells)	total activity (U/mL)
parent (medium A*)	0.72	0.40
↓		
parent (medium R*)	66	363
↓ MNNG treatment		
Am 3	65	465
↓ MNNG treatment		
Am 324	125	952
↓ feeding of methacrylamide		
Am 324	141	1260

*medium A: dextrin, K_2HPO_4, NaCl, MgSO and isobutyronitrile
medium R: sucrose, K_2HPO_4, KH_2PO_4, $MgSO_4$, $FeSO_4$, soybean hydrolyzate and methacrylamide

Fig. 4.2.1.84 – 3

- The third generation of industrial strains is the *Rhodococcus rhodochrous* J1 that produces two kinds of nitrile-converting enzymes, the nitrilase and nitrile hydratase. The latter one was found after optimization of fermentation medium.

- Addition of cobalt ions greatly increases nitrilase hydratase activity in comparison to Fe ions for the *Pseudomonas chlororaphis* strain.

- The difference in metal-ion cofactors can be ascribed to a small number of amino acids at their ligand-binding sites, resulting in higher stability of *Rhodococcus rhodochrous* J1 strain against reducing and oxidizing agents. Although the association of 20 subunits depresses the flexibility of the protein, it increases the stability.

- The strains forms two kinds of nitrile hydratases with different molecular weights and characteristics. The following table compares the different enyzmes and shows the advantages of the high-molecular mass hydratase:

Nitrile hydratase
Rhodococcus rhodochrous

| | *Pseudomonas chloraphis* B23 | *Rhodococcus rhodochrous* J1 | |
| | | low molecular weigh | high molecular weigh |
	(L-NHaseT)	(L-NHaseT)	(H-NHaseT)
molecular weight	100,000	130,000	520,000
subunit molecular weight	α25,000	α26,000	α26,000
	β25,000	β29,000	β29,000
number of subunits	4	4-5	18-20
absorption maxima (nm)	280,720	280,415	280,415
(415/280) / (720/280)	0.014	0.031	0.016
optimum temperature (°C)	20	40	35
heat stability	20	30	50
optimum pH	7.5	8.8	6.5
pH stability	6.0-7.5	6.5-8.0	6.0-8.5
V_{max} at 20°C (U·mg^{-1} protein)	1,470		1,760
K_M at 20°C (mM)	34.6		1.89

Fig. 4.2.1.84 – 4

- As inducer urea is used, which is much cheaper than methacrylamide for the *Pseudomonas chlororaphis*. This allows an increase in the amount of L-NHase in the cell free extract to more than 50 % of the total soluble protein.

- The nitrile hydratases act also on other nitriles with yields of 100 %. The most impressive example is the conversion of 3-cyanopyridine to nicotinamide. The product concentration is about 1,465 g·L^{-1}. This conversion (1.17 g·L^{-1} dry cell mass) can be named 'pseudocrystal enzymation' since at the start of the reaction the educt is solid and with ongoing reaction it is solubilized. The same is valid for the product which crystallizes at higher conversions so that at the end of the reaction the medium is solid again (see also Lonza, page 361).

- The following table shows some examples and the end concentrations of possible products for *Rhodococcus rhodocrous* J1 induced by crotonamide:

product	concentration
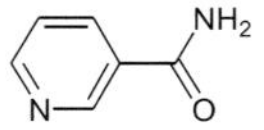 (pyridine-3-carboxamide)	1,465 g·L^{-1}
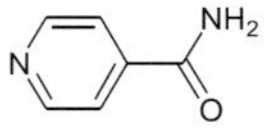 (pyridine-4-carboxamide)	1,099 g·L^{-1}
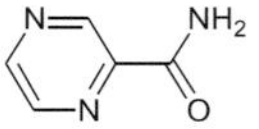 (pyrazine-2-carboxamide)	985 g·L^{-1}
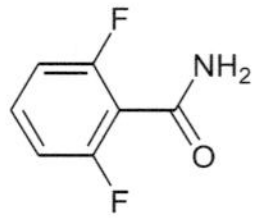 (2,6-difluorobenzamide)	306 g·L^{-1}
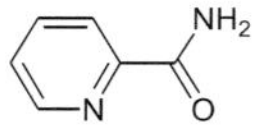 (pyridine-2-carboxamide)	977 g·L^{-1}
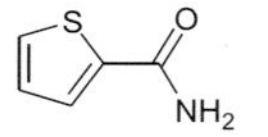 (thiophene-2-carboxamide)	210 g·L^{-1}
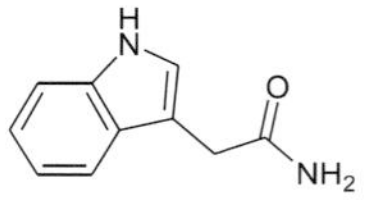 (indole-3-acetamide)	1,045 g·L^{-1}
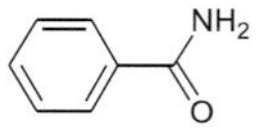 (benzamide)	489 g·L^{-1}
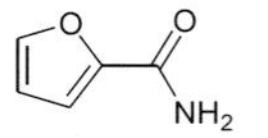 (furan-2-carboxamide)	522 g·L^{-1}

Fig. 4.2.1.84 – 5

- Since acrylamide is unstable and polymerizes easily, the process is carried out at low temperatures (about 5 °C).

- Although the cells, which are immobilized on polyacrylamide gel, and the contained enzyme is very stable towards acrylonitrile, the educt has to be fed continuously to the reaction mixture due to inhibition effects at higher concentrations.

- The following table summarized important production data for the discussed strains:

	Rhodococcus sp. NT774	*Pseudomonas chloraphis* B23	*Rhodococcus rhodochrous* J1
tolarence to acrylamide (%)	27	40	50
arylic acid formation	very little	barely detected	barely detected
cultivation time (h)	48	45	72
activity of culture (U·mL^{-1})	900	1,400	2,100
specific activit (U·mg^{-1} cells)	60	85	76
cell yield (g·L^{-1})	15	17	28
acrylamide productivity (g·g^{-1} cells)	500	850	>7,000
total amount of production (t·a^{-1})	4,000	6,000	>30,000
final concentration of acrylamide (%)	20	27	40
first year of production scale	1985	1988	1991

Fig. 4.2.1.84 – 6

3) Flow scheme

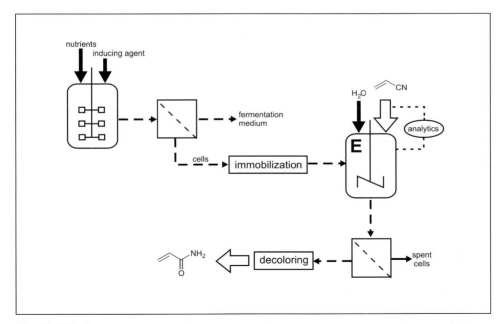

Fig. 4.2.1.84 – 7

4) Process parameters

conversion:	> 99.99 %
yield:	> 99.99 %
selectivity:	> 99.99 %
reactor type:	fed batch
capacity:	> 30,000 t $\cdot$ a^{-1}
residence time:	5 h
space-time-yield:	1,920 g $\cdot$ L^{-1} $\cdot$ d^{-1}
down stream processing:	filtration and decoloring
enzyme activity:	76,000 U $\cdot$ g$_{cells}$; 2,100,000 U $\cdot$ L^{-1}
start-up date:	1991
production site:	Japan
company:	Nitto Chemical Industry Co., Ltd., Japan

5) Product application

- Acrylamide is an important bulk chemical used in coagulators, soil conditioners and stock additives for paper treatment and paper sizing, and for adhesives, paints and petroleum recovering agents.

6) Literature

- Nagasawa, T., Shimizu, H., Yamada, H. (1993) The superiority of the third-generation catalyst, *Rhodococcus rhodochrous* J1 nitrile hydratase, for industrial production of acrylamide, Appl. Microb. Biotechnol. **40**, 189–195

- Shimizu, H., Fujita, C., Endo, T., Watanabe, I. (1993) Process for preparing glycine from glycinonitrile, Nitto Chemical Industry Co., Ltd., US 5238827

- Shimizu, H., Ogawa, J., Kataoka, M., Kobayashi, M. (1997) Screening of novel microbial enzymes for the production of biologically and chemically usesful commpounds, in: New Enzymes for Organic Synthesis; Adv. Biochem. Eng. Biotechnol. **58** (Ghose, T. K., Fiechter, A., Blakebrough, N. eds.), pp. 56–59

- Yamada, H., Tani, Y. (1982) Process for biologically producing amide, EP 093782

- Yamada, H., Kobayashi, M (1996) Nitrile hydratase and its application to industrial production of acrylamide, Biosci. Biotech. Biochem. **60** (9), 1391–1400

- Yamada, H., Tani, Y. (1987) Process for biological preparation of amides, Nitto Chemical Industry Co., Ltd., US 4637982

Carnitine dehydratase
Escherichia coli

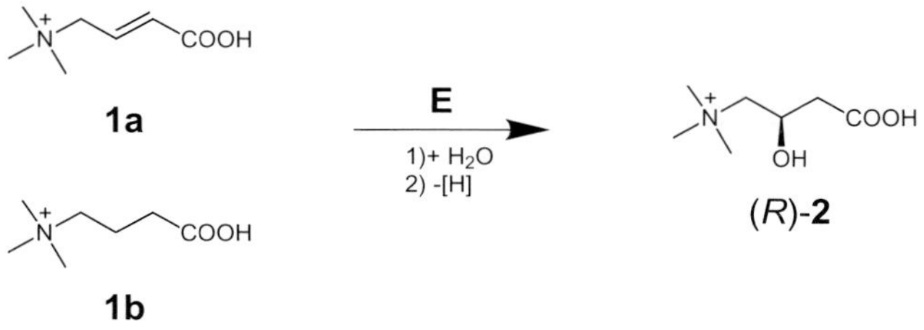

1a = crotonobetaine
1b = 4-butyrobetaine
2 = L-carnitine (3-hydroxy-4-(trimethylamino)butanoate)

Lonza AG

Fig. 4.2.1.89 – 1

1) Reaction conditions

[**1a**]:	0.69 M, 70 g · L^{-1} [102.11 g · mol^{-1}]
[**1b**]:	0.67 M, 70 g · L^{-1} [104.13 g · mol^{-1}]
[**2**]:	> 0.58 M, 70 g · L^{-1} [120.13 g · mol^{-1}]
pH:	7.0
T:	30 °C
medium:	aqueous
reaction type:	C-O bond cleavage by elimination of water
catalyst:	whole cell
enzyme:	carnitine: NAD$^+$ 3-oxidoreductase (L-carnitine 3-dehydrogenase, L-carnitine hydrolyase, carnitine dehydratase)
strain:	mutant strain of *Agrobacterium/Rhizobium* HK 1331-b, microorganisms DSM 3225 (HK1331-b), *Pseudomonas putida* T1
CAS (enzyme):	[104382–17–4]

2) Remarks

- *Agrobacterium* produces only L-carnitine.

- 4-Butyrobetaine is degraded via a reaction sequence that might be compared with the β-oxidation of fatty acids:

369

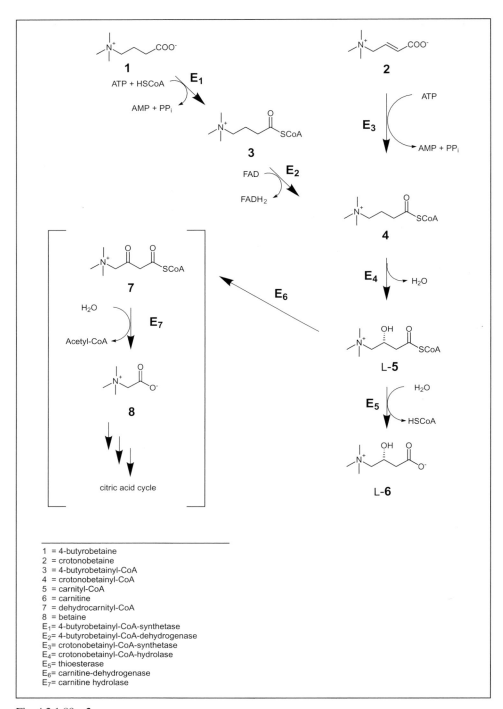

1 = 4-butyrobetaine
2 = crotonobetaine
3 = 4-butyrobetainyl-CoA
4 = crotonobetainyl-CoA
5 = carnityl-CoA
6 = carnitine
7 = dehydrocarnityl-CoA
8 = betaine
E_1= 4-butyrobetainyl-CoA-synthetase
E_2= 4-butyrobetainyl-CoA-dehydrogenase
E_3= crotonobetainyl-CoA-synthetase
E_4= crotonobetainyl-CoA-hydrolase
E_5= thioesterase
E_6= carnitine-dehydrogenase
E_7= carnitine hydrolase

Fig. 4.2.1.89 – 2

- The mutant strain blocks the L-carnitine dehydrogenase **E6** and excretes the accumulated product.

- Only growing cells are active.

- The purified enzymes could not be used for the biotransformation due to their high instability.

- At high product concentration most tested strains show an inhibition and strains with lower inhibition are not so selective.

- The product tolerance and selectivity could be optimized in a strain development.

- Apart from usual batch fermentations, continuous production is also feasable since the cells go into a 'maintenance state' with high metabolic activity and low growth rate. The cells can be recycled after separation from the fermentation broth by filtration.

- In the continuous process only 92 % conversion is reached. For almost complete conversions, fed batch processes are used. As a result, product purification becomes easier.

- Duration of process development: 30 months.

- A chemical resolution process that was developed at Lonza was not competitive with the biotechnological route:

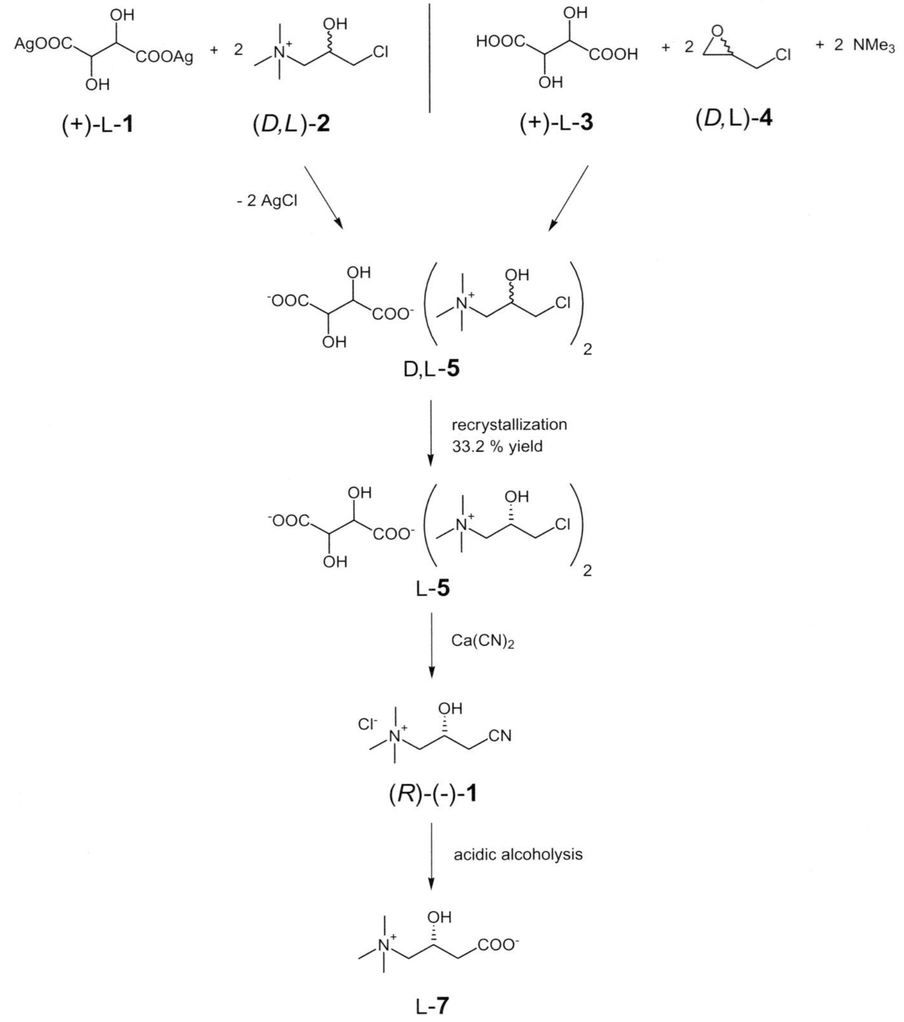

Carnitine dehydratase
Escherichia coli

Fig. 4.2.1.89 – 3

3) Flow scheme

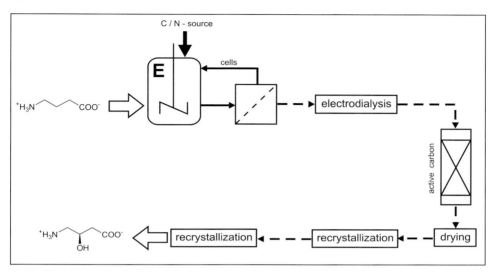

C / N - source

cells

E

^+H_3N～～COO⁻

electrodialysis

active carbon

^+H_3N～～COO⁻
OH

recrystallization ◄ – – – recrystallization ◄ – – drying

Fig. 4.2.1.89 – 4

4) Process parameters

conversion:	> 99.5 %
yield:	99.5 %
selectivity:	very high (only 3-dehydrocarnitine is accepted for transformation to L-carnitine)
ee:	99.9 %
chemical purity:	99.8 %
reactor type:	fed batch
reactor volume:	50,000 L
capacity:	$300 \, t \cdot a^{-1}$
space-time-yield:	$> 130 \, g \cdot L^{-1} \cdot d^{-1}$
down stream processing:	cross-flow filtration, ion exchange, crystallization
start-up date:	1993
production site:	Czech Republic
company:	Lonza AG, Switzerland

5) Product application

- L-Carnitine is used in infant, health sport and geriatric nutrition.

- The physiological functions are:

  Transport of long-chain fatty acids through the mitrochondrial membrane and their oxidation play an important role in energy metabolism.

  It is involved in regulating the level of blood lipids.

  L-Carnitine is utilized as a drug to increase cardiac output, improve myocardial function and treat cartinine deficiency (especially after hemodialysis).

6) Literature

- Hoeks, F.W.J.M.M.(1991) Verfahren zur diskontinuierlichen Herstellung von L-Carnitin auf mikrobiologischem Weg, Lonza AG, EP 0 410 430 A2

- Kieslich, K. (1991), Biotransformations of industrial use, 5th Leipziger Biotechnologiesymposium 1990, Acta Biotechnol. **11** (6), 559–570

- Kitamura, M., Ohkuma, T., Takaya, H., Noyori, R. (1988) A practical asymmetric synthesis of carnitine, Tetrahedron Lett. **29**, 1555–1556

- Kulla, H. (1991) Enzymatic hydroxylations in industrial application, CHIMICA **45**, 81–85

- Kulla, H., Lehky, P., Squaratti, A. (1991) Verfahren zur kontinuierlichen Herstellung von L-Carnitin, Lonza AG, EP 0 195 944 B1

- Kulla, H., Lehky, P. (1991) Verfahren zur Herstellung von L-Carnitin auf mikrobiologischem Wege, Lonza AG, EP 0 158 194 B1

- Macy, J., Kulla, H., Gottschalk, G. (1976) H_2-dependent anaerobic growth of *Escherichia coli* on L-malate: succinate formation, J. Bacteriol. **125**, 423–428

- Seim, H., Ezold, R., Kleber, H.-P., Strack, E. (1980) Stoffwechsel des L-Carnitins bei Enterobakterien Z. Allg. Mikrobiol. **20**, 591–594

- Vandecasteele, J.-P. (1980) Enzymatic synthesis of L-carnitine by reduction of an achiral precursor: the problem of reduced nicotinamide adenine dinucleotide recycling, Appl. Environ. Microbiol. **39**, 327–334

- Voeffray, R., Perlberger, J.-C., Tenud, L., Gosteli, J. (1987) L-Carnitine. Novel synthesis and determination of the optical purity, Helv. Chim. Acta. **70**, 2058–2064

- Zhou, B.-N., Gopalan, A.S., van Middleworth, F., Shieh, W.-R., Sih, C.J. (1983) Stereochemical control of yeast reductions. 1. Asymmetric synthesis of L-carnitine, J. Am. Chem. Soc. **105**, 5925–5926

- Zimmermann, Th.P., Robins, K.T., Werlen, J., Hoeks, F.W.(1997) Bio-transformation in the production of L–carnitine, in: Chirality in Industry (Collins, A.N., Sheldrake, G.N., Crosby, J., eds.), pp. 287–305, John Wiley and Sons Ltd, New York

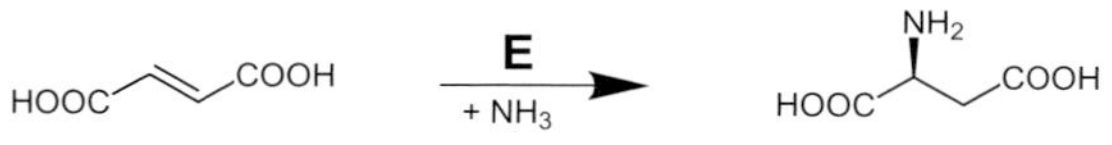

1 = fumaric acid
2 = aspartic acid

BioCatalytics

Fig. 4.3.1.1 – 1

1) Reaction conditions

[**1**]:	1.5 M, 174.1 g · L^{-1} [116.07 g · mol^{-1}]
[**2**]:	1.5 M, 199.5 g · L^{-1} [133.10 g · mol^{-1}]
pH:	8.5
T:	inlet 27 °C; outlet 37 °C
medium:	aqueous
reaction type:	C-N-bond cleavage
catalyst:	immobilized enzyme
enzyme:	L-aspartate-ammonia lyase (aspartase)
strain:	*Escherichia coli*
CAS (enzyme):	[9027–30–9]

2) Remarks

- The presence of 1 mM MgCl$_2$ enhances the activity and stability of the enzyme.

- By using isolated, on silica-based support immobilized enzyme, a higher productivity was achieved than the comparable process using immobilized cells.

- Tanabe Seiyaku uses for the same synthesis an immobilized whole cell system since 1973 (see page 381), in contrast to Mitsubishi Petrochemical Co. Ltd., which uses suspended whole cells (see page 379). Kyowa Hakko Kogyo Co, Ltd., Japan, also uses an immobilized enzyme (see page 377).

- The product solution is acidified to pH 2.8, chilled and the product precipitate is filtered off.

3) Flow scheme

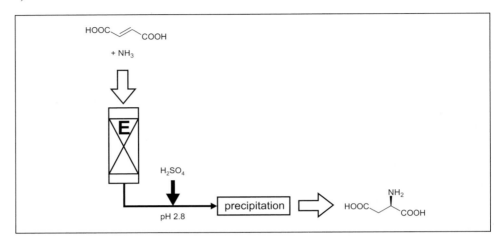

Fig. 4.3.1.1 – 2

4) Process parameters

conversion: 99 %
yield: 95 %
selectivity: 96 %
optical purity: >99.9 %
chemical purity: 99 %
reactor type: plug flow reactor
reactor volume: 75 L
residence time: 0.07 h
space-time-yield: $3 \ kg \cdot L^{-1} \cdot h^{-1}$
down stream processing: precipitation
enzyme consumption: $t_{1/2}$ = 6 months
company: BioCatalytics, USA

5) Product application

- The product is the precursor for the synthesis of aspartame (see page 270).

- Alternatively it can be used for parental nutrition, food additive or as chiral synthon in organic synthesis.

6) Literature

- Rozzel, D. (1998) BioCatalytics, personal communication

Aspartase
Escherichia coli

EC 4.3.1.1

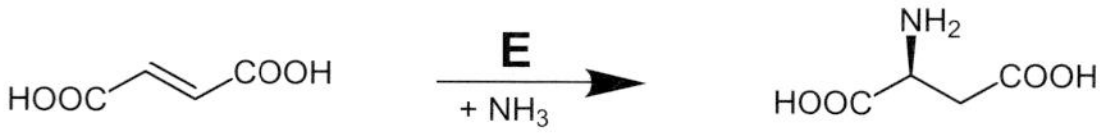

1 L-2

1 = fumaric acid
2 = aspartic acid

Kyowa Hakko Kogyo Co, Ltd.

Fig. 4.3.1.1 – 1

1) Reaction conditions

[**1**]:	2 M, 232.14 g · L^{-1} [116.07 g · mol^{-1}]
T:	37 °C
medium:	aqueous
reaction type:	C-N bond cleavage
catalyst:	immobilized enzyme
enzyme:	L-aspartate ammonia-lyase (fumaric aminase, aspartase)
strain:	*Escherichia coli*
CAS (enzyme):	[9027–30–9]

2) Remarks

- As source of the amine 4 M NH_4OH is added.
- Isolated enzyme is immobilized on Duolite A-7, a weakly basic anion-exchange resin.
- The column reactor is operated for over 3 month at over 99 % conversion.
- Tanabe Seiyaku uses for the same syntheses an immobilized whole cell system since 1973 (see page 381), instead of suspended whole cells used by Mitsubishi Petrochemical Co. Ltd. (see page 379). Biocatalytics uses an immobilized enzym (see page 375).

3) Flow scheme

Not published.

4) Process parameters

conversion:	> 99 %
ee:	> 99.9 %
reactor type:	plug flow reactor
residence time:	0.75 h
enzyme consumption:	$t_{1/2}$ = 18 d
start-up date:	1974
company:	Kyowa Hakko Kogyo Co, Ltd., Japan

5) Product application

- L-Aspartic acid is used as an intermediate for aspartame, an artificial sweetener (see page 270).

- It can also be used as acidulant in pharmaceuticals and foods.

6) Literature

- Tanaka, A., Tosa, T., Kobayashi, T. (1993) Industrial Application of Immobilized Biocatalysts, Marcel Dekker Inc., New York

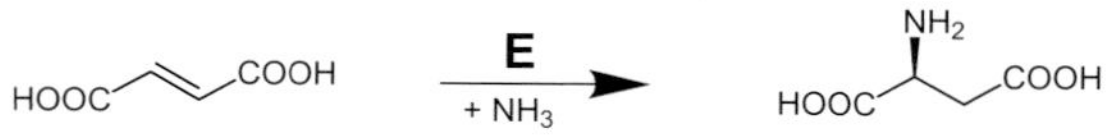

1 = fumaric acid
2 = aspartic acid

Mitsubishi

Fig. 4.3.1.1 – 1

1) Reaction conditions

[**1**]:	1.3 M, 150.9 g·L^{-1} [116.10 g·mol^{-1}]
pH:	9–10
T:	54 °C
medium:	aqueous
reaction type:	C-N bond cleavage
catalyst:	suspendend whole cells
enzyme:	L-aspartate ammonia-lyase (aspartase)
strain:	*Brevibacterium flavum*
CAS (enzyme):	[9027–30–9]

2) Remarks

- As source of the amine 4 M NH_4OH is added.

- L-Malic acid is formed as a side product by an intracellular fumarase. But the fumarase can be inactivated by thermal effects. As optimal conditions for the suppression of fumarase activity the following parameters were established: Incubation of the cells (3 % w/v) at 45°C for 5 hours in the presence of 2 M NH_4OH, 0.75 M L-aspartic acid, 0.0075 M $CaCl_2$ and 0.08 % (w/v) of the nonionic detergent Tween 20. L-Aspartic acid and $CaCl_2$ act as protectors against thermal inactivation. By the addition of Tween 20 the production of L-aspartic acid is increased by 40 %.

- L-Aspartic acid is produced from fumaric acid stoichiometrically.

- The bacterial cells are retained by ultrafiltration.

- The industrial production of L-aspartic acid has been carried out since 1958.

- Kyowa Hakko Kogyo Co. Ltd. uses an immobilized enzyme (see page 377) and Tanabe Seiyaku uses immobilized whole cells (see page 381).

3) Flow scheme

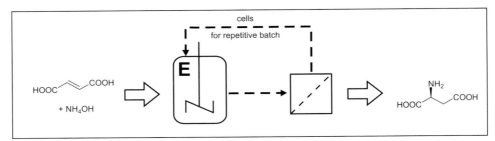

Fig. 4.3.1.1 – 2

4) Process parameters

yield: > 99.99 %
selectivity: > 99.99 %
reactor type: repetitive batch
start-up date: 1986
company: Mitsubishi Petrochemical Co. Ltd., Japan

5) Product application

- L-Aspartic acid is used as a precursor for the synthesis of the low calorie synthetic sweetener aspartame (see page 270).

- It is also an intermediate for several pharmaceuticals.

6) Literature

- Tanaka, A., Tosa, T., Kobayashi, T. (1993) Industrial Application of Immobilized Biocatalysts, Marcel Dekker Inc., New York

- Terasawa, M., Yukawa, H, Takayama, Y. (1985) Production of L-aspartic acid from *Brevibacterium* by cell re-using process, Proc. Biochem. **20**, 124–128

- Yamagata, H., Teresawa, M., Yukawa, H. (1994) A novel industrial process for L-aspartic acid production using an ultrafiltration-membrane, Cat. Today **22**, 621–627

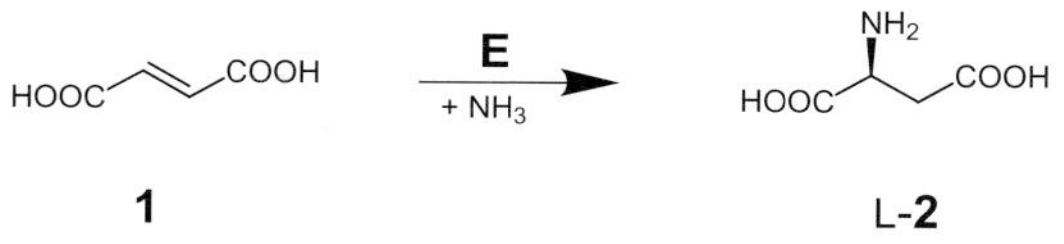

Fig. 4.3.1.1 – 1

1) Reaction conditions

[**1**]:	1 M, 150.14 g·L⁻¹ [150.14 g·mol⁻¹] (ammonium fumarate)
pH:	8.5
T:	37 °C
medium:	aqueous
reaction type:	C-N bond cleavage
catalyst:	immobilized whole cells
enzyme:	L-aspartate ammonia-lyase (fumaric aminase)
strain:	*Escherichia coli*
CAS (enzyme):	[9027–30–9]

where the subscripts are rendered as $[\mathbf{1}]$: 1 M, 150.14 g$\cdot$L^{-1} [150.14 g$\cdot$mol^{-1}] (ammonium fumarate)

2) Remarks

- L-Aspartic acid is produced batchwise via fermentation or enzymatic synthesis since 1953.

- The stability of isolated and immobilized aspartase is not satisfactory. Therefore the cells are immobilized on polyacrylamide or, preferably on κ-carrageenan.

- This process is the first example of the application of immobilized whole cells.

- It is one of the rare examples where the synthesis of an amino acid via an enzymatic route is economically more attractive than usual fermentation methods.

- The costs of the process are reduced to two thirds in comparison to batchwise operation.

- The activity of the cells is increased 10-fold by immobilization.

- The half-life of the cells is about 12 days. By addition of about 1 mM Mg^{2+}, Mn^{2+} or Ca^{2+} it can be increased to more than 120 days.

- A heat exchanger (multiple, small isothermic pipes) is used, because the reaction is exothermic and the use of a fixed bed reactor is advantageous.

- The product is isolated by titration to the isoelectric point (pH 2.8) with H_2SO_4 and filtration of the precipitate.

- This synthesis step is often combined with the synthesis of L-alanine in a two step biotransformation (see page 334).

3) Flow scheme

See page 336

4) Process parameters

yield:	> 95 %
reactor type:	plug-flow reactor
reactor volume:	1,000 L
capacity:	700 t · a^{-1}
down stream processing:	isoelectric precipitation
enzyme activity:	3,200 U/g$_{cells}$
start-up date:	1973
company:	Tanabe Seiyaku Co., Ltd., Japan

5) Product application

- The product is used as a precursor for the synthesis of the artificial sweetener aspartame, pharmaceuticals and food additives.

6) Literature

- Adelberg, E.A., Mandel M. Chen G.C.C. (1965) Optimal conditions for mutagenesis by *N*-methyl-*N*-nitrosoguanidine in *Escherichia coli* K-12, Biochem. Biophys. Res. Commun. **18**, 788–795

- Cheetham, P.S.J. (1994) Case studies in applied biocatalysis, in: Applied Biocatalysis (Cabral, J.M.S., Best, D., Boross, L., Tramper, J., eds.) pp. 77–78, Chur, Switzerland, Harwood Academic Publishers

- Chibata, I., Tosa, T., and Sato, T. (1974) Immobilized aspartase containing microbial cells: Preparation and enzymatic properties, Appl. Microbiol. **27**, 878–885

- Chibata, I. (1978) Immobilized Enzymes, Research and Development, John Wiley & Sons, New York

- Furui, M. (1985) Heat-exchange column with horizontal tubes for immobilized cell reactions with generation of heat, J. Ferment. Technol. **63**, 4, 371–375

- Nishimura, N., Kisumi, M (1984), Aspartase-hyperproducing mutants of *Escherichia coli* B, Appl. Environ. Microbiol. **48**, 1072–1075

- Sato, T., Tetsuya, T. (1993) Production of L-aspartic acid, in: Industrial Applications of Immobilized Biocatalysts (Tanaka, A., Tosa, T., Kobayashi, T., eds.), pp. 15–24, Marcel Dekker Inc., New York

- Takata, I., Tosa, T., and Chibata, I. (1977) Screening of matrix suitable for immobilization of microbial cells, J. Solid-Phase Biochem. **2**, 225–236

- Tosa, T., Shibatani, T. (1995) Industrial application of immobilized biocatalysts in Japan, Ann. N. Y. Acad. Sci. **750**, 364–375

L-Phenylalanine ammonia-lyase
Rhodotorula rubra

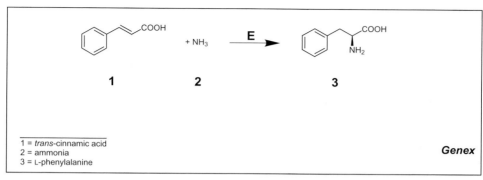

1 = *trans*-cinnamic acid
2 = ammonia
3 = L-phenylalanine

Genex

Fig. 4.3.1.5 – 1

1) Reaction conditions

[**1**]:	0.088 M, 13.02 g·L^{-1} [148.16 g·mol^{-1}]
[**2**]:	9.307 M, 158.5 g·L^{-1} [17.03 g·mol^{-1}]
[**3**]:	0.258 M, 42.7 g·L^{-1} [165.19 g·mol^{-1}]
pH:	10.6
T:	25 °C
medium:	aqueous
reaction type:	C-N bond formation
catalyst:	suspended whole cells
enzyme:	L-phenylalanine ammonia-lyase (PAL, L-phenylalanine deaminase, tyrase)
strain:	*Rhodococcus rubra* (Genex 1983), *Rhodotorula rubra* (Genex 1986), wild type
CAS (enzyme):	[9024–28–6]

2) Remarks

- The PAL-producing microorganisms are initially cultivated under aerobic, growth-promoting conditions.

- Due to the instability of the enzyme towards oxygen, the biotransformation is performed under anaerobic, static conditions.

- An aqueous solution of *trans*-cinnamic acid is mixed with 29 % aqueous ammonia and the pH is adjusted by addition of carbon dioxide.

- As biocatalyst 5.88 g·L^{-1} (dry weight) *Rhodotorula rubra* cells are added.

- The reaction is performed in fed batch mode with periodical addition of concentrated ammonium cinnamate solution.

- The enzyme is deactivated by oxygen and by agitation. Therefore the reaction medium is sparged with nitrogen before the addition of the cells. Instead of stirring, the bioreactor contents are mixed after each addition of substrate solution by sparging with nitrogen.

- Instead of starting from *trans*-cinnamic acid, the fermentation process now starts from glucose. The yields of this *de novo* process are high and up to 25 g·L^{-1} of L-phenylalanine are obtained.

- Prior to this and related processes, L-phenylalanine was mainly obtained by extraction from human hair and other non-microbiological sources.

3) Flow scheme

Not published.

4) Process parameters

yield:	85.7 %
reactor type:	fed batch
down stream processing:	centrifugation, evaporation, crystallization
company:	Genex Corporation, USA

5) Product application

- L-Phenylalanine is used in the manufacture of the artificial sweetener aspartame and in parenteral nutrition.

- The product is used as a building block for the synthesis of the macrolide antibiotic rutamycin B:

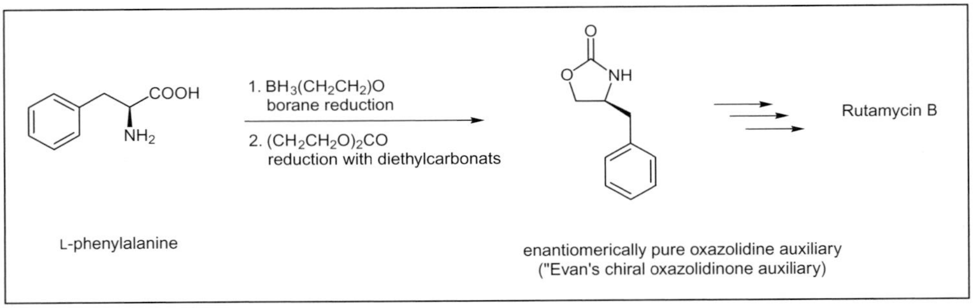

Fig. 4.3.1.5 – 2

- L-Phenylalanine ammonia lyase is effective in the therapy of transplantable tumors in mice.

6) Literature

- Kirk-Othmer (1991–1997) Encyclopedia of Chemical Technology, John Wiley, New York

- Sheldon, R. A. (1993) Chirotechnology, Marcel Dekker-Verlag, New York

- Crosby, J. (1991) Synthesis of optically active compounds: A large scale perspective, Tetrahedron **47**, 4789–4846

- Vollmer, P.J., Montgomery, J.P., Schruber, J.J., Yang, H.-H. (1985) Method for stabilizing the enzymatic activity of phenylalanine ammonia lyase during L-phenylalanine production, Genex Corporation, US 4584269

- Wiseman, A. (1995) Handbook of Enzyme and Biotechnology, Ellis Horwood, Chichester

GlcNAc 2-epimerase
Escherichia coli

1 = *N*-acetyl-D-glucosamine
2 = *N*-acetyl-D-mannosamine

Marukin Shoyu

Fig. 5.1.3.8 – 1

1) Reaction conditions

[**1**]:	0.8 M, 177 g·L⁻¹ [221.21 g·mol⁻¹]
pH:	7.2
T:	30 °C
medium:	aqueous
reaction type:	epimerization
catalyst:	solubilized enzyme
enzyme:	*N*-acyl-D-glucosamine 2-epimerase (*N*-acylglucosamine 2-epimerase)
strain:	*Escherichia coli*
CAS (enzyme):	[37318–34–6]

2) Remarks

- This biotransformation is integrated into the production of *N*-acetylneuraminic acid (Neu5Ac), see page 340.

- By application of the *N*-acylglucosamine 2-epimerase it is possible to start with the inexpensive *N*-acetyl-D-glucosamine instead of *N*-acetyl-D-mannosamine.

- The epimerase is used for the *in situ* synthesis of *N*-acetyl-D-mannosamine (ManNAc). Since the equilibrium is on the side of the educt, the reaction is driven by the subsequent transformation of ManNAc and pyruvate to Neu5Ac.

- The *N*-acylglucosamine 2-epimerase is cloned from porcine kidney and transformed and overexpressed in *Escherichia coli*.

- To reach maximal axctivitiy ATP and Mg²⁺ need to be added.

- Since the whole synthesis is reversible high GlcNAc concentrations are used.

- The chemical epimerization of GlcNAc is employed by Glaxo (page 338).

3) Flow scheme

Not published.

4) Process parameters

conversion:	84 %
reactor type:	batch
reactor volume:	200 L
capacity:	multi kg
down stream processing:	subsequent, *in situ* biotransformation to *N*-acetylneuraminic acid
enzyme supplier:	Marukin Shoyu Co., Ltd., Japan
company:	Marukin Shoyu Co., Ltd., Japan

5) Product application

- *N*-Acetyl-D-mannosamine serves as *in situ* generated substrate for the synthesis of *N*-acetylneuraminic acid.

6) Literature

- Kragl, U., Gygax, D., Ghisalba, O., Wandrey, C. (1991) Enzymatic process for preparing *N*-acetylneuraminic acid, Angew. Chem. Int. Ed. Engl. **30**, 827–828

- Maru, I., Ohnishi, J., Ohta, Y., Tsukada, Y. (1998) Simple and large-scale production of *N*-acetylneuraminic acid from *N*-acetyl-D-glucosamine and pyruvate using *N*-acyl-D-glucosamine 2-epimerase and *N*-acetylneuraminate lyase, Carbohydrate Res. **306**, 575–578

- Maru, I., Ohta, Y., Murata, K., Tsukada, Y. (1996) Molecular cloning and identification of *N*-acyl-D-glucosamine 2-epimerase from porcine kidney as a renin-binding protein, Biol. Chem. **271**, 16294–16299

Xylose isomerase

Bacillus coagulans/Streptomyces rubiginosus/Streptomyces phaechromogenes **EC 5.3.1.5**

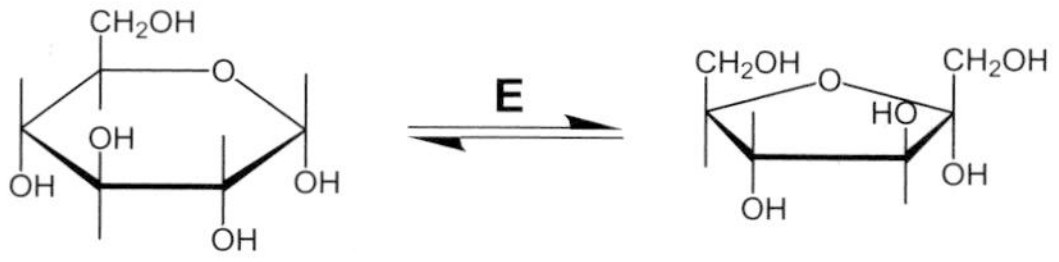

1 **2**

Novo-Nordisk
Gist-brocardes
Miles Kali-Chemie
Finnsugar
Nagase

1 = glucose
2 = fructose

Fig. 5.3.1.5 – 1

1) Reaction conditions

[**1**]:	> 95 % dry matter
pH:	7.5 – 8.0
T:	50 – 60 °C
medium:	aqueous
reaction type:	isomerization
catalyst:	immobilized whole cells or isolated enzyme
enzyme:	D-xylose ketol-isomerase (xylose-isomerase, glucose-isomerase)
strain:	several, see remarks
CAS (enzyme):	[9023–82–9]

2) Remarks

- Glucose isomerase is produced by several microorganisms as an intracellular enzyme. The following table shows some examples:

Trade name	**Microorganism**	**Company**	**Country**
Sweetzyme	*Bacillus coagulans*	Novo-Nordisk	Denmark
Maxazyme	*Actinplanes missouriensis*	Gist-Brocades	The Netherlands
Optisweet	*Streptomyces rubiginosus*	Miles Kali-Chemie	Germany
Sweetase	*Streptomyces phaechromogenes*	Nagase	Japan

Fig. 5.3.1.5 – 2

- The commercially important varieties show superior affinity to xylose and are therefore classified as xylose-isomerases.
- Since the isolation of the intracellular enzyme is very expensive, whole cells are used instead. In almost all cases the enzymes or cells are immobilized using different techniques depending on strain and supplier.
- The educt is purified glucose (dextrose) syrup from the saccharification stage.

Xylose isomerase
Bacillus coagulans/Streptomyces rubiginosus/Streptomyces phaechromogenes

EC 5.3.1.5

- Since these isomerases belong to the group of metalloenzymes, Co^{2+} and Mg^{2+} are required.

- The reaction enthalpy is slightly endothermic and reversible. The equilibrium conversion is about 50 % at 55 °C.

- To limit byproduct formation, the reaction time must be minimized. This can be done economically only by using high concentrations of immobilized isomerase.

- Several reactors are operated in parallel or in series, containing enzymes of different ages. The feed to a single reactor is controlled by the conversion of this reactor.

- The educt has to be highly purified (filtration, adsorption on charcoal, ion exchange) to prevent fast deactivation and clogging of the catalyst bed (for first part of process see page 231).

- Plants producing more than 1,000 t of HFCS (high fructose corn syrup) (based on dry matter) per day typically use at least 20 individual reactors.

- The product HFCS contains 42 % fructose (53 % glucose) or 55 % fructose (41 % glucose) (as dry matter).

- Glucose isomerases have half-lives of more than 100 days. To maintain the necessary activity the enzyme is replaced after deactivation of about 12.5 %.

- The reaction temperature is normally above 55°C to prevent microbial infection although enzyme stability is lowered.

388

3) Flow scheme

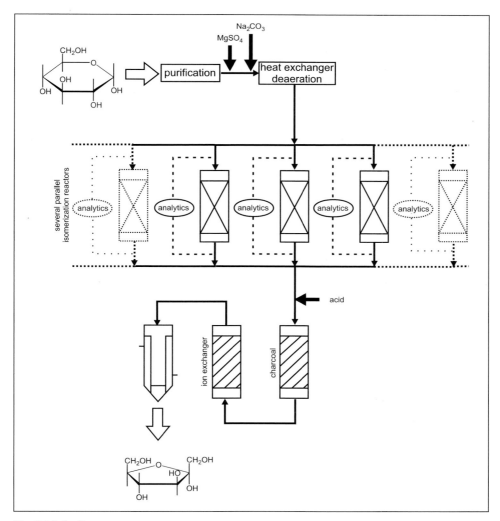

Fig. 5.3.1.5 – 3

4) Process parameters

reactor type:	continuous, fixed bed
reactor volume:	variable
capacity:	$> 7 \cdot 10^6 \, \text{t} \cdot \text{a}^{-1}$
residence time:	0.17 – 0.33 h
down stream processing:	55 % fructose: chromatography; 42 % fructose: no down stream processing
enzyme consumption:	see remarks
start-up date:	1967 by Clinton Corn Processing Co. (USA); 1974 with immobilized enzyme

389

Xylose isomerase
Bacillus coagulans/Streptomyces rubiginosus/Streptomyces phaechromogenes

EC 5.3.1.5

production site: Denmark, the Netherlands, Germany, Finland, Japan

company: Novo Nordisk, Gist-Brocades, Miles Kali-Chemie, Finnsugar, Nagase and others

5) Product application

- The product is named high-fructose corn syrup (HFCS) or ISOSIRUP.

- It is an alternative sweetener to sucrose or invert sugar in the food and beverage industries.

- The chromatographically enriched form (55 % fructose) is used for sweetening alcoholic beverages.

- 42 % HFCS obtained directly by enzymatic isomerization is used mainly in the baking and dairy industries.

6) Literature

- Antrim, R.L., Colilla, W., Schnyder, B.J. and Hemmingesen, S.H., in: Applied Biochemistry and Bioengineering 2 (Wingard, L.B., Katchalski-Katzir, E., Goldstein, L., eds.), pp. 97–183, Academic Press, New York

- Blanchard, P.H., Geiger, E.O. (1984) Production of high-fructose corn syrup in the USA, **11**, Sugar Technol. Rev. 1–94

- Gerhartz, W.(1990) Enzymes in Industry: Production and Application, Verlag Chemie, Weinheim

- Hupkes, J.V., Tilburg, R. van (1976) [Industrial] applications of the catalytic power of enzymes, Neth. Chem. Weekbl. **69**, K14-K17

- Ishimatsu, Y., Shigesada, S., Kimura, S., (1975) Immobilized enzymes, Denki Kaguku Kogyo K.K. Japan, US 3915797

- Landis, B.H., Beery, K.E. (1984) Developments in soft drink technology, **3**, Elsevier Applied Science Publishers, London, pp. 85–120

- Oestergaard, J., Knudsen, S.L. (1976) Use of Sweetenzyme in industrial continuous isomerization. Various process alternatives and corresponding product types, Stärke **28**, 350–356

- Straatsma, J., Vellenga, K., Witt, H.G.J. de, Joosten, G.E. (1983) Isomerization of glucose to fructose. 1. The stability of a whole cell immobilized glucose isomerase catalyst, Ind. Eng. Chem. Process Des. Dev. **22**, 349–356

- Straatsma, J., Vellenga, K., Witt, H.G.J. de, Joosten, G.E. (1983) Isomerization of glucose to fructose. 2. Optimization of reaction conditions in the production of high fructose syrup by isomerization of glucose catalyzed by a whole cell immobilized glucose isomerase catalyst, Ind. Eng. Chem. Process Des. Dev. **22**, 356–361

- Tewari, Y.B., Goldberg, R.N. (1984) Thermodynamics of the conversion of aqueous glucose to fructose, J. Solution Chem., **13**, 523–547

- Weidenbach, G., Bonse, D., Richter, G. (1984) Glucose isomerase immobilized on silicon dioxide-carrier with high productivity, Stärke **36**, 412–416

- White, J.S., Parke, W. (1989) Fructose adds variety to breakfast, Cereals Foods World **34**, 392–398

390

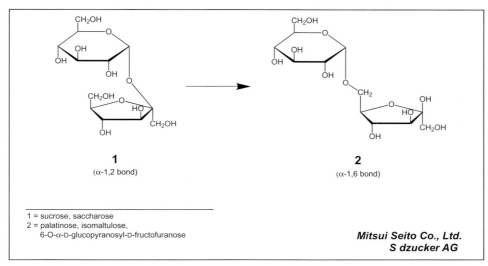

1
(α-1,2 bond)

2
(α-1,6 bond)

1 = sucrose, saccharose
2 = palatinose, isomaltulose,
    6-O-α-D-glucopyranosyl-D-fructofuranose

Mitsui Seito Co., Ltd.
S dzucker AG

Fig. 5.4.99.11 – 1

1) Reaction conditions

pH:	5.8
T:	40 °C
medium:	aqueous
reaction type:	isomerization
catalyst:	immobilized, whole cells
enzyme:	Sucrose glucosylmutase (isomaltulose synthetase, sucrose α-glucosyltransferase)
strain:	*Protaminobacter rubrum*
CAS (enzyme):	[159940–49–5]

2) Remarks

- Palatinose occurs naturally in low amounts in honey and sugar-cane extract.

- Palatinose is derived from palatin, the Latin name for the Pfalz region in Germany.

- α-glucosyltransferase simultaneously produces isomaltulose and smaller amounts of trehalulose (1-O-α-D-glucopyranosyl-β-D-fructose) from sucrose.

- Palatinose (isomaltulose) is the kinetically preferred product, while trehalulose is the thermodynamically preferred product.

- The commercial product contains about 5 % water.

α-Glucosyl transferase
Protaminobacter rubrum

EC 5.4.99.11

3) Process parameters

conversion:	> 99.5 %
yield:	85 %
selectivity:	85 %
capacity:	> 4,000 t · a^{-1}
start-up date:	1985
production site:	Europe, Japan
company:	Südzucker AG, Germany; Mitsui Seito Co. Ltd., Japan

4) Product application

- Used (also hydrogenated: Palatinit, isomaltit) as sweetener with a similiar taste as sucrose but only 42 % of the sweetness of sucrose and with only half of the calorific value.

- It is used as a substitute for sucrose because of the low insulin stimulation.

- The advantage of palatinose or its hydrogenated derivates is that it is decomposed only slightly by *Streptomyces* mutants and dental plaque suspensions resulting in a low cariogenic potential. The level of acid and glucan production is decreased compared to sucrose.

5) Literature

- Cabral, J., Best, D., Boross, L., Tramper, J. (1993) Applied Biocatalysis, Harwood Academic Publishers, Chur, Switzerland

- McAllister, M., Kelly, T., Doyle, E., Fogarty (1990) The isomaltulose synthesizing enzyme of *Serratia plymuthica*, Biotechnol. Lett. **12**, 667–672

- Park, Y.K.; Uekane, R.T., Pupin, A.M. (1992) Conversion of sucrose to isomaltose by microbial glucosyltransferase, Biotechnol. Lett. **14**, 547–551

- Takazoe, I., Frostell, G., Ohta, K., Topitsoglou, Sasaki, N. (1985) Palatinose – a sucrose substitute, Swed. Dent. J. **9**, 81–87

- Tosa, T., Shibatani, T. (1995) Industrial application of immobilized biocatalysts in Japan, Ann. N. Y. Acad. Sci. **750**, 365–375

Dear reader,

there are many processes that are not included in our book.
If you know of any processes which you find worth to be shown in our book, please do not hesitate to contact us and to help us to increase the number of interesting processes in the field of biotechnology.

We will add your suggestion in the next edition of this book.

There are several possibilities how you can get in contact with us:

1) Address

Dr. Andreas Liese
Institute for Biotechnology
Forschungszentrum Jülich GmbH
D-52425 Jülich, Germany

Tel.: +49(0)2461 / 61-6044
Fax.: +49(0)2461 / 61-3870

Dr. Karsten Seelbach
Corporate Process Technology
Degussa-Hüls AG
D-45764 Marl, Germany

Tel.: +49(0)2365 / 49-4972
Fax.: +49(0)2365 / 49-5580

2) Email

industrial.biotransformations@web.de

3) Please send us the following completed form

On the following pages you will find a blank form with all requested parameters. After completion please fax it to one of the above given fax numbers.

Thank you very much for your help!

Data Sheet

Data Sheet for your process

enzyme:
strain:
EC-number:

Reaction scheme (if more complex please enclose an extra page)

1) Reaction conditions

concentration of starting material/product 1:
concentration of starting material/product 2:
concentration of starting material/product 3:
pH:
T:
medium:
catalyst (immobilizes, solubilized etc.):

2) Remarks (if necessary please enclose an extra page)

3) Flow scheme (if necessary please enclose an extra page)

Data Sheet

4) Process parameters

conversion:
yield:
selectivity:
optical purity:
chemical purity:
reactor type:
reactor volume:
capacity:
residence time:
space-time-yield:
down stream processing:
enzyme activity:
enzyme consumption:
enzyme supplier:
start-up date:
closing date:
production site:
company:
contact address:

5) Product application (if necessary please enclose an extra page)

6) Literature (if necessary please enclose an extra page)

Index of enzyme name

enzyme name	EC number	strain	company	page
Acetolactate decarboxylase	4.1.1.5	*Bacillus brevis*	Novo Nordisk, Denmark	332
N-Acetyl-D-neuraminic acid aldolase	4.1.3.3	*Escherichia coli*	Glaxo Wellcome, UK	338
Alcohol dehydrogenase	1.1.1.1	*Acinetobacter calcoaceticus*	Bristol-Myers Squibb, USA	107
		Neurospora crassa	Zeneca Life Science Molecules, UK	99
		Rhodococcus erythropolis	Forschungszentrum Jülich GmbH, Germany	103
Amidase	3.5.1.4	*Comamonas acidovorans*	Lonza AG, Switzerland	274
		Klebsiella terrigena	Lonza AG, Switzerland	276
D-Aminoacid oxidase	1.4.3.3	*Trijonopsis variabilis*	Hoechst Marion Roussel, Germany	129
D-Aminoacid transaminase	2.6.1.21	*Bacillus* sp.	Monsanto, USA	175
			NSC Technologies, USA	175
Aminoacylase	3.5.1.14	*Aspergillus niger*	Chiroscience Ltd., UK	296
		Aspergillus oryzae	Degussa-Hüls AG, Germany	300
Aminopeptidase	3.4.11.1	*Pseudomonas putida*	DSM, Netherlands	236
α-Amylase/ Amyloglucosidase	3.2.1.1/ 3.2.1.3	*Bacillus licheniformis* / *Aspergillus niger*	Several companies	231
Aspartase	4.3.1.1	*Brevibacterium flavum*	Mitsubishi Petrochemical Co., Japan	379
		Escherichia coli	BioCatalytics, USA	375
			Kyowa Hakko Kogyo, Japan	377
			Tanabe Seiyaku Co., Japan	381
Aspartate β-decarboxylase	4.1.1.12	*Pseudomonas dacunhae*	Tanabe Seiyaku Co., Japan	334
Benzoate dioxygenase	1.14.12.10	*Pseudomonas putida*	ICI, UK	143
		Pseudomonas solanacearum	Chiroscience Ltd., UK	312
Carbamoylase	3.5.1.77	*Pseudomonas* sp.	Dr. Vig Medicaments, India	308
Carboxypeptidase B	3.4.17.2	*Pig Pancreas*	Eli Lilly, USA	243
			Hoechst Marion Roussel, Germany	245
Carnitine dehydratase	4.2.1.89	*Escherichia coli*	Lonza AG, Switzerland	369
Catalase	1.11.1.6	*Microbial source*	Novartis, Switzerland	135
Cyclodextrin glycosyltransferase	2.4.1.19	*Bacillus circulans*	Mercian Co. Ltd., Japan	172
Dehalogenase	3.8.1.2	*Pseudomonas putida*	Zeneca Life Science Molecules, UK	322
	1.1.1.1	*Zygosaccharomyces rouxii*	Eli Lilly, USA	110
	1.1.X.X	*Candida sorbophila*	Merck Research Laboratories, USA	121
Dehydrogenase	1.1.X.X	*Geotrichum candidum*	Bristol-Myers Squibb, USA	119
Desaturase	1.X.X.X	*Rhodococcus* sp.	KAO Corp., Japan	165
Enoyl-CoA hydratase	4.2.1.17	*Candida rugosa*	Kanegafuchi, Japan	349, 351
Fumarase	4.2.1.2	*Brevibacterium flavum*	Tanabe Seiyaku Co., Japan	347
		Corynebacterium glutamicum	AMINO GmbH, Germany	344
β-Galactosidase	3.2.1.23	*Saccharomyces lactis*	Central del Latte, Italy	215
			Snow Brand Milk Products Co., Ltd., Japan	215
			Sumitomo Chemical Industries, Japan	215
GlcNAc 2-epimerase	5.1.3.8	*Escherichia coli*	Marukin Shoyu Co., Japan	385
α-Glucosyl transferase	5.4.99.11	*Protaminobacter rubrum*	Mitsui Seito Co., Japan	391

397

398

enzyme name	EC number	strain	company	page
Penicillin amidase	3.5.1.11	*Bacillus megaterium*	Asahi Chemical Industry Co., Ltd., Japan	286
			Toyo Jozo Inc., Japan	286
		Escherichia coli	Fujisawa Pharmaceutical Co., Japan	283
			Gist-Brocades/DSM, The Netherlands	283
			Novo-Nordisk, Denmark	283
			Pfizer, USA	283
			Unifar, Turkey	283
		Escherichia coli	Eli Lilly, USA	288
		Escherichia coli / Arthrobacter viscosus	Dr. Vig. Medicaments, India	281
L-Phenylalanine ammonia-lyase	4.3.1.5	*Rhodotorula rubra*	Genex Corp., USA	383
Phosphorylase	2.4.2.2	*Erwinia carotovora*	Yamasa, Japan	234
Pyruvate decarboxylase	4.1.1.1	*Saccharomyces cerevisiae*	Knoll (BASF), Germany	330
			Krebs Biochemicals Ltd., India	330
			Malladi Drugs, India	330
Racemase	5.1.1.15	*Achromobacter obae*	Toray Industries, Japan	314
Reductase	1.X.X.X	*Saccharomyces cerevisiae*	Hoffmann La-Roche, Switzerland	161
D-Sorbitol dehydrogenase	1.1.99.21	*Gluconobacter oxydans*	Bayer AG, Germany	116
Subtilisin	3.4.21.62	*Bacillus licheniformis*	Coca-Cola, USA	256
	3.4.21.62	*Bacillus licheniformis*	Hoffmann La-Roche, Switzerland	259, 263
		Bacillus sp.	Hoffmann La-Roche, Switzerland	266
Thermolysin	3.4.24.27	*Bacillus proteolicus*	DSM, The Netherlands	270
			Holland Sweetener Company, The Netherlands	270
			Toso, Japan	270
Transaminase	2.6.1.X	*Escherichia coli*	Celgene, USA	179
Trypsin	3.4.21.4	*Pig Pancreas*	Eli Lilly, USA	249
			Hoechst Marion Roussel, Germany	251
			Novo Nordisk, Denmark	253
Tryptophan synthase	4.2.1.20	*Escherichia coli*	AMINO GmbH, Germany	353
Tyrosine phenol lyase	4.1.99.2	*Erwinia herbicola*	Ajinomoto Co., Japan	342
Urease	3.5.1.5.	*Lactobacillus fermentum*	Asahi Chemical Industry, Japan	279
			Toyo Jozo, Japan	279
Xylose isomerase	5.3.1.5	*Bacillus coagulans/Streptomyces rubiginosus/Streptomyces phaechromogenes*	Finnsugar, Finland	387
			Gist-Brocades, the Netherlands	387
			Miles Kali-Chemie, Germany	387
			Nagase, Japan	387
			Novo-Nordisk, Denmark	387

Index of strain

strain	enzyme name	EC number	company	page
Achromobacter obae	Racemase	5.1.1.15	Toray Industries, Japan	314
Achromobacter xylosoxidans	Nicotinic acid hydroxylase	1.5.1.13	Lonza AG, Switzerland	131
Acinetobacter calcoaceticus	Alcohol dehydrogenase	1.1.1.1	Bristol-Myers Squibb, USA	107
Agrobacterium sp.	Nitrilase	3.5.5.1	Lonza AG, Switzerland	317
Alcaligenes faecalis	Nitrilase	3.5.5.1	Lonza AG, Switzerland	320
Alcaligenes sp.	Haloalkane dehalogenase	3.8.1.5	Daiso Co. Ltd., Japan	326
Arthrobacter sp.	Lipase	3.1.1.3	Sumitomo Chemical Co., Japan	208
	Oxygenase	1.13.11.1	Mitsubishi Kasei Corp., Japan	137
Arthrobacter viscosus	Penicillin amidase	3.5.1.11	Dr. Vig. Medicaments, India	281
Aspergillus niger	Aminoacylase	3.5.1.14	Chiroscience Ltd., UK	296
	Amyloglucosidase	3.2.1.3	Several companies	231
Aspergillus oryzae	Aminoacylase	3.5.1.14	Degussa-Hüls AG, Germany	300
Aureobacterium sp.	β-Lactamase	3.5.2.6	Chiroscience Ltd., UK	310
Bacillus brevis	Acetolactate decarboxylase	4.1.1.5	Novo Nordisk, Denmark	332
	D-Hydantoinase	3.5.2.2	Kanegafuchi, Japan	304
Bacillus circulans	Cyclodextrin glycosyltransferase	2.4.1.19	Mercian Co. Ltd., Japan	172
Bacillus coagulans	Xylose isomerase	5.3.1.5	Finnsugar, Finland	387
			Gist-Brocades, the Netherlands	387
			Miles Kali-Chemie, Germany	387
			Nagase, Japan	387
			Novo-Nordisk, Denmark	387
Bacillus licheniformis	α-Amylase	3.2.1.1	Several companies	231
	Subtilisin	3.4.21.62	Coca-Cola, USA	256
	Subtilisin	3.4.21.62	Hoffmann La-Roche, Switzerland	259, 263
Bacillus megaterium	Penicillin amidase	3.5.1.11	Asahi Chemical Industry Co., Ltd., Japan	286
			Toyo Jozo Inc., Japan	286
Bacillus proteolicus	Thermolysin	3.4.24.27	DSM, The Netherlands	270
			Holland Sweetener Company, The Netherlands	270
			Toso, Japan	270
Bacillus sphaericus	Leucine dehydrogenase	1.4.1.9	Degussa-Hüls AG, Germany	125
Bacillus sp.	D-Amino acid transaminase	2.6.1.21	Monsanto, USA	175
			NSC Technologies, USA	175
	Subtilisin	3.4.21.62	Hoffmann La-Roche, Switzerland	266
Beauveria bassiana	Oxidase	1.X.X.X	BASF, Germany	169
Brevibacterium flavum	Aspartase	4.3.1.1	Mitsubishi Petrochemical Co., Japan	379
	Fumarase	4.2.1.2	Tanabe Seiyaku Co., Japan	347
Burkholderia plantarii	Lipase	3.1.1.3	BASF, Germany	181
Candida antarctica	Lipase	3.1.1.3	Schering Plough, USA	205
			UNICHEMA Chemie BV, The Netherlands	217
Candida cylindracea	Lipase	3.1.1.3	Sepracor, USA	202
Candida rugosa	Enoyl-CoA hydratase	4.2.1.17	Kanegafuchi, Japan	349, 351

strain	enzyme name	EC number	company	page
Candida sorbophila	Dehydrogenase	1.1.X.X	Merck Research Laboratories, USA	121
Comamonas acidovorans	Amidase	3.5.1.4	Lonza AG, Switzerland	274
Corynebacterium glutamicum	Fumarase	4.2.1.2	AMINO GmbH, Germany	344
Cryptococcus laurentii	Lactamase	3.5.2.11	Toray Industries, Japan	314
Erwinia carotovora	Nucleosidase	3.2.2.1	Yamasa, Japan	234
	Phosphorylase	2.4.2.2	Yamasa, Japan	234
Erwinia herbicola	Tyrosine phenol lyase	4.1.99.2	Ajinomoto Co., Japan	342
Escherichia coli	Aspartase	4.3.1.1	BioCatalytics, USA	375
			Kyowa Hakko Kogyo, Japan	381
			Tanabe Seiyaku Co., Japan	377
	Carnitine dehydratase	4.2.1.89	Lonza AG, Switzerland	369
	GlcNAc 2-epimerase	5.1.3.8	Marukin Shoyu Co., Japan	385
	Glutaryl amidase	3.1.1.41	Hoechst Marion Roussel, Germany	225
	N-Acetyl-D-neuraminic acid aldolase	4.1.3.3	Glaxo Wellcome, UK	338
			Marukin Shoyu Co., Ltd., Japan	340
			Research Center Jülich GmbH, Germany	340
	Oxygenase	1.14.13.44	Fluka, Switzerland	145
	Penicillin acylase	3.5.1.11	Chemferm, The Netherlands	290
	Penicillin amidase	3.5.1.11	Dr. Vig. Medicaments, India	281
			Eli Lilly, USA	288
			Fujisawa Pharmaceutical Co., Japan	283
			Gist-Brocades/DSM, The Netherlands	283
			Novo-Nordisk, Denmark	283
			Pfizer, USA	283
			Unifar, Turkey	283
	Transaminase	2.6.1.X	Celgene, USA	179
	Tryptophan synthase	4.2.1.20	AMINO GmbH, Germany	353
Fusarium oxysporum	Lactonase	3.1.1.25	Fuji Chemicals Industries, Japan	220
Geotrichum candidum	Dehydrogenase	1.1.X.X	Bristol-Myers Squibb, USA	119
Gluconobacter oxydans	D-Sorbitol dehydrogenase	1.1.99.21	Bayer AG, Germany	116
Klebsiella terrigena	Amidase	3.5.1.4	Lonza AG, Switzerland	276
Lactobacillus fermentum	Urease	3.5.1.5	Asahi Chemical Industry, Japan	279
			Toyo Jozo, Japan	279
Microbial source	Catalase	1.11.1.6	Novartis, Switzerland	135
Mucor miehei	Lipase	3.1.1.3	Chiroscience Ltd., UK	192
Neurospora crassa	Alcohol dehydrogenase	1.1.1.1	Zeneca Life Science Molecules, UK	99
Nocardia autotropica	Oxygenase	1.14.14.1	Merck Sharp & Dohme, USA	151
Nocardia corallina	Monooxygenase	1.14.14.1	Nippon Mining, Japan	153
Pig Pancreas	Carboxypeptidase B	3.4.17.2	Eli Lilly, USA	243
			Hoechst Marion Roussel, Germany	245
	Lipase	3.1.1.3	DSM, The Netherlands	196
	Trypsin	3.4.21.4	Eli Lilly, USA	249
			Hoechst Marion Roussel, Germany	251
			Novo Nordisk, Denmark	253

Index of company

company	strain	enzyme name	EC number	page
Ajinomoto Co., Japan	*Erwinia herbicola*	Tyrosine phenol lyase	4.1.99.2	342
AMINO GmbH, Germany	*Corynebacterium glutamicum*	Fumarase	4.2.1.2	344
	Escherichia coli	Tryptophan synthase	4.2.1.20	353
Asahi Chemical Industry Co., Ltd., Japan	*Bacillus megaterium*	Penicillin amidase	3.5.1.11	286
	Lactobacillus fermentum	Urease	3.5.1.5.	279
	Pseudomonas sp.	Glutaryl amidase	3.1.1.41	229
BASF, Germany	*Beauveria bassiana*	Oxidase	1.X.X.X	169
	Burkholderia plantarii	Lipase	3.1.1.3	181
Bayer AG, Germany	*Gluconobacter oxydans*	D-Sorbitol dehydrogenase	1.1.99.21	116
BioCatalytics, USA	*Escherichia coli*	Aspartase	4.3.1.1	375
Bristol-Myers Squibb, USA	*Acinetobacter calcoaceticus*	Alcohol dehydrogenase	1.1.1.1	107
	Geotrichum candidum	Dehydrogenase	1.1.X.X	119
	Pseudomonas cepacia	Lipase	3.1.1.3	185, 189
Celgene, USA	*Escherichia coli*	Transaminase	2.6.1.X	179
Central del Latte, Italy	*Saccharomyces lactis*	β-Galactosidase	3.2.1.23	215
Chemferm, The Netherlands	*Escherichia coli*	Penicillin acylase	3.5.1.11	290
Chiroscience Ltd., UK	*Aspergillus niger*	Aminoacylase	3.5.1.14	296
	Aureobacterium sp.	β-Lactamase	3.5.2.6	310
	Mucor miehei	Lipase	3.1.1.3	192
	Pseudomonas fluorescens	Lipase	3.1.1.3	199
	Pseudomonas solanacearum	β-Lactamase	3.5.2.6	312
Ciba-Geigy, Switzerland	*Staphylococcus epidermidis*	Lactate dehydrogenase	1.1.1.28	113
Coca-Cola, USA	*Bacillus licheniformis*	Subtilisin	3.4.21.62	256
Daiso Co. Ltd., Japan	*Alcaligenes* sp. or *Pseudomonas* sp.	Haloalkane dehalogenase	3.8.1.5	326
Degussa-Hüls AG, Germany	*Aspergillus oryzae*	Aminoacylase	3.5.1.14	300
	Bacillus sphaericus	Leucine dehydrogenase	1.4.1.9	125
Dr. Vig Medicaments, India	*Arthrobacter viscosus*	Penicillin amidase	3.5.1.11	281
	Escherichia coli	Penicillin amidase	3.5.1.11	281
	Pseudomonas sp.	Hydantoinase	3.5.2.4	308
	Pseudomonas sp.	Carbamoylase	3.5.1.77	308
DSM, The Netherlands	*Bacillus proteolicus*	Thermolysin	3.4.24.27	270
	Porcine pancreas	Lipase	3.1.1.3	196
	Pseudomonas pseudoalcaligenes	Malease	4.2.1.31	355
	Pseudomonas putida	Aminopeptidase	3.4.11.1	236
	Serratia marescens	Lipase	3.1.1.3	210
DuPont, USA	*Pseudomonas chlororaphis*	Nitrile hydratase	4.2.1.84	359
Eli Lilly, USA	*Escherichia coli*	Penicillin amidase	3.5.1.11	288
	Pig Pancreas	Carboxypeptidase B	3.4.17.2	243
		Trypsin	3.4.21.4	249
	Zygosaccharomyces rouxii	Dehydrogenase	1.1.1.1	110
Finnsugar, Finland	*Bacillus coagulans/Streptomyces rubiginosus/Streptomyces phaechromogenes*	Xylose isomerase	5.3.1.5	387
Fluka, Switzerland	*Escherichia coli*	Oxygenase	1.14.13.44	145
Forschungszentrum Jülich GmbH, Germany	*Rhodococcus erythropolis*	Alcohol dehydrogenase	1.1.1.1	103
Fuji Chemicals Industries, Japan	*Fusarium oxysporum*	Lactonase	3.1.1.25	220

company	strain	enzyme name	EC number	page
Fujisawa Pharmaceutical Co., Japan	*Escherichia coli*	Penicillin amidase	3.5.1.11	283
Genencor, USA	*Pseudomonas putida*	Naphthalene dioxygenase	1.13.11.11	139
Genex Corp., USA	*Rhodotorula rubra*	L-Phenylalanine ammonia-lyase	4.3.1.5	383
Gist-Brocades/DSM, The Netherlands	*Escherichia coli*	Penicillin amidase	3.5.1.11	283
Gist-Brocades, The Netherlands	*Bacillus coagulans/Streptomyces rubiginosus/Streptomyces phaechromogenes*	Xylose isomerase	5.3.1.5	387
Glaxo Wellcome, UK	*Escherichia coli*	*N*-Acetyl-D-neuraminic acid aldolase	4.1.3.3	338
Hoechst Marion Roussel, Germany	*Escherichia coli*	Glutaryl amidase	3.1.1.41	225
	Pig Pancreas	Carboxypeptidase B	3.4.17.2	245
		Trypsin	3.4.21.4	251
	Trijonopsis variabilis	D-Aminoacid oxidase	1.4.3.3	129
Hoffmann La-Roche, Switzerland	*Bacillus licheniformis*	Subtilisin	3.4.21.62	259, 263
	Bacillus sp.	Subtilisin	3.4.21.62	266
	Baker's yeast	Reductase	1.X.X.X	161
Holland Sweetener Company, The Netherlands	*Bacillus proteolicus*	Thermolysin	3.4.24.27	270
ICI, UK	*Pseudomonas putida*	Benzoate dioxygenase	1.14.12.10	143
International BioSynthetics	*Rhodococcus erythropolis*	Oxidase	1.X.X.X	163
Kanegafuchi, Japan	*Bacillus brevis*	D-Hydantoinase	3.5.2.2	304
	Candida rugosa	Enoyl-CoA hydratase	4.2.1.17	349, 351
KAO Corp., Japan	*Rhodococcus* sp.	Desaturase	1.X.X.X	165
Knoll (BASF), Germany	*Saccharomyces cerevisiae*	Pyruvate decarboxylase	4.1.1.1	330
Krebs Biochemicals Ltd., India	*Saccharomyces cerevisiae*	Pyruvate decarboxylase	4.1.1.1	330
Kyowa Hakko Kogyo, Japan	*Escherichia coli*	Aspartase	4.3.1.1	377
Lonza AG, Switzerland	*Achromobacter xylosoxidans*	Nicotinic acid hydroxylase	1.5.1.13	131
	Agrobacterium sp.	Nitrilase	3.5.5.1	317
	Alcaligenes faecalis	Nitrilase	3.5.5.1	320
	Comamonas acidovorans	Amidase	3.5.1.4	274
	Escherichia coli	Carnitine dehydratase	4.2.1.89	369
	Klebsiella terrigena	Amidase	3.5.1.4	276
	Pseudomonas putida	Monooxygenase	1.14.13.X	148
	Rhodococcus rhodochrous	Nitrile hydratase	4.2.1.84	361
Malladi Drugs, India	*Saccharomyces cerevisiae*	Pyruvate decarboxylase	4.1.1.1	330
Marukin Shoyu Co., Ltd., Japan	*Escherichia coli*	*N*-Acetyl-D-neuraminic acid aldolase	4.1.3.3	340
		GlcNAc 2-epimerase	5.1.3.8	385
Mercian Co. Ltd., Japan	*Bacillus circulans*	Cyclodextrin glycosyltransferase	2.4.1.19	172
Merck Research Laboratories, USA	*Candida sorbophila*	Dehydrogenase	1.1.X.X	121
Merck Sharp & Dohme, USA	*Nocardia autotropica*	Oxygenase	1.14.14.1	151
Miles Kali-Chemie, Germany	*Bacillus coagulans/Streptomyces rubiginosus/Streptomyces phaechromogenes*	Xylose isomerase	5.3.1.5	387
Mitsubishi Kasei Corp., Japan	*Arthrobacter* sp.	Oxygenase	1.13.11.1	137

company	strain	enzyme name	EC number	page
Mitsubishi Petrochemical Co., Japan	*Brevibacterium flavum*	Aspartase	4.3.1.1	379
Mitsui Seito Co., Japan	*Protaminobacter rubrum*	α-Glucosyl transferase	5.4.99.11	391
Monsanto, USA	*Bacillus* sp.	D-Amino acid transaminase	2.6.1.21	175
Nagase, Japan	*Bacillus coagulans/Streptomyces rubiginosus/Streptomyces phaechromogenes*	Xylose isomerase	5.3.1.5	387
Nippon Mining, Japan	*Nocardia corallina*	Monooxygenase	1.14.14.1	153
Nitto Chemical Industry, Japan	*Rhodococcus rhodochrous*	Nitrile hydratase	4.2.1.84	362
Novartis, Switzerland	*Microbial source*	Catalase	1.11.1.6	135
Novo Nordisk, Denmark	*Bacillus brevis*	Acetolactate decarboxylase	4.1.1.5	332
	Bacillus coagulans	Xylose isomerase	5.3.1.5	387
	Streptomyces rubiginosus	Xylose isomerase	5.3.1.5	387
	Streptomyces phaechromogenes	Xylose isomerase	5.3.1.5	387
	Escherichia coli	Penicillin amidase	3.5.1.11	283
	Pig Pancreas	Trypsin	3.4.21.4	253
NSC Technologies, USA	*Bacillus* sp.	D-Amino acid transaminase	2.6.1.21	175
Pfizer, USA	*Escherichia coli*	Penicillin amidase	3.5.1.11	283
Research Center Jülich GmbH, Germany	*Escherichia coli*	N-Acetyl-D-neuraminic acid aldolase	4.1.3.3	340
Schering Plough, USA	*Candida antarctica*	Lipase	3.1.1.3	205
Sepracor, USA	*Candida cylindracea*	Lipase	3.1.1.3	202
Shell-Brocades, USA	*Pseudomonas oleovorans*	Oxidase	1.14.X.X	159
Snow Brand Milk Products Co., Ltd., Japan	*Saccharomyces lactis*	β-Galactosidase	3.2.1.23	215
Südzucker AG, Germany	*Protaminobacter rubrum*	α-Glucosyl transferase	5.4.99.11	391
Sumitomo Chemical Co., Japan	*Arthrobacter* sp.	Lipase	3.1.1.3	208
	Saccharomyces lactis	β-Galactosidase	3.2.1.23	215
Tanabe Seiyaku Co., Japan	*Brevibacterium flavum*	Fumarase	4.2.1.2	347
	Escherichia coli	L-Aspartase	4.3.1.1	381
	Pseudomonas dacunhae	Aspartate β-decarboxylase	4.1.1.12	334
Toray Industries, Japan	*Cryptococcus laurentii*	Lactamase	3.5.2.11	314
	Achromobacter obae	Racemase	5.1.1.15	
Toso, Japan	*Bacillus proteolicus*	Thermolysin	3.4.24.27	270
Toyo Jozo Inc., Japan	*Bacillus megaterium*	Penicillin amidase	3.5.1.11	286
	Lactobacillus fermentum	Urease	3.5.1.5.	279
	Pseudomonas sp.	Glutaryl amidase	3.1.1.41	229
UNICHEMA Chemie BV, The Netherlands	*Candida antarctica*	Lipase	3.1.1.3	217
Unifar, Turkey	*Escherichia coli*	Penicillin Amidase	3.5.1.11	283
Yamasa, Japan	*Erwinia carotovora*	Nucleosidase	3.2.2.1	234
	Erwinia carotovora	Phosphorylase	2.4.2.2	234
Zeneca Life Science Molecules, UK	*Neurospora crassa*	Alcohol dehydrogenase	1.1.1.1	99
	Pseudomonas putida	Dehalogenase	3.8.1.2	322

Index of starting material

starting material	enzyme name	EC number	company	page
7-ACCA (= 7-aminodeacetoxy methyl-3-chlorocephalosporanic acid)	Penicillin acylase	3.5.1.11	Chemferm, The Netherlands	290
acetaldehyde	Pyruvate decarboxylase	4.1.1.1	Knoll (BASF), Germany	330
			Krebs Biochemicals Ltd., India	330
			Malladi Drugs, India	330
acetic acid 2-methyl-4-oxo-3-prop-2-ynyl-cyclopent-2-enyl ester	Lipase	3.1.1.3	Sumitomo Chemical Co., Japan	208
acetolactate, α-	Acetolactate decarboxylase	4.1.1.5	Novo Nordisk, Denmark	332
acetyl-D-glucosamine; N-	GlcNAc 2-epimerase	5.1.3.8	Marukin Shoyu Co., Japan	385
	N-Acetyl-D-neuraminic acid aldolase	4.1.3.3	Marukin Shoyu Co., Ltd., Japan	340
			Research Center Jülich GmbH, Germany	340
acetyl-D-glucosamine; N- (GlcNAc)	N-Acetyl-D-neuraminic acid aldolase	4.1.3.3	Glaxo Wellcome, UK	338
acetyl-D,L-3-(4-thiazolyl)alanine; N-	Aminoacylase	3.5.1.14	Chiroscience Ltd., UK	296
acetyl-D,L-methionine; N-	Aminoacylase	3.5.1.14	Degussa-Hüls AG, Germany	300
acetyl-D-mannosamine; N-	N-Acetyl-D-neuraminic acid aldolase	4.1.3.3	Marukin Shoyu Co., Ltd., Japan	340
			Research Center Jülich GmbH, Germany	340
acetyl-D-mannosamine; N- (ManNAc)	N-Acetyl-D-neuraminic acid aldolase	4.1.3.3	Glaxo Wellcome, UK	338
acrylonitrile	Nitrile hydratase	4.2.1.84	Nitto Chemical Industry, Japan	362
7-ADCA (= 7-aminodeacetoxy methylcephalosporanic acid)	Penicillin acylase	3.5.1.11	Chemferm, The Netherlands	290
adenosine	Nucleosidase	3.2.2.1	Yamasa, Japan	234
	Phosphorylase	2.4.2.2	Yamasa, Japan	234
adiponitrile	Nitrile hydratase	4.2.1.84	DuPont, USA	359
alkene	Monooxygenase	1.14.14.1	Nippon Mining, Japan	153
amino acid amide; α-H or α-substituted	Aminopeptidase	3.4.11.1	DSM, Netherlands	236
amino acid; α-H or α-substituted	Aminopeptidase	3.4.11.1	DSM, Netherlands	236
amino acid; L-	D-Amino acid transaminase	2.6.1.21	Monsanto, USA	175
			NSC Technologies, USA	175
amino-D-sorbitol; 1-	D-Sorbitol dehydrogenase	1.1.99.21	Bayer AG, Germany	116
amino-D-sorbitol; 1- (N-protected)	D-Sorbitol dehydrogenase	1.1.99.21	Bayer AG, Germany	116
amino methyl ester; α-H or α-substituted	Aminopeptidase	3.4.11.1	DSM, Netherlands	236

starting material	enzyme name	EC number	company	page
ammonia	Aspartase	4.3.1.1	BioCatalytics, USA	375
	GlcNAc 2-epimerase	5.1.3.8		385
	L-Phenylalanine ammonia-lyase	4.3.1.5	Genex Corp., USA	383
6-APA (= 6-aminopenicillanic acid)	Penicillin acylase	3.5.1.11	Chemferm, The Netherlands	290
arylallyl ether	Oxidase	1.14.X.X	Shell-Brocades, USA	159
aspartic acid; D,L-	Aspartate β-decarboxylase	4.1.1.12	Tanabe Seiyaku Co., Japan	334
aspartic acid; L-	Aspartate β-decarboxylase	4.1.1.12	Tanabe Seiyaku Co., Japan	334
	Thermolysin	3.4.24.27	DSM, The Netherlands	270
			Holland Sweetener Company, The Netherlands	270
			Toso, Japan	270
azabicyclo[2.2.1]hept-5-en-3-one; 2-	β-Lactamase	3.5.2.6	Chiroscience Ltd., UK	310
azabicyclo[2.2.1]hept-5-en-3-one; 2-	β-Lactamase	3.5.2.6	Chiroscience Ltd., UK	312
azetidinone acetate; racemic cis-	Lipase	3.1.1.3	Bristol-Myers Squibb, USA	185
azetidinone; cis-[(2R,3S),(2S,3R)]-	Penicillin amidase	3.5.1.11	Eli Lilly, USA	288
butyl-2-phenyl-4H-oxazol-5-one; 4-*tert*	Lipase	3.1.1.3	Chiroscience Ltd., UK	192
benzaldehyde	Pyruvate decarboxylase	4.1.1.1	Knoll (BASF), Germany	330
			Krebs Biochemicals Ltd., India	330
			Malladi Drugs, India	330
benzene; and derivatives thereof	Benzoate dioxygenase	1.14.12.10	ICI, UK	143
benzoic acid	Oxygenase	1.13.11.1	Mitsubishi Kasei Corp., Japan	137
benzyl-3-[[1-methyl-1-((morpholino-4-yl)carbonyl)ethyl]sulfonyl]propionic acid ethyl ester; (R,S)-2-	Subtilisin	3.4.21.62	Hoffmann La-Roche, Switzerland	259
benzyl-3-(*tert*-butylsulfonyl)propionic acid ethyl ester; (R,S)-2-	Subtilisin	3.4.21.62	Hoffmann La-Roche, Switzerland	263
benzyloxy-3,5-dioxo-hexanoic acid ethyl ester; 6-	Alcohol dehydrogenase	1.1.1.1	Bristol-Myers Squibb, USA	107
butanedioic acid diethyl ester; (R,S)-(2-methylpropyl)-	Subtilisin	3.4.21.62	Hoffmann La-Roche, Switzerland	266
butyric acid	Enoyl-CoA hydratase	4.2.1.17	Kanegafuchi, Japan	349
2-oxo-butyric acid; 3,3-dimethyl-2-	D-Amino acid transaminase	2.6.1.21	Monsanto, USA	175
			NSC Technologies, USA	175
butyrobetaine; 4-	Carnitine dehydratase	4.2.1.89	Lonza AG, Switzerland	369
caprolactam; α-amino-ε	Lactamase	3.5.2.11	Toray Industries, Japan	314
	Racemase	5.1.1.15	Toray Industries, Japan	314

starting material	enzyme name	EC number	company	page
carbamoyl-D-hydroxyphenyl glycine; D-*N*-	Hydantoinase	3.5.2.4	Dr. Vig Medicaments, India	308
	Carbamoylase	3.5.1.77	Dr. Vig Medicaments, India	308
cephalosporin C	D-Aminoacid oxidase	1.4.3.3	Hoechst Marion Roussel, Germany	129
cephalosporin G	Penicillin amidase	3.5.1.11	Dr. Vig. Medicaments, India	281
chloro-1,2-propanediol; (*R/S*)-3-	Haloalkane dehalogenase	3.8.1.5	Daiso Co. Ltd., Japan	326
chloro-3-oxobutanoic acid methyl ester; 4-	Dehydrogenase	1.1.X.X.	Bristol-Myers Squibb, USA	119
chloropropionic acid; (*R/S*)-	Dehalogenase	3.8.1.2	Zeneca Life Science Molecules, UK	322
cinnamic acid; *trans*-	L-Phenylalanine ammonia-lyase	4.3.1.5	Genex Corp., USA	383
coumaranone; 2-	Lactonase	3.1.1.25	Fuji Chemicals Industries, Japan	224
crotonobetaine	Carnitine dehydratase	4.2.1.89	Lonza AG, Switzerland	369
cyanopyrazine; 2-	Nitrilase	3.5.5.1	Lonza AG, Switzerland	317
cyanopyridine; 2-	Nitrilase	3.5.5.1	Lonza AG, Switzerland	320
cyanopyridine; 3-	Nitrile hydratase	4.2.1.84	Lonza AG, Switzerland	361
dextrose	Xylose isomerase	5.3.1.5	Finnsugar, Finland	387
			Gist-Brocades, the Netherlands	387
			Miles Kali-Chemie, Germany	387
			Nagase, Japan	387
			Novo-Nordisk, Denmark	387
dichloro-1-propanol; (*R/S*)-2,3-	Haloalkane dehalogenase	3.8.1.5	Daiso Co. Ltd., Japan	326
difluoro-phenyl)-allyl]-propane-1,3-diol; 2-[2-(2,4-	Lipase	3.1.1.3	Schering Plough, USA	205
dihydrocoumarin	Lactonase	3.1.1.25	Fuji Chemicals Industries, Japan	220
dihydro-*(6S)*-methyl-4H-thieno[2,3b]-thiopyran-4-one-7,7-dioxide; 5,6-	Alcohol dehydrogenase	1.1.1.1	Zeneca Life Science Molecules, UK	99
dimethyl-1,3-dioxolane-4-methanol; (*R,S*)-2,2-	Oxidase	1.X.X.X	International BioSynthetics	163
dimethylcyclopropanecarboxamide; 2,2-	Amidase	3.5.1.4	Lonza AG, Switzerland	274
dimethylpyrazine; 2,5-	Monooxygenase	1.14.13.X	Lonza AG, Switzerland	148
erythrono-γ-lactone; D,L-	Lactonase	3.1.1.25	Fuji Chemicals Industries, Japan	224
ethylmethoxyacetate	Lipase	3.1.1.3	BASF, Germany	181
bis(4-fluorophenyl])-3-(1-methyl-1H-tetrazol-5-yl)-1,3-butadienyl]-tetrahydro-4-hydroxy-2H-pyran-2-one; [4-[4α,6β(*E*)]]-6-[4,4-	Lipase	3.1.1.3	Bristol-Myers Squibb, USA	189

starting material	enzyme name	EC number	company	page
fumaric acid	Aspartase	4.3.1.1	BioCatalytics, USA	375
			Kyowa Hakko Kogyo, Japan	377
			Mitsubishi Petrochemical Co., Japan	379
	Fumarase	4.2.1.2	AMINO GmbH, Germany	344
			Tanabe Seiyaku Co., Japan	347
	L-Aspartase	4.3.1.1	Tanabe Seiyaku Co., Japan	381
galactono-γ-lactone; D,L-	Lactonase	3.1.1.25	Fuji Chemicals Industries, Japan	220
GlcNAc = N-acetyl-D-glucosamine	GlcNAc 2-epimerase	5.1.3.8	Marukin Shoyu Co., Japan	385
			Research Center Jülich GmbH, Germany	340
glucono-δ-lactone; D,L-	Lactonase	3.1.1.25	Fuji Chemicals Industries, Japan	220
glucooctanoic-γ-lactone; α,β-	Lactonase	3.1.1.25	Fuji Chemicals Industries, Japan	220
glucose	Xylose isomerase	5.3.1.5	Finnsugar, Finland	387
			Gist-Brocades, the Netherlands	387
			Miles Kali-Chemie, Germany	387
			Nagase, Japan	387
			Novo-Nordisk, Denmark	387
glucose; D-	D-Sorbitol dehydrogenase	1.1.99.21	Bayer AG, Germany	116
glutaryl-7-aminocephalo-sporanic acid	Glutaryl amidase	3.1.1.41	Asahi Chemical Industry Co., Ltd., Japan	229
			Hoechst Marion Roussel, Germany	225
			Toyo Jozo, Japan	229
glycero-D-gulo-heptono-γ-lactone; D,L-	Lactonase	3.1.1.25	Fuji Chemicals Industries, Japan	220
glycero-D-manno-heptono-γ-lactone; D,L-	Lactonase	3.1.1.25	Fuji Chemicals Industries, Japan	220
glycidate; (R,S)-	Lipase	3.1.1.3	DSM, The Netherlands	196
glyoxylic acid	D-Hydantoinase	3.5.2.2	Kanegafuchi, Japan	304
guanosine	Nucleosidase	3.2.2.1	Yamasa, Japan	234
	Phosphorylase	2.4.2.2	Yamasa, Japan	234
gulono-γ-lactone; D,L-	Lactonase	3.1.1.25	Fuji Chemicals Industries, Japan	220
Hydrogen peroxide, H$_2$O$_2$	Catalase	1.11.1.6	Novartis, Switzerland	135
hydrogen peroxide	Catalase	1.11.1.6	Novartis, Switzerland	135
hydroxy-2-methyl-3-oxobutanoate, 2-	Acetolactate decarboxylase	4.1.1.5	Novo Nordisk, Denmark	332

starting material	enzyme name	EC number	company	page
ManNAc = *N*-acetyl-D-mannosamine	*N*-Acetyl-D-neuraminic acid aldolase	4.1.3.3	Glaxo Wellcome, UK	338
		4.1.3.3	Marukin Shoyu Co., Ltd., Japan	340
			Research Center Jülich GmbH, Germany	340
mannono-γ-lactone; D,L-	Lactonase	3.1.1.25	Fuji Chemicals Industries, Japan	220
methoxyphenyl)glycidic acid methyl ester; *trans*-3-(4-	Lipase	3.1.1.3	DSM, The Netherlands	210
methoxypropanone	Transaminase	2.6.1.X	Celgene, USA	179
methylenedioxyacetophenone; 3,4-	Dehydrogenase	1.1.1.1	Eli Lilly, USA	110
methyl phenoxyacetate	Penicillin amidase	3.5.1.11	Eli Lilly, USA	288
methyl phenylacetate	Penicillin amidase	3.5.1.11	Eli Lilly, USA	288
molasses	Pyruvate decarboxylase	4.1.1.1	Knoll (BASF), Germany	330
			Krebs Biochemicals Ltd., India	330
MPGM = 3-(4-methoxyphenyl)-glycidic acid methyl ester	Lipase	3.1.1.3	DSM, The Netherlands	210
myristic acid	Lipase	3.1.1.3	UNICHEMA Chemie BV, The Netherlands	217
niacin	Nicotinic acid hydroxylase	1.5.1.13	Lonza AG, Switzerland	131
nicotinic acid	Nicotinic acid hydroxylase	1.5.1.13	Lonza AG, Switzerland	131
nitro-phenyl)-*N*-(2-oxo-2-pyridin-3-yl-ethyl)-acetamide; 2-(4-	Dehydrogenase	1.1.X.X	Merck Research Laboratories, USA	121
nitrotoluene	Catalase	1.11.1.6	Novartis, Switzerland	135
OPBA = 2-oxo-4-phenylbutyric acid	Lactate dehydrogenase	1.1.1.28	Ciba-Geigy, Switzerland	113
orotidine	Nucleosidase	3.2.2.1	Yamasa, Japan	234
	Phosphorylase	2.4.2.2	Yamasa, Japan	234
oxazolinone, 5-(4H)-	Lipase	3.1.1.3	Chiroscience Ltd., UK	192
oxiranyl-methanol	Lipase	3.1.1.3	DSM, The Netherlands	196
oxo-4-phenylbutyric acid; 2- (= OPBA)	Lactate dehydrogenase	1.1.1.28	Ciba-Geigy, Switzerland	113
oxo-butyric acid; 2-	D-Amino acid transaminase	2.6.1.21	Monsanto, USA	175
			NSC Technologies, USA	175
oxoisophorone	Reductase	1.X.X.X	Hoffmann La-Roche, Switzerland	161
oxo-pentanedioic acid; 2-	D-Amino acid transaminase	2.6.1.21	Monsanto, USA	175
			NSC Technologies, USA	175
oxo-propionic acid; 3-(4-hydroxy-phenyl)-2-	D-Amino acid transaminase	2.6.1.21	Monsanto, USA	175
			NSC Technologies, USA	175

Index of starting material

starting material	enzyme name	EC number	company	page
palmitic acid	Lipase	3.1.1.3	UNICHEMA Chemie BV, The Netherlands	217
pantolactone; D,L-	Lactonase	3.1.1.25	Fuji Chemicals Industries, Japan	220
penicillin-G	Penicillin amidase	3.5.1.11	Asahi Chemical Industry Co., Ltd., Japan	286
			Dr. Vig. Medicaments, India	281
			Fujisawa Pharmaceutical Co., Japan	283
			Gist-Brocades/DSM, The Netherlands	283
			Novo-Nordisk, Denmark	283
			Pfizer, USA	283
			Unifar, Turkey	283
			Toyo Jozo Inc., Japan	286
phenol	D-Hydantoinase	3.5.2.2	Kanegafuchi, Japan	304
phenoxypropionic acid; (R)-2- (POPS)	Oxidase	1.X.X.X	BASF, Germany	169
phenyl-2-propanone; 1-	Alcohol dehydrogenase	1.1.1.1	Research center Jülich GmbH, Germany	103
phenyl acrylic acid; 3-	L-Phenylalanine ammonia-lyase	4.3.1.5	Genex Corp., USA	383
phenylalanine-isopropylester; (R,S)-	Subtilisin	3.4.21.62	Coca-Cola, USA	256
phenylalanine methyl ester	Thermolysin	3.4.24.27	DSM, The Netherlands	270
			Holland Sweetener Company, The Netherlands	270
			Toso, Japan	270
phenyl-butyric acid; 2-	D-Amino acid transaminase	2.6.1.21	Monsanto, USA	175
			NSC Technologies, USA	175
phenylethylamine; 1-	Lipase	3.1.1.3	BASF, Germany	181
phenylglycineamide; D-(–)- = PGA)	Penicillin acylase	3.5.1.11	Chemferm, The Netherlands	290
phenylglycinemethylester; D-(–)- (= PGM)	Penicillin acylase	3.5.1.11	Chemferm, The Netherlands	290
phenylphenol; 2-	Oxygenase	1.14.13.44	Fluka, Switzerland	145
picolinic acid	Nitrilase	3.5.5.1	Lonza AG, Switzerland	320
piperazine-2-carboxamide	Amidase	3.5.1.4	Lonza AG, Switzerland	276
propanol; 2-	Lipase	3.1.1.3	UNICHEMA Chemie BV, The Netherlands	217
pyrazine-2-carboxamide	Amidase	3.5.1.4	Lonza AG, Switzerland	276
pyrazine-2-carboxylic acid	Nitrilase	3.5.5.1	Lonza AG, Switzerland	317
pyridine-3-carboxylate	Nicotinic acid hydroxylase	1.5.1.13	Lonza AG, Switzerland	131

starting material	enzyme name	EC number	company	page
pyrocatechol	Tyrosine phenol lyase	4.1.99.2	Ajinomoto Co., Japan	342
pyruvic acid	*N*-Acetyl-D-neuraminic acid aldolase	4.1.3.3	Marukin Shoyu Co., Ltd., Japan	340
			Research Center Jülich GmbH, Germany	340
			Glaxo Wellcome, UK	338
	Tyrosine phenol lyase	4.1.99.2	Ajinomoto Co., Japan	342
ribono-γ-lactone; D,L-	Lactonase	3.1.1.25	Fuji Chemicals Industries, Japan	220
ribose-1-phosphate	Nucleosidase	3.2.2.1	Yamasa, Japan	234
	Phosphorylase	2.4.2.2	Yamasa, Japan	234
saccharose	α-Glucosyl transferase	5.4.99.11	Mitsui Seito Co., Japan	391
			Südzucker AG, Germany	391
serine; L-	Tryptophan synthase	4.2.1.20	AMINO GmbH, Germany	353
simvastatin	Oxygenase	1.14.14.1	Merck Sharp & Dohme, USA	151
starch	Cyclodextrin glycosyltransferase	2.4.1.19	Mercian Co. Ltd., Japan	172
	α-Amylase	3.2.1.1	Several companies	231
	Amyloglucosidase	3.2.1.3	Several companies	231
sucrose	α-Glucosyl transferase	5.4.99.11	Mitsui Seito Co., Japan	391
			Südzucker AG, Germany	391
trimethylpyruvic acid	Leucine dehydrogenase	1.4.1.9	Degussa-Hüls AG, Germany	125
tryptophan, L-	Naphthalene dioxygenase	1.13.11.11	Genencor, USA	139
urea	D-Hydantoinase	3.5.2.2	Kanegafuchi, Japan	304
	Urease	3.5.1.5.	Asahi Chemical Industry, Japan	279
			Toyo Jozo, Japan	279
uridine	Nucleosidase	3.2.2.1	Yamasa, Japan	234
	Phosphorylase	2.4.2.2	Yamasa, Japan	234

Index of product

product	enzyme name	EC number	company	page
7-ACA = 7-aminocephalosporanic acid	Glutaryl amidase	3.1.1.41	Asahi Chemical Industry Co., Ltd., Japan	229
			Hoechst Marion Roussel, Germany	225
			Toyo Jozo, Japan	229
ACE-inhibitors	Aminopeptidase	3.4.11.1	DSM, Netherlands	236
	Lactate dehydrogenase	1.1.1.28	Ciba-Geigy, Switzerland	113
acetaldehyde	Pyruvate decarboxylase	4.1.1.1	Knoll (BASF), Germany	330
			Krebs Biochemicals Ltd., India	330
			Malladi Drugs, India	330
acetic acid 2-methyl-(4S)-oxo-3-prop-2-ynyl-cyclopent-2-enyl ester	Lipase	3.1.1.3	Sumitomo Chemical Co., Japan	208
acetic acid 4-(2,4-difluoro-phenyl)-2-hydroxymethyl-pent-4-enyl ester	Lipase	3.1.1.3	Schering Plough, USA	205
acetoin, (R)-2-	Acetolactate decarboxylase	4.1.1.5	Novo Nordisk, Denmark	332
acetone	Transaminase	2.6.1.X	Celgene, USA	179
acetyl-D-mannosamine; N-	GlcNAc 2-epimerase	5.1.3.8	Marukin Shoyu Co., Japan	385
acetyl-D-methionine; N-	Aminoacylase	3.5.1.14	Degussa-Hüls AG, Germany	300
acetyl-L-3-(4-thiazolyl)alanine; N-	Aminoacylase	3.5.1.14	Chiroscience Ltd., UK	296
acetylneuraminic acid; N-	N-Acetyl-D-neuraminic acid aldolase	4.1.3.3	Marukin Shoyu Co., Ltd., Japan	340
			Research Center Jülich GmbH, Germany	340
acetylneuraminic acid; N- (Neu5Ac)	N-Acetyl-D-neuraminic acid aldolase	4.1.3.3	Glaxo Wellcome, UK	338
acrylamide	Nitrile hydratase	4.2.1.84	Nitto Chemical Industry, Japan	362
ADCA; 7-	Penicillin amidase	3.5.1.11	Dr. Vig. Medicaments, India	281
adipamide	Nitrile hydratase	4.2.1.84	DuPont, USA	359
alanine; D-	D-Amino acid transaminase	2.6.1.21	Monsanto, USA	175
			NSC Technologies, USA	175
alanine; L-	Aminoacylase	3.5.1.14	Degussa-Hüls AG, Germany	300
alanine; L-	Aspartate β-decarboxylase	4.1.1.12	Tanabe Seiyaku Co., Japan	334
alkyl glycidyl ethers; (R)-	Monooxygenase	1.14.14.1	Nippon Mining, Japan	153
allylglycine, L-2-	Aminoacylase	3.5.1.14	Chiroscience Ltd., UK	296
alprenolol	Lipase	3.1.1.3	DSM, The Netherlands	196
amino-3-phenyl-propionic acid; 2-	L-Phenylalanine ammonia-lyase	4.3.1.5	Genex Corp., USA	383
amino acid; D-	D-Amino acid transaminase	2.6.1.21	Monsanto, USA	175
			NSC Technologies, USA	175

product	enzyme name	EC number	company	page
aminobutyric acid; L-	Aminoacylase	3.5.1.14	Degussa-Hüls AG, Germany	300
aminocephalosporanic acid; 7- (7-ACA)	Glutaryl amidase	3.1.1.41	Asahi Chemical Industry Co., Ltd., Japan	229
			Hoechst Marion Roussel, Germany	225
			Toyo Jozo, Japan	229
amino-cyclopent-2-ene carboxylic acid; 2-	β-Lactamase	3.5.2.6	Chiroscience Ltd., UK	312
	Lactamase	3.5.2.11	Toray Industries, Japan	314
	Racemase	5.1.1.15	Toray Industries, Japan	314
amino-cyclopent-2-enecarboxylic acid; 4-	β-Lactamase	3.5.2.6	Chiroscience Ltd., UK	310
amino deacetoxy cephalosporinic acid; 7-	Penicillin amidase	3.5.1.11	Dr. Vig. Medicaments, India	281
amino-ε-caprolactam; D-α-	Lactamase	3.5.2.11	Toray Industries, Japan	314
	Racemase	5.1.1.15	Toray Industries, Japan	314
aminoindole-3-propionic acid; α-	Tryptophan synthase	4.2.1.20	AMINO GmbH, Germany	353
amino-L-sorbose; 6- (N-protected)	D-Sorbitol dehydrogenase	1.1.99.21	Bayer AG, Germany	116
amino penicillanic acid; 6-	Penicillin amidase	3.5.1.11	Asahi Chemical Industry Co., Ltd., Japan	286
			Dr. Vig. Medicaments, India	281
			Fujisawa Pharmaceutical Co., Japan	283
			Gist-Brocades/DSM, The Netherlands	283
			Novo-Nordisk, Denmark	283
			Pfizer, USA	283
			Unifar, Turkey	283
			Toyo Jozo Inc., Japan	286
ammonia	Urease	3.5.1.5.	Asahi Chemical Industry, Japan	279
			Toyo Jozo, Japan	279
amoxicillin	Penicillin acylase	3.5.1.11	Chemferm, The Netherlands	290
ampicillin	Aminopeptidase	3.4.11.1	DSM, Netherlands	236
	Penicillin acylase	3.5.1.11	Chemferm, The Netherlands	290
APA; 6-	Penicillin amidase	3.5.1.11	Asahi Chemical Industry Co., Ltd., Japan	281
			Dr. Vig. Medicaments, India	286
			Fujisawa Pharmaceutical Co., Japan	283
			Gist-Brocades/DSM, The Netherlands	283
			Novo-Nordisk, Denmark	283
			Pfizer, USA	283

product	enzyme name	EC number	company	page
			Unifar, Turkey	283
			Toyo Jozo Inc., Japan	286
arylglycidyl ether	Oxidase	1.14.X.X	Shell-Brocades, USA	159
aspartame (α-L-aspartyl-L-phenyl-alanine methyl ester, APM)	Aminopeptidase	3.4.11.1	DSM, Netherlands	236
	Thermolysin	3.4.24.27	DSM, The Netherlands	270
			Holland Sweetener Company, The Netherlands	270
			Toso, Japan	270
aspartic acid; D-	Aspartate β-decarboxylase	4.1.1.12	Tanabe Seiyaku Co., Japan	334
aspartic acid; L-	Aspartase	4.3.1.1	BioCatalytics, USA	375
aspartic acid; L-	Aspartase	4.3.1.1	Kyowa Hakko Kogyo, Japan	377
			Mitsubishi Petrochemical Co., Japan	379
			Tanabe Seiyaku Co., Japan	381
atenolol	Lipase	3.1.1.3	DSM, The Netherlands	196
	Oxidase	1.14.X.X	Shell-Brocades, USA	159
azabicyclo[2.2.1]hept-5-en-3-one; 2-	β-Lactamase	3.5.2.6	Chiroscience Ltd., UK	310
azabicyclo[2.2.1]hept-5-en-3-one; (–)-2-	β-Lactamase	3.5.2.6	Chiroscience Ltd., UK	312
	Lactamase	3.5.2.11	Toray Industries, Japan	314
	Racemase	5.1.1.15	Toray Industries, Japan	314
azetidinone; (4R,3R)-	Lipase	3.1.1.3	Bristol-Myers Squibb, USA	185
azetidinone acetate; (3R,4S)-	Lipase	3.1.1.3	Bristol-Myers Squibb, USA	185
azetidinone; cis-(2R,3S)			Eli Lilly, USA	288
benzapril	Aminopeptidase	3.4.11.1	DSM, Netherlands	236
benzodiazepine	Dehydrogenase	1.1.1.1	Eli Lilly, USA	110
benzyl-3-[[1-methyl-1-((morpho-lino-4-yl)carbonyl)ethyl]sulfonyl]-propionic acid ethyl ester; (R)-2-	Subtilisin	3.4.21.62	Hoffmann La-Roche, Switzerland	259
benzyl-3-[[1-methyl-1-((morpho-lino-4-yl)carbonyl)ethyl]sulfonyl]-propionic acid; (S)-2-	Subtilisin	3.4.21.62	Hoffmann La-Roche, Switzerland	259
benzyl-3-(tert-butylsulfonyl)propionic acid ethyl ester; (R)-2-	Subtilisin	3.4.21.62	Hoffmann La-Roche, Switzerland	263
benzyl-3-(tert-butylsulfonyl)propionic acid; (S)-2-	Subtilisin	3.4.21.62	Hoffmann La-Roche, Switzerland	263
benzyloxy-(3R,5S)-dihydroxy-hexanoic acid ethyl ester; 6-	Alcohol dehydrogenase	1.1.1.1	Bristol-Myers Squibb, USA	107
benzylserine; O-L-	Aminoacylase	3.5.1.14	Degussa-Hüls AG, Germany	300
bis(4-fluorophenyl])-3-(1-methyl-1H-tetrazol-5-yl)-1,3-butadienyl]-tetra-hydro-4-hydroxy-2H-pyran-2-one; [4R-[4α,6β(E)]]-6-[4,4-	Lipase	3.1.1.3	Bristol-Myers Squibb, USA	189
Boc-Taz = N-boc-L-3-(4-thiazolyl)-alanine	Aminoacylase	3.5.1.14	Chiroscience Ltd., UK	296
bromocatechol; 3-	Oxygenase	1.14.13.44	Fluka, Switzerland	145
butanedioic acid 4-ethyl ester; (R)-(2-methylpropyl)-	Subtilisin	3.4.21.62	Hoffmann La-Roche, Switzerland	266

416

product	enzyme name	EC number	company	page
dopa; L- (= L-3,4-dihydroxyphenyl-alanine)	Tyrosine phenol lyase	4.1.99.2	Ajinomoto Co., Japan	342
enalapril	Aminopeptidase	3.4.11.1	DSM, Netherlands	236
ephedrine, L-	Pyruvate decarboxylase	4.1.1.1	Knoll (BASF), Germany	330
			Krebs Biochemicals Ltd., India	330
			Malladi Drugs, India	330
epichlorohydrin	Haloalkane dehalogenase	3.8.1.5	Daiso Co. Ltd., Japan	326
epoxide; 1,2-	Monooxygenase	1.14.14.1	Nippon Mining, Japan	153
epoxyalkanes; (R)-1,2-	Monooxygenase	1.14.14.1	Nippon Mining, Japan	153
epoxydec-9-ene; (R)-1,2-	Monooxygenase	1.14.14.1	Nippon Mining, Japan	153
erythrono-γ-lactone; D-	Lactonase	3.1.1.25	Fuji Chemicals Industries, Japan	220
ethyl-3-methylpropane glycine, L-3-	Leucine dehydrogenase	1.4.1.9	Degussa-Hüls AG, Germany	125
ethylcatechol; 3-	Oxygenase	1.14.13.44	Fluka, Switzerland	145
4-exo-hydroxy-2-oxybicyclo[3.3.0]-oct-7-en-3-one butyrate ester	Lipase	3.1.1.3	Chiroscience Ltd., UK	199
fluvalinate	Aminopeptidase	3.4.11.1	DSM, Netherlands	236
fructofuranose; 6-O-α-D-gluco-pyranosyl-D-	α-Glucosyl transferase	5.4.99.11	Mitsui Seito Co., Japan	391
			Südzucker AG, Germany	391
fructose	Xylose isomerase	5.3.1.5	Finnsugar, Finland	387
			Gist-Brocades, the Netherlands	387
			Miles Kali-Chemie, Germany	387
			Nagase, Japan	387
			Novo-Nordisk, Denmark	387
galactono-γ-lactone; D-	Lactonase	3.1.1.25	Fuji Chemicals Industries, Japan	220
galactose	β-Galactosidase	3.2.1.23	Central del Latte, Italy	215
			Snow Brand Milk Products Co., Ltd., Japan	215
			Sumitomo Chemical Industries, Japan	215
glucono-δ-lactone; D-	Lactonase	3.1.1.25	Fuji Chemicals Industries, Japan	220
glucooctanoic-γ-lactone; α,β-	Lactonase	3.1.1.25	Fuji Chemicals Industries, Japan	220
glucose	α-Amylase	3.2.1.1	Several companies	231
	Amyloglucosidase	3.2.1.3	Several companies	231
	β-Galactosidase	3.2.1.23	Central del Latte, Italy	215
			Snow Brand Milk Products Co., Ltd., Japan	215
			Sumitomo Chemical Industries, Japan	215
glutanic acid, D-	D-Amino acid transaminase	2.6.1.21	Monsanto, USA	175
			NSC Technologies, USA	175

product	enzyme name	EC number	company	page
glutaryl-7-aminocephalosporanic acid	D-Aminoacid oxidase	1.4.3.3	Hoechst Marion Roussel, Germany	129
glycero-D-gulo-heptono-γ-lactone; D-	Lactonase	3.1.1.25	Fuji Chemicals Industries, Japan	220
glycero-D-manno-heptono-γ-lactone; D-	Lactonase	3.1.1.25	Fuji Chemicals Industries, Japan	220
glycidate; (R)-	Lipase	3.1.1.3	DSM, The Netherlands	196
glycidol	Haloalkane dehalogenase	3.8.1.5	Daiso Co. Ltd., Japan	326
gulono-γ-lactone; D-	Lactonase	3.1.1.25	Fuji Chemicals Industries, Japan	220
high fructose corn syrup (HFCS)	Xylose isomerase	5.3.1.5	Novo-Nordisk, Denmark	387
H₂O	Catalase	1.11.1.6	Novartis, Switzerland	135
homophenylalanine; L-	Aminoacylase	3.5.1.14	Degussa-Hüls AG, Germany	300
HPBA = (R)-2-hydroxy-4-phenylbutyric acid	Lactate dehydrogenase	1.1.1.28	Ciba-Geigy, Switzerland	113
4-hydroxy-2-oxabicyclo[3.3.0]oct-7-en-3-one (endo)	Lipase	3.1.1.3	Chiroscience Ltd., UK	199
hydroxy-2-pyridin-3-yl-ethyl)-2-(4-nitro-phenyl)-acetamide; (R)-N-(2-	Dehydrogenase	1.1.X.X	Merck Research Laboratories, USA	121
hydroxy-3-methyl-2-prop-2-ynyl-cyclopent-2-enone; (R)-4-	Lipase	3.1.1.3	Sumitomo Chemical Co., Japan	208
hydroxy-4-phenylbutyric acid; (R)-2- (= HPBA)	Lactate dehydrogenase	1.1.1.28	Ciba-Geigy, Switzerland	113
hydroxy-4-(trimethylamino)butanoate; (R)-3-	Carnitine dehydratase	4.2.1.89	Lonza AG, Switzerland	369
hydroxybutan-2-one, (R)-3-	Acetolactate decarboxylase	4.1.1.5	Novo Nordisk, Denmark	332
hydroxy-isobutyric acid; (R)-β-	Enoyl-CoA hydratase	4.2.1.17	Kanegafuchi, Japan	351
hydroxy-n-butyric acid; (R)-β-	Enoyl-CoA hydratase	4.2.1.17	Kanegafuchi, Japan	349
hydroxy-methyl-simvastatin; 6-β-	Oxygenase	1.14.14.1	Merck Sharp & Dohme, USA	151
hydroxynicotinate; 6-	Nicotinic acid hydroxylase	1.5.1.13	Lonza AG, Switzerland	131
hydroxyphenoxy)propionic acid; (R)-2-(4- (HPOPS)	Oxidase	1.X.X.X	BASF, Germany	169
hydroxyphenyl glycine; D-p-	Hydantoinase / Carbamoylase	3.5.2.4/ 3.5.1.77	Dr. Vig Medicaments, India	308
	D-Hydantoinase	3.5.2.2	Kanegafuchi, Japan	304
hydroxypicolinic acid; 6-	Nitrilase	3.5.5.1	Lonza AG, Switzerland	320
hydroxypyrazine-2-carboxylic acid; 5-	Nitrilase	3.5.5.1	Lonza AG, Switzerland	317
hydroxy-pyridine-3-carboxylate; 6-	Nicotinic acid hydroxylase	1.5.1.13	Lonza AG, Switzerland	131
ibuprofen; (R)	Lipase	3.1.1.3	Sepracor, USA	202
ibuprofen, methoxyethyl ester; (R)	Lipase	3.1.1.3	Sepracor, USA	202
indigo	Naphthalene dioxygenase	1.13.11.11	Genencor, USA	139
indole	Naphthalene dioxygenase	1.13.11.11	Genencor, USA	139
indole-2,3-dihydrodiol; cis-	Naphthalene dioxygenase	1.13.11.11	Genencor, USA	139
indole-3-ol; 1H-	Naphthalene dioxygenase	1.13.11.11	Genencor, USA	139
indoxyl	Naphthalene dioxygenase	1.13.11.11	Genencor, USA	139
insulin, di-arg-	Trypsin	3.4.21.4	Eli Lilly, USA	249
			Hoechst Marion Roussel, Germany	251
insulin, human	Carboxypeptidase B	3.4.17.2	Eli Lilly, USA	243
			Hoechst Marion Roussel, Germany	245

product	enzyme name	EC number	company	page
ManNAc = *N*-acetyl-ᴅ-mannosamine	GlcNAc 2-epimerase	5.1.3.8	Marukin Shoyu Co., Japan	385
mannono-γ-lactone; ʟ-	Lactonase	3.1.1.25	Fuji Chemicals Industries, Japan	220
methionine; ʟ-	Aminoacylase	3.5.1.14	Degussa-Hüls AG, Germany	300
methoxyisopropylamine; (*S*)-	Transaminase	2.6.1.X	Celgene, USA	179
methoxyphenyl)glycidic acid; 3-(4-	Lipase	3.1.1.3	DSM, The Netherlands	210
methoxyphenyl)glycidic acid methyl ester; *trans-(2R,3S)*-(4-	Lipase	3.1.1.3	DSM, The Netherlands	210
methyl-1,2-epoxyalkanes; (*R*)-2-	Monooxygenase	1.14.14.1	Nippon Mining, Japan	153
methylamino-1-phenyl-1-propanyl, (+)-threo-2-	Pyruvate decarboxylase	4.1.1.1	Knoll (BASF), Germany	330
			Krebs Biochemicals Ltd., India	330
			Malladi Drugs, India	330
methylenedioxyphenyl)-2-propanol; (*S*)-4-(3,4-	Dehydrogenase	1.1.1.1	Eli Lilly, USA	110
methylpyrazine-2-carboxylic acid; 5-	Monooxygenase	1.14.13.X	Lonza AG, Switzerland	148
metroprolol	Lipase	3.1.1.3	DSM, The Netherlands	196
	Oxidase	1.14.X.X	Shell-Brocades, USA	159
miglitol (1,5.didesoxy-1,5-(2-hydroxyethylimino)-ᴅ-gluctiol; *N*-(2-hydroxyethyl)-1-desoxynojirimycin)	ᴅ-Sorbitol dehydrogenase	1.1.99.21	Bayer AG, Germany	116
MPGM = 3-(4-methoxyphenyl)glycidic acid methyl ester	Lipase	3.1.1.3	DSM, The Netherlands	210
cis,cis-muconic acid	Oxygenase	1.13.11.1	Mitsubishi Kasei Corp., Japan	137
naphthylalanine, ʟ-2-	Aminoacylase	3.5.1.14	Chiroscience Ltd., UK	296
neopentylglycine; ʟ-	Leucine dehydrogenase	1.4.1.9	Degussa-Hüls AG, Germany	125
Neu5Ac = *N*-acetylneuraminic acid	*N*-Acetyl-ᴅ-neuraminic acid aldolase	4.1.3.3	Glaxo Wellcome, UK	338
			Marukin Shoyu Co., Ltd., Japan	340
			Research Center Jülich GmbH, Germany	340
nicotinamide	Nitrile hydratase	4.2.1.84	Lonza AG, Switzerland	361
norleucine; ʟ-	Aminoacylase	3.5.1.14	Degussa-Hüls AG, Germany	300
norvaline; ʟ-	Aminoacylase	3.5.1.14	Degussa-Hüls AG, Germany	300
oxygen, O_2	Catalase	1.11.1.6	Novartis, Switzerland	135
Paclitaxel	Lipase	3.1.1.3	Bristol-Myers Squibb, USA	185
palatinose	α-Glucosyl transferase	5.4.99.11	Mitsui Seito Co., Japan	391
			Südzucker AG, Germany	391
pantoic acid; ᴅ-	Lactonase	3.1.1.25	Fuji Chemicals Industries, Japan	220
pantolactone; ᴅ-	Lactonase	3.1.1.25	Fuji Chemicals Industries, Japan	220
pentafluorostyrene oxide; (*R*)-	Monooxygenase	1.14.14.1	Nippon Mining, Japan	153
phenoxy-1-isopropylamino-2-propanol; (–)-3-	Oxidase	1.14.X.X	Shell-Brocades, USA	159
phenyl-2-propanol; (*S*)-1-	Alcohol dehydrogenase	1.1.1.1	Research Center Jülich GmbH, Germany	103

product	enzyme name	EC number	company	page
rutamycin B	L-Phenylalanine ammonia-lyase	4.3.1.5	Genex Corp., USA	383
Schiff base of α-H or α-substituted amino acid \ amide	Aminopeptidase	3.4.11.1	DSM, Netherlands	236
spirapril	Aminopeptidase	3.4.11.1	DSM, Netherlands	236
taxol	Lipase	3.1.1.3	Bristol-Myers Squibb, USA	185
thiazolyl)alanine; D-3-(4-	Aminoacylase	3.5.1.14	Chiroscience Ltd., UK	296
trandolapril	Aminopeptidase	3.4.11.1	DSM, Netherlands	236
trifluoro-1,2-epoxypropane; (R)-3,3,3-	Monooxygenase	1.14.14.1	Nippon Mining, Japan	153
trimethylcyclohexane-1,4-dione; (6R)-2,2,6-	Reductase	1.X.X.X	Hoffmann La-Roche, Switzerland	161
tryosine; D-	D-Amino acid transaminase	2.6.1.21	Monsanto, USA	175
			NSC Technologies, USA	175
tryptophan; L-	Aminoacylase	3.5.1.14	Degussa-Hüls AG, Germany	300
	Tryptophan synthase	4.2.1.20	AMINO GmbH, Germany	353
tyrosine; L-	Aminoacylase	3.5.1.14	Degussa-Hüls AG, Germany	300
valine; L-	Aminoacylase	3.5.1.14	Degussa-Hüls AG, Germany	300
vitamin B_3	Nitrile hydratase	4.2.1.84	Lonza AG, Switzerland	361